FLUID MECHANICS

FLUID MECHANICS

RUTH H. ROGERS
Reader in Mathematics
School of Engineering and Applied Science
University of Sussex

ROUTLEDGE & KEGAN PAUL
London, Henley and Boston

First published in 1978
by Routledge & Kegan Paul Ltd
39 Store Street,
London WC1E 7DD,
Broadway House,
Newtown Road,
Henley-on-Thames,
Oxon RG9 1EN and
9 Park Street,
Boston, Mass. 02108, USA
Printed in Great Britain by
Thomson Litho Ltd
East Kilbride, Scotland

British Library Cataloguing in Publication Data

Rogers, Ruth H
Fluid mechanics.
1. Fluid mechanics
I. Title
532   QC145.2   78-40013

ISBN 0 7100 8681 4

PREFACE

This textbook is intended as an introduction to fluid
mechanics, suitable for students of mathematics, physics
and engineering. To cover the whole book could take
thirty or forty lectures: in many universities and
colleges less time than this will be available and it
will be necessary to omit some chapters. At various
times and places I have used different parts of the book
as an introduction to the subject: now I strongly favour
presenting the subject matter in the order I have used
here, and suggest that at least the first seven chapters
of the book should be covered if time is not available
for the whole. In this way the student is given a
physical insight into what is happening to the fluid and
how each term in the mathematical equations corresponds
to a particular feature in its behaviour. I have
included an account of the basic mathematical methods
commonly used in the subject (although I have assumed
familiarity with the elements of vector calculus) but
have been content, in places, to use physical intuition
or experimental evidence to justify some of the
approximations which have been made. In doing this I
have laid myself open to criticism by engineers that I
use too much mathematics and to criticism by mathemat-
icians that I use too little rigour. It is a matter of
personal judgment where to draw the line and I feel that,
as the subject stands at present, the approach I have
used is a valid one. I hope I have made it intelligible.

I have begun by discussing the molecular structure of
fluids and showing briefly (at least for a gas) how this
can be incorporated into continuum theories such as those
described in the rest of the book. Some readers may
prefer to skip this at first and begin with the discussion
of boundary conditions in Section 1.6. If they do,

however, I hope they will think it worth while to return
to this discussion at a later stage.

The subject is one in which the student can readily
relate, or try to relate, the theory to experiment,
because he so often observes moving fluids in everyday
life. After discussing the way in which the theory is
most easily related to visualization of flows, I have
discussed a number of one-dimensional flows in which
viscous forces are negligible; these flows can easily be
understood by the student and should help him to develop
a feel for the subject.

An understanding of the role of viscosity is central
to the study of fluid mechanics. To emphasize the fact
that viscous (internal friction) forces are always
present in any flow, I have treated a number of simple
flows in which viscosity is of prime importance. This is
followed by a simplified account of boundary-layer theory
and I discuss where and when viscous forces may be
neglected. It is only after doing this that I derive
Euler's equations and introduce the concepts of vorticity
and circulation. The special case of two-dimensional
irrotational flow is discussed and a number of examples
of the use of conformal mappings is considered.

Finally I have discussed the problem of waves on the
surface of a liquid. This involves the idea of a
dispersive medium and the concept of group velocity is
introduced.

I am grateful to many friends and colleagues who have
made helpful suggestions from time to time while I have
been writing the book. In particular, I should like to
thank Dr A.H. Craven and Mr G.B. Trustrum who read the
typescript; they made many constructive criticisms, most
of which I have used.

I also wish to thank the Director of the Imperial War
Museum for permission to reproduce Plate 3(b), and the
Controller of Her Majesty's Stationery Office for the use
of Crown Copyright material in Figure 9.13
(adapted from the original). Plate 2 is reproduced by
kind permission of Dr D.H. Peregrine. Most of the data
quoted have been taken from *Tables of Physical and
Chemical Constants* (13th edition) by Kaye and Laby.

Finally, I wish to thank Mrs Jill Foster for her care
and patience in preparing the typescript.

Ruth H. Rogers
University of Sussex

THE PHYSICAL BACKGROUND

1.1 THE MACROSCOPIC VIEW OF MATTER

It is the purpose of the author of this book, and indeed that of anyone who speaks of fluid mechanics, to discuss the motion of gases and liquids on a *macroscopic* scale. This means that the distances involved are much greater than the mean free path of the molecules of the fluid, and that the times involved are much greater than the average time between successive collisions of a particular molecule. On this scale the fluid appears to be a continuum, and an 'infinitesimally' small volume (say $10^{-11} cm^3$) is still sufficiently large to contain many millions of molecules. Similarly, an 'infinites- imally' small time is large enough (say $10^{-9} s$) for most molecules to have been involved in eight or nine collis- ions. This means that we can talk about an average velocity of the molecules: here the average is taken either over all the molecules in the 'infinitesimal' volume or over an 'infinitesimal' time for one molecule.* This average velocity of the molecules is referred to as the *velocity of the fluid* at a point in the 'infinites- imal' volume at an instant during the 'infinitesimal' time. Similarly, the mass of the molecules contained in the 'infinitesimal' volume is essentially unchanged during the 'infinitesimally' small time (in spite of the fact that molecules are entering and leaving the region all the time, the number entering is approximately equal to the number leaving during this small time interval) and the ratio of this mass to the 'infinitesimal' volume

---

* The two averages are usually assumed to be the same. This is known as the ergodic theorem in statistical mechanics, and has considerable philosophical implications.

itself is called the *density* of the fluid at a point in the 'infinitesimal' volume at an instant during the 'infinitesimal' time. We shall in future omit the inverted commas round the word 'infinitesimal', but the reader should always bear in mind their implied presence.
    Although we are not concerned with the details of the molecular motion, it is necessary for us to discover what effect they have on the large-scale motion of the fluid. In any medium there are forces of attraction between individual molecules, and this means that there will be similar forces between infinitesimally small regions of the continuous medium. In a solid the molecules are so close together that these internal forces are very large and prevent any large relative motion of the molecules (except when very large external forces are applied, and then the solid may break); in this case any two molecules which are close together at one instant will always be close together. At the other extreme, the molecules of a gas are so far apart that (except during collisions, and these take a negligibly small time) the attractions between them can be ignored; the molecules are therefore free to move about independently, and we shall see in the next section that this molecular motion produces the appearance of internal forces on a macroscopic scale. A liquid behaves in a way which is more complicated; the molecular attractions are not negligible, but they are not so great as to prevent large relative motion between individual molecules.

1.2 THE KINETIC THEORY OF GASES

First of all we shall consider the comparatively simple case of a gas in which the intermolecular forces are negligible. Suppose we consider a region V which is bounded by the surface ∂V. Inside V the molecules are rushing about (in air on the surface of the earth, their average speed is about 500 m/s), and some of them pass out of V through ∂V; with them they carry momentum out of V. Also, the molecules of the gas outside V are rushing about and some of them pass into V through ∂V; these carry momentum into V. We shall see that, when the mean velocity* of the molecules is zero (that is, when the gas is in equilibrium), then the two effects add up to a uni-directional flow of momentum across an element δS of the surface ∂V.

---

* The mean velocity as distinct from the mean speed.

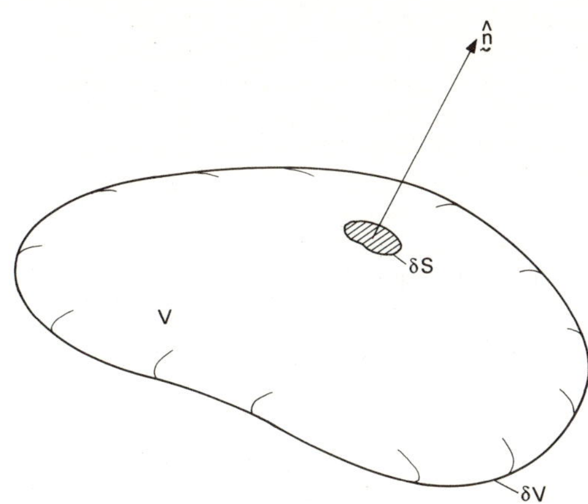

FIGURE 1.1

Consider first the molecules which are leaving V
through δS (Figure 1.1): these carry out of V momentum
all of which has a positive component in the direction $\hat{\underset{\sim}{n}}$,
the unit vector which is normal to δS and which points
out of V. But those which are entering V through δS
carry momentum into V all of which has a negative
component in the direction $\hat{\underset{\sim}{n}}$, and this is exactly
equivalent to a loss from V of momentum having a
positive component in the direction $\hat{\underset{\sim}{n}}$. Thus all the
transfer of momentum across δS by molecular motion is in
the same direction, and is indistinguishable (on the
macroscopic scale) from the effect of an internal force
acting normally inwards on the gas in V due to the
presence of the gas outside V; as long as the gas is in
equilibrium, it is clear that the apparent force is
normal, since the tangential components of momentum
carried across δS cancel out on averaging over
sufficiently many molecules. Evidently there is an equal
and opposite force apparently acting through δS on the
gas outside V due to the gas inside V, since the
transfer of molecular momentum into this region across
δS from V is equal and opposite to the transfer from
this region across δS into V. So this apparent force
has all the usual properties of a normal internal force,
and is indistinguishable from one. On the macroscopic
scale we shall therefore refer to it as an internal
force; its magnitude per unit area of the surface on
which it acts is called the *pressure*.

The situation when the mean velocity of the molecules is not zero is very similar. Suppose that a velocity $\underset{\sim}{u}$ is superposed on all the molecules in an infinitesimal volume so that the velocity of a typical molecule at some instant is $\underset{\sim}{u} + \underset{\sim}{v}$, and that the mean value of $\underset{\sim}{v}$ is zero. In this case the number of molecules leaving V through δS is different from the number entering V through δS: for example, if $\underset{\sim}{u}$ has a positive component in the direction of $\underset{\sim}{\hat{n}}$, more molecules leave than enter V through δS. The amount of momentum carried out of V per unit area per unit time by molecules leaving V is $\overline{\rho(u_n + v_n)(\underset{\sim}{u} + \underset{\sim}{v})}$, and this has normal component $\overline{\rho(u_n + v_n)^2}$ and tangential component $\overline{\rho(u_n + v_n)(\underset{\sim}{u}_t + \underset{\sim}{v}_t)}$, where the suffix n denotes the normal component and the suffix t the tangential component. When $\underset{\sim}{u}$ is uniform, the normal component becomes $\rho(u_n^2 + \overline{v_n^2})$ and the tangential component $\rho u_n \underset{\sim}{u}_t$ since v_n, $\underset{\sim}{v}_t$ and $v_n \underset{\sim}{v}_t$ are all zero in the mean. In this case, therefore, the only part of the momentum which is invisible on the macroscopic scale is given by the term $\overline{\rho v_n^2}$ in the normal component. The corresponding amount of momentum leaving V due to the molecules entering V is, as before, exactly the same and so there is a total momentum in the direction $\underset{\sim}{\hat{n}}$ of magnitude $\overline{\rho v_n^2}\, \delta S \delta t$ transported through δS out of V in time δt which is invisible on the macroscopic scale. This is indistinguishable from an internal force, as for a gas in equilibrium.

When $\underset{\sim}{u}$ is not constant, this is not quite true, since the value of $\underset{\sim}{u}$ (and of u_n) in the expression $\overline{\rho(u_n + v_n)(\underset{\sim}{u} + \underset{\sim}{v})}$ is not precisely that on the surface itself, but the value at a distance from the surface equal approximately to one mean free path. Further, for the molecules leaving V the appropriate value is that at a distance of the mean free path from outside V. It is not too difficult to show* (and it is shown in standard textbooks on the kinetic theory of gases) that, to the first order in the mean free path, this introduces extra terms which must be included in the internal force

---

* Other methods of obtaining this result are given in
  Chapter 5.

and that these terms are linear functions of the spatial
derivatives of the velocity $\underset{\sim}{u}$. In most gases, however,
it happens that the magnitude of this extra force (which
has both normal and tangential components) is very much
less than that of the normal component discussed in the
previous paragraph, unless the velocity gradients are
very large. It is convenient to think of the apparent
internal force across δS as being made up of these two
parts: a normal force due to the $\overline{\rho v_n^2}$ term (which we
still call the pressure), and a so-called *viscous* force
with components in all directions which is zero when the
velocity is uniform. It is clear that the value of
$\overline{\rho v_n^2}$ is independent of the direction of $\overset{\wedge}{\underset{\sim}{n}}$ as in the
equilibrium case; but the viscous force (which depends,
as we saw above, on the first spatial derivatives of $\underset{\sim}{u}$)
depends very strongly on the direction of $\overset{\wedge}{\underset{\sim}{n}}$ relative to
the motion of the fluid.

The mean energy of a molecule of mass m is made up of
three parts: its kinetic energy $\frac{1}{2}m\underset{\sim}{u}^2$ due to the mean
motion of the gas of which it is an element, its
potential energy due to its position in any external
force field which may be present, and its internal energy.
This internal energy is, in turn, of several kinds:
examples are its translational energy $\frac{1}{2}m\overline{\underset{\sim}{v}^2}$, its
rotational energy if it is not a spherically symmetric
molecule (such as in a monatomic gas), and its vibrat-
ional energy due to internal oscillations within the
molecule. At very high temperatures ionization or
dissociation may occur and this will cause changes in the
internal energy, but in this book we shall not be
concerned with situations when this occurs. In fact we
shall restrict our attention to those problems where the
translational and the rotational energy are the only
types of internal energy which vary: this restriction
merely means that we shall not consider flow in which the
temperature varies by several hundred degrees.

The *temperature* T of a gas is a measure of its
translational internal energy with a constant of
proportionality k given by $\frac{3}{2}kT = \frac{1}{2}m\,\overline{\underset{\sim}{v}^2}$. This
constant is known as Boltzmann's constant and its
numerical value is $k = 1.38 \times 10^{-26}$ kJ/$^\circ$K. It is one of
the principles of statistical mechanics that the internal
energy in a gas of the simple kind under discussion is
partitioned equally between the degrees of freedom of a
molecule: it follows that if n_f is the number of

degrees of freedom of a molecule, its internal energy is $\frac{1}{2}n_f kT$. The internal energy of the gas per unit mass is therefore $E = n_f kT/2m$. So, for a monatomic molecule, $E = 3kt/2m$ and for a diatomic molecule (which has two rotational degrees of freedom as well as three translational ones) $E = E_0 + 5kT/2m$.* It is evident that, for a gas of this kind, E is an increasing linear function of the temperature T. Further, if the total energy of a molecule remains constant and we ignore viscous forces, $E + \frac{1}{2}u^2$ remains constant as long as there is no change in potential energy; hence if the magnitude of the mean velocity of the gas increases, then its internal energy and therefore its temperature decreases - and conversely.

It is easy to see that for a gas which behaves in this way

$$p = \rho\,\overline{v_n^2} = \tfrac{1}{3}\rho\,\overline{\underset{\sim}{v}^2} = R\rho T,$$

where $R = k/m$ is a constant for a given gas.† Typical values of R for some gases are shown in Table 1.1. The equation of state

$$p = R\rho T \tag{1}$$

is used throughout this book for a gas. A gas satisfying this equation and having in addition its internal energy a linear function of the temperature alone (at least over a wide range of temperature) is called a *perfect gas*. These two properties serve as a definition on the macroscopic scale although, from the point of view of kinetic theory, they are (as we have seen) consequences of assumptions about the molecules of a gas. We note that equation (1) is true because of the way in which we have defined p: if we had chosen p to be the normal component of the internal force (including that component of the

---

* For very low temperatures diatomic molecules have no rotational degrees of freedom, and so $E = 3kT/2m$ when T is small: at temperatures high enough for rotational effects to be important, the constant E_0 is chosen so that E is a continuous function of T. Since we shall be concerned only with this range of values of the temperature, we quote this form. The numerical value of E_0 is immaterial since we shall be concerned only with changes in E.

† Sometimes R is taken to be the universal constant $16k/m_0$ where m_0 is the mass of an atom of oxygen, and then $p = R(m_0/16m)\rho T$, but this is not usual in fluid mechanics.

TABLE 1.1   Values of the gas constant R, for various gases

Gas	$R(kJ/kg°K)$
Hydrogen	4.130
Helium	2.080
Nitrogen	0.297
Dry air	0.287
Nitric oxide	0.277
Oxygen	0.260
Argon	0.208
Carbon dioxide	0.188
Nitrous oxide	0.188

viscous force), the equation would not hold. A perfect gas in which the viscous force is negligible (and we shall see later that this often happens) is called an *ideal gas*.

1.3 THE INTERNAL FORCES IN A LIQUID

It is as yet impossible to produce a similarly detailed account of the molecular structure of a liquid. The effects described for a gas are still present in a liquid; but of equal importance are the intermolecular attractions. We are unable, therefore, to use the equation of state (1) for a liquid. Fortunately, however, the molecules of a liquid are so closely packed together that its density is not easily changed by the application of a force, and this enables us to construct easily an equation of state for a liquid. (An increase of pressure due to the atmosphere increases the density of water by less than 0.005 per cent, and that of other liquids by the same order of magnitude.) A change in temperature does, however, change the density and so we have for a liquid in general

$$\rho = \rho(T). \tag{2}$$

This equation of state is usually linear, but a notable exception is water, whose variation of density with temperature is shown in Figure 1.2.

   Viscous effects in many liquids are of the same form as in a gas: that is, they produce internal forces which are linear functions of the spatial derivatives of the velocity of the fluid. This enables us to use the same

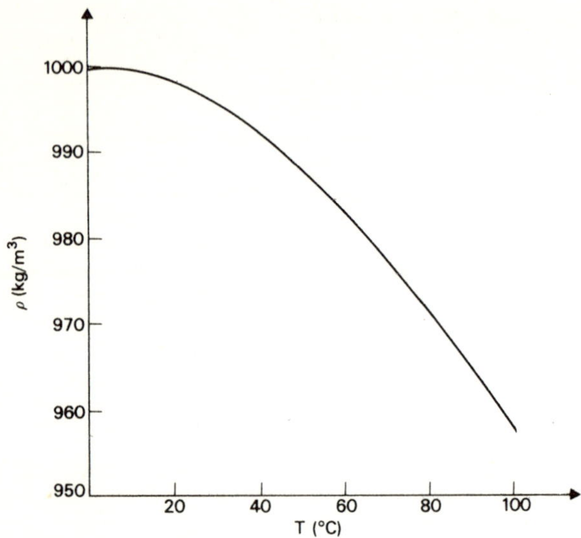

FIGURE 1.2

theory for both liquids and gases in many problems. (We shall not be concerned in this book with 'non-Newtonian' liquids: these liquids contain exceptionally long molecules and viscous effects are more complicated.

1.4 INCOMPRESSIBILITY

An *incompressible fluid* is one whose density may be treated as constant. For a liquid whose temperature does not vary, this condition holds to a very good approximation as we saw in the previous section. On the other hand it is quite obvious that the density of a gas does not in general remain constant, and so it is apparently not incompressible. However, we shall show that, for mean velocities whose magnitudes are sufficiently small and for certain types of motion, it is sometimes justifiable to neglect the changes in density of an element of the gas as it moves about. In such cases we can treat the gas as if it were incompressible, and we refer to the motion of an *incompressible gas*. Conventionally, it has been common to refer to the motion of incompressible fluids as 'hydrodynamics' (as if all such fluids were liquid, like water) and to that of compressible fluids as 'aerodynamics' (as if all the motion of a gas must be treated as compressible). There is, of course, a difference between the behaviour of a liquid and an incompressible gas, because a liquid can have a free surface whereas a gas expands to fill its container, and so the boundary conditions are different. But the equations governing the motion are the same.

To justify the statement made above we consider first the temperature T of the gas. We have seen in Section 1.3 that a change in $\underset{\sim}{u}$ corresponds to a change in T. Suppose we have an element of gas whose speed* varies between zero and 100 m/s. This corresponds to a change in temperature of the order 7°K (if the change is at constant volume) which is a percentage change of the order of $2\frac{1}{2}$ per cent for temperatures of about 280°K. When work is also done in changing the volume of the fluid (as for a change at constant pressure), the change in temperature is even less. Here we have neglected the possible change in internal energy due to any external forces which act on the fluid. The most common of such forces is that due to gravity; a change of height of 10 m corresponds to a change in kinetic energy per unit mass of 100 m^2/s^2 and this, in turn, is equivalent to a change in temperature of 0.1°K. This effect may there- fore be neglected unless very large-scale vertical motions take place (as in some atmospheric problems, for example).

A similar argument applies to the pressure p, for we shall show that the maximum change in pressure when u changes from zero to $\overline{100}$ m/s is $\frac{1}{2}\rho u^2$, and this is to be compared with ρv^2, the actual pressure. The percentage change is again of the order of 4 per cent: As in the case of temperature, the change in pressure due to change in vertical height is negligible compared with $\overline{\rho v^2}$.

We now use the equation of state (1) which holds for any perfect gas. Throughout its motion, an element of gas does not change either its pressure or its temper- ature by more than about 4 per cent when its speed changes from zero to about 100 m/s. But this means that the density cannot change by more than about 8 per cent: so, to this order of approximation, the density changes can be neglected.

Any disturbances to the gas are propagated by the molecules as they rush about and have collisions with their neighbours. The speed at which these disturbances are propagated is called the *speed of sound* (since sound is nothing more than a small, rapidly varying disturbance

---

* By the speed of an element of fluid we mean the magni- tude of the mean velocity of its molecules: that is, $|\underset{\sim}{u}|$

in a fluid); we shall denote this speed by the letter a.
Since the propagation occurs by means of the molecular
motion of the molecules, it is clear that speed a must
be of the same order of magnitude as v. So the above
argument holds only when the speed u of the gas is very
much less than the speed of sound in the gas.*

There is another restriction to the idea of an
incompressible gas. When we are concerned with the
motion inside a container (for example, in a pipe), there
will be so many molecular collisions with the walls that
the above arguments will no longer hold. In fact they
refer only to what is sometimes called 'external aerodynamics' - that is, to the flow of a gas past an
obstacle. In this case a given element of gas may have
its energy changed slightly by collisions with the
obstacle, but it is quickly swept past and so does not go
on losing energy in this way.

A further restriction is that there should be no large
applied temperature differences at the boundaries.†
Evidently any large temperature change in a fluid due to
contact with a boundary whose temperature is significantly
different from that of the fluid itself will cause a
change in density: and this is true whether the fluid is
a gas which satisfies the equation of state (1) or a
liquid which satisfies the equation of state (2). We
shall not consider situations of this kind in this book.

1.5 ENTROPY AND ADIABATIC MOTION

The kinetic energy per unit mass of the random part of
the molecular motion of a gas is, as we have seen, the
internal energy E of the gas. For a perfect gas, E
is a function of the temperature T only. If a small
quantity $M\delta Q$ of heat is given to a small volume Mv of
the fluid whose mass is M (and where $v = 1/\rho$ is the
specific volume), this energy can be used both to change
the volume and to increase the internal energy. If the
fluid is in equilibrium, the work needed to change the

---

* The errors involved in making this approximation will
   be discussed in Chapter 2.
† It is worth noting here that when this condition holds
   for a liquid, the same arguments apply (qualitatively
   at least) and so we can then regard the temperature,
   and therefore the density, of the liquid as constant
   throughout the liquid.

volume from Mv to M$(v + \delta v)$ is $\iint p \; x \; dS$ where x

is the displacement of the element δS of the volume Mv (see Figure 1.3). Since the volume is small, we can

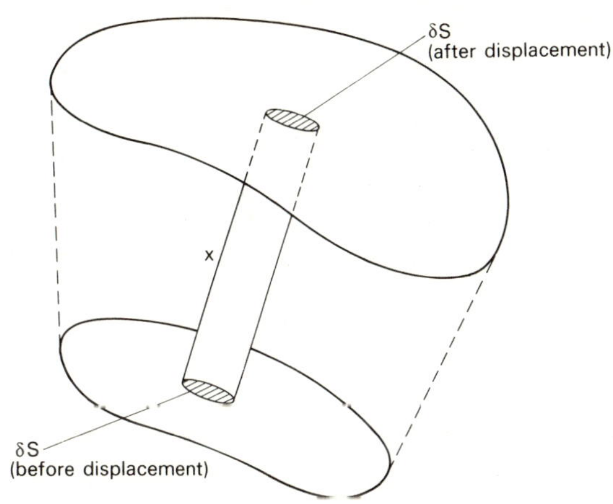

FIGURE 1.3

regard p as uniform throughout, at least to first order. Then the double integral is $p\iint x \; dS = p \; M \; \delta v$. Hence

$$\delta Q = p\delta v + \delta E = R\delta T - v\delta p + \delta E,$$

or, in terms of the density ρ ,

$$\delta Q = p\delta(1/\rho) + \delta E = R\delta T - \frac{1}{\rho}\delta p + \delta E \qquad (3)$$

to the first order, using the equation of state (1). The specific heat of the gas is the ratio of the quantities δQ and δT in equation (3); but this ratio depends on the conditions under which the heat is applied. If the volume is not allowed to change, it is called the specific heat at constant volume:

$$c_v = \lim_{\delta T \to 0} \left(\frac{\delta Q}{\delta T}\right)_{\delta v=0} = \frac{dE}{dT}, \qquad (4a)$$

where an ordinary (instead of a partial) derivative is used since $E = E(T)$. If the pressure is maintained constant during the process, then the ratio is the specific heat at constant pressure:

$$c_p = \lim_{\delta T \to 0} \left(\frac{\delta Q}{\delta T}\right)_{\delta p = 0} = R + \frac{dE}{dT} \, . \tag{4b}$$

This gives immediately the important result that

$$c_p - c_v = R \, . \tag{5}$$

If we write $\gamma = c_p/c_v$ (usually γ is simply referred to as the ratio of the specific heats of a gas, reference to which specific heats being omitted), we can find expressions for c_p and c_v in terms of γ and R. In fact

$$\gamma = \frac{c_p}{c_v} \; ; \quad c_v = \frac{R}{\gamma - 1} \, , \quad c_p = \frac{\gamma R}{\gamma - 1} \, . \tag{6}$$

For a perfect gas, the quantities c_v and c_p (and therefore γ) are constant for a wide range of values of T. In fact their values depend on the number of degrees of freedom of a molecule of the gas: we have seen in Section 1.2 that $E = n_f kT/2m = \frac{1}{2} n_f RT$. It follows that

$c_v = \frac{1}{2} n_f R$, $c_p = \frac{1}{2}(n_f + 2)R$ and $\gamma = (n_f + 2)/n_f$.

Hence, as long as the temperature change is such that n_f does not change, none of the quantities varies with the temperature. Table 1.2 shows how the value of γ for air (a real gas, not a perfect one) varies with temperature when the pressure is equal to that of one atmosphere: it is clear from the Table that at temperatures between about 300°K (27°C) and 750°K (477°C), a change in temperature of 100°K causes a variation in γ of only about 0.6 per cent. It is therefore permissible to treat γ as a constant at these temperatures. Typical values of γ for other gases are also shown in Table 1.2.

It is useful, at this stage, to introduce the idea of a *reversible process*. This, as its name implies, is a process which could be reversed. An example of such a process would be the bouncing of a perfectly elastic ball dropped from a height h above the ground; the ball falls to the ground and then rises again to a height h and is momentarily at rest there again. We can regard a reversible process as one which is a succession of states such that the variables in successive states differ from

TABLE 1.2 Values of the ratio of specific heats γ, for gases at various temperatures, at a pressure of 1 atmosphere.

Gas	T(°C)	γ
Argon	0	1.667
Helium	0	1.63
Carbon monoxide	− 85	1.40
	1,800	1.30
Hydrogen	4 −17	1.41
Nitrogen	20	1.401
Oxygen	5 −14	1.400
Dry air	0	1.400
	100	1.397
	200	1.390
	500	1.355
	1,000	1.319
	2,000	1.298
Carbon dioxide	4 −11	1.300
	300	1.22
	500	1.20
Water vapour	100	1.334

each other by an infinitesimal amount. In practice, no process is truly reversible (the ball is not perfectly elastic and will never quite reach the height h again), but many processes are approximately so. For example, it often happens that frictional or other dissipative forces have a negligible effect over the periods of time we are interested in, and the process may be treated as reversible to a good approximation. A process which is not reversible is said to be *irreversible*.

Consider what happens to a perfect gas which starts off with density ρ_1 and temperature T_1 (and pressure $p_1 = R\rho_1 T_1$). After a time, during which some heat may be given to the gas, it has density ρ_2 and temperature T_2 (and pressure $p_2 = R\rho_2 T_2$). It is found experimentally that the amount of heat required to implement this change depends on how the change is made. For example, if the temperature is first held constant at T_1 while the density is changed from ρ_1 to ρ_2, and then the density is held constant at ρ_2 while the temperature is changed from T_1 to T_2, the amount of heat used up is $c_v(T_2 - T_1) + R\,T_1\log_e(\rho_1/\rho_2)$. On the other hand, if the temperature is first changed at constant density ρ_1 and then the density is changed at constant temperature T_2, the heat used up is $c_v(T_2 - T_1) + R\,T_2\log_e(\rho_1/\rho_2)$;

and, of course, if the change is carried out in a different way, yet a different amount of heat is used. If, however, we divide equation (3) by T and write the resulting small quantity as δS, we obtain

$$\delta S = \frac{\delta Q}{T} = \frac{p}{T}\,\delta\left(\frac{1}{\rho}\right) + \frac{1}{T}\,\delta E = R\rho\,\delta\left(\frac{1}{\rho}\right) + \frac{c_v}{T}\,\delta T,$$

since $p = R\rho T$ and $dE/dT = c_v$ for a perfect gas. Hence

$$\delta S = -\frac{R}{\rho}\,\delta\rho + \frac{c_v}{T}\,\delta T$$

$$= \delta(-R\log_e\rho + c_v\log_e T).$$

It is clear from this expression that the change that occurs in S when the state of a perfect gas is changed in the way described in the previous paragraph, is

$$S_2 - S_1 = c_v\log_e(T_2/T_1) + R\log_e(\rho_1/\rho_2)$$

and that the value of this expression is the same however the change of state occurs. Using equation (6) we can write this expression as

$$S_2 - S_1 = c_v\left[\log_e(T_2/T_1) + (\gamma - 1)\log_e(\rho_1/\rho_2)\right]$$

$$= c_v\,\log_e\left[\frac{T_2}{T_1}\left(\frac{\rho_1}{\rho_2}\right)^{\gamma-1}\right]$$

and in terms of pressure and density, this is

$$S_2 - S_1 = c_v\,\log_e\left[\frac{p_2}{p_1}\left(\frac{\rho_1}{\rho_2}\right)^{\gamma}\right].$$

We define *entropy* as the quantity

$$S = S_0 + c_v\log_e(p\,\rho^{-\gamma}), \tag{7}$$

where S_0 is a suitable reference value.

CONTENTS

PLATES

Between pp.144–5
1 Laminar and turbulent flow in steam emerging from
   a kettle
2 The Severn bore
3 Bow waves (a) behind swimming ducks and (b) behind
   minesweepers

It sometimes happens that there is no heat given to or
taken from a system during a process: the process is then
said to be *adiabatic*. In such a situation, the increase
(or decrease) in internal energy is exactly balanced by
the work done by (or against) the pressure when the
density changes. Another situation which is of some
importance is that of a process during which the entropy
S remains constant: such a process is said to be
isentropic and it is clear from equation (7) that, for
an isentropic process

$$p = K \rho^{\gamma}, \qquad (8)$$

where K is a constant. When an adiabatic process is
reversible it is clear from the above discussion that it
is also isentropic, and that a reversible isentropic
process is adiabatic. The two words are often used
interchangeably, but this is legitimate only for
reversible processes. In Chapter 3 we shall discuss the
motion of a gas through a shock wave: there is no heat
exchange during this process, so the motion is adiabatic,
but we shall find that there is a change of entropy.
When the entropy does change in an irreversible
process, it is found that it always increases as long as
the system is isolated from its surroundings. For a
reversible process in an isolated system, the entropy
remains constant. This is expressed in the Second Law
of Thermodynamics which says that the entropy of an
isolated system can never decrease.
So far in this section we have supposed that equation
(3) always holds for a reversible process. When there is
an average motion of the fluid, we might expect that some
of the heat put into the system may be used to change,
say, the average kinetic energy, the potential energy,
the chemical energy (if chemical reactions take place) or
the magnetohydrodynamic energy (if a magnetic field is
present and the fluid is electrically conducting). It
will do some work as the fluid moves against the pressure
forces and also against any viscous forces which may be
present. We need, therefore, to replace equation (3) by
one of the form

$$\delta Q = p\delta(1/\rho) + \delta E + \delta(\text{kinetic energy}) +$$
$$+ \delta(\text{potential energy}) + \delta W_{v},$$

where we have supposed that there are no changes in the
chemical or magnetohydrodynamic energy, and where δW_{v} is

the work done against the viscous forces. In practice
the changes in the kinetic energy and the potential
energy often exactly balance out; also δW_V is often

negligible and is identically zero in a perfect gas.
When this happens equation (3) does, after all, hold.

1.6 BOUNDARY CONDITIONS

In practice there are two main types of surface which can
bound a fluid: a rigid boundary and a free surface. A
rigid boundary is one whose position is determined by
some mechanism other than the motion of the fluid itself,
the most common type being a solid surface. The body
itself may move (as, for example, when a steel ball falls
through a liquid), and this motion may be affected by the
motion of the fluid but, relative to a frame of reference
fixed in the body, the shape of the body is unchanged.
A free surface is one between two immiscible fluids:*
its shape depends strongly on the motion of the fluid.
Of course, it is possible to have a boundary which is
somewhere between these two (for example, a thin rubber
sheet), but we shall not consider this here.
   Before discussing what happens to a fluid when it is
near a boundary, we note the special (but commonly
occurring) case of a fluid which is in a vessel very
large compared with the region we are considering. Here
we introduce the usual mathematical friction of an
infinite fluid. We refer to conditions 'at infinity', by
which we mean the limit as r (the distance from the
origin) approaches infinity. This approximation is
justifiable as along as the vessel is large enough for
the actual conditions within it, but far from the region
we are concerned with (which is assumed to include the
origin), to be indistinguishable from the conditions 'at
infinity' in the model. (We shall not usually use the
inverted commas in the phrase 'at infinity', but it
should be remembered that they are always implied.) It
is then unnecessary to bother with the true boundaries
of the vessel as long as we know what values the
variables (such as pressure, temperature, density,
velocity) have a long way from the origin.

---

* It may well happen that the density of one of these
  fluids is negligibly small compared with that of the
  other; in this case, the less-dense fluid can be
  treated as if it were a vacuum.

Now we consider what happens at a rigid boundary.
Except in gases whose densities are very low indeed, it
seems that molecules which reach the surface adhere to it
for a time which depends on, amongst other things, the
temperature. Unless the temperature is very high, this
time is approximately the same as that for two or three
collisions, and there is time for the molecules to adjust
to the equilibrium conditions of the surface itself.
When the molecules leave the surface, therefore, they will
do so in random directions. It follows that, relative to
the surface, the mean velocity of the fluid is zero, and
this is the correct boundary condition to take when all
the internal forces are included in full. When, however,
we try to neglect tangential forces because (as we saw
before) they are usually very small compared with the
normal ones, we get into difficulties. This is hardly
surprising since the tangential forces are proportional
to the derivative of the velocity, and this may well be
quite large near the boundary.

The remarkable thing is that it is still worth while
considering motions in which tangential forces are
neglected - but if we do this, the boundary conditions
have to be modified. It is still necessary to have the
normal component of the relative velocity zero, because
if it were not so there would be either a vacuum formed
(if there were a component of relative velocity away from
the body) or a penetration by the fluid into the surface
(which would involve a change of shape of the surface).
This is the only condition we can apply, and this means
that we allow the fluid to slip past the surface: the
tangential component of the relative velocity is called
the *slip velocity*. From a molecular point of view, what
is happening is that the molecules are regarded as
bouncing off the boundary as in a perfectly elastic
collision. Thus the boundary has no direct effect on the
tangential component of momentum of the fluid relative to
the boundary, which is the same as saying that we are
neglecting tangential forces. There is now less likeli-
hood of large velocity derivatives occurring than if the
velocity were reduced completely to zero. We shall see
in Chapter 6 under what circumstances it is reasonable to
neglect tangential forces and how we can then satisfy the
true boundary condition instead of the modified one we
have discussed here. When tangential forces may be
neglected, the fluid is said to be *inviscid*.

For a free surface the situation is more complicated,
since we cannot know the shape of the surface in advance.
When molecules of immiscible fluids collide they will
interchange momentum, although the fluids do not mix.

It follows that, as for a rigid boundary, the true
condition for a free surface between immiscible fluids is
that the relative velocity between them is zero. For the
free surface between a fluid and a vacuum, of course,
there is no such condition on the velocity. We have,
however, extra boundary conditions. One is that the
internal forces are, when surface tension effects are
neglected, the same on both sides of the boundary: in
the case of fluid in contact with a vacuum, the internal
forces in the fluid are zero at the surface. (The effect
of surface tension on a liquid is illustrated in
Chapter 10.) The other is that the normal component of
velocity of the fluid at the free surface is equal to the
speed at which the surface itself is moving; for if the
motion of the fluid in the direction from fluid I to
fluid II is greater than that of the surface itself, then
some of fluid I will be carried into fluid II which (by
definition) means that it carries the surface between
them with it.

If tangential forces are neglected, we have to make
similar modifications to the boundary conditions. It is
now possible for the fluids to slip past each other, but
their normal components of velocity are still equal to
each other and to the speed at which the surface itself
moves. Further, the pressure (that is, the normal
component of the internal stress) is continuous across
the boundary when surface-tension effects are neglected.

Although they are not, strictly speaking, boundaries
it is convenient to mention here the possibility of
singularities in the flow. Again this is a mathematical
idealization of certain physical phenomena which turns
out to be useful. There are two main types of singularity
which have to be considered - sources and vortices. A
source (or sink) is a source of fluid mass: although we
know that mass cannot be created (or destroyed) in the
middle of the fluid there are some occasions when it is
useful to suppose that it is. For example, when a narrow
tube emits fluid into (or sucks fluid out of) a large
bulk of fluid, the effect is the same (except very near
the source or sink of fluid) to all intents and purposes
as that of a point source (or sink) of fluid at the mouth
of the tube. The condition near such a point, of course,
is that the mass of fluid emerging from it per unit time
is the given value.

The other basic type of singularity, the vortex, is
concerned with circulatory motion. We shall discuss this
in more detail in Chapter 7. Other types of singularity
can occur, but these are usually combinations of sources
or vortices.

1.7 THE BASIC PRINCIPLES GOVERNING FLUID MOTION

We shall, throughout the rest of this book, concern ourselves with motion on a macroscopic scale. To do this we
have to use the ideas discussed above, and this chapter
has been devoted to a brief account of the basic explanation of these ideas. We shall derive equations
involving the velocity, density, pressure and temperature
of the fluid and solve them in certain cases using
boundary conditions of the appropriate kind as discussed
in the previous section.

To obtain the equations governing the motion, we use
the same principles as for ordinary Newtonian mechanics.
(We shall exclude from our discussion relativistic
effects: all the speeds with which we shall be concerned
are very much less than that of light.) We will finish
this chapter with a brief discussion of these principles.

First of all, we use the principle of conservation of
mass. This means that if we have some region V which
contains fluid, then the rate of increase of mass within
V is equal to the rate at which fluid flows in through
∂V, the boundary of V. There are certain exceptions to
this law. One is the point source which we discussed in
Section 1.6 - the mathematical idealization corresponding
to the emission of fluid into V from outside at (or in
the neighbourhood of) an isolated point. Another occurs
in the atmosphere when water vapour, originally part of
the gaseous fluid contained in the atmosphere, condenses
out and falls out of the atmosphere as liquid rain. It
turns out that it is possible to treat the gas as a fluid
whose mass is being partially destroyed.

Next there is the principle of conservation of energy.
Here we have the usual principle that the rate of
increase of energy in the volume V is equal to the rate
of inflow of energy carried by the fluid through ∂V,
together with the work done on the fluid by any external
or internal forces which may be present. The difficulty
here is always that of being sure that all types of
energy have been included: thus we have to include
kinetic energy, potential energy, internal (or thermal)
energy and also, when they are relevant, chemical energy
and electrical energy.

Finally, we consider the momentum of the fluid.
Newton's second law of motion still applies, but again we
have to be a little careful how we apply it. The reason
for this is that the momentum of the fluid inside V can
change for two reasons - the first is that any forces
(either internal or external) acting on the fluid tend to
change its momentum, and the second is that fluid which

flows into (or out of) V carries momentum with it.

When we neglect the tangential component of the internal forces, we would expect (from the preceding physical argument) the magnitude of the normal stress on a surface point to be independent of the direction of the surface through the point. This is easily verified by considering the momentum of the fluid in a tetrahedron ABCD (Figure 1.4) where AB, AC, AD are parallel to

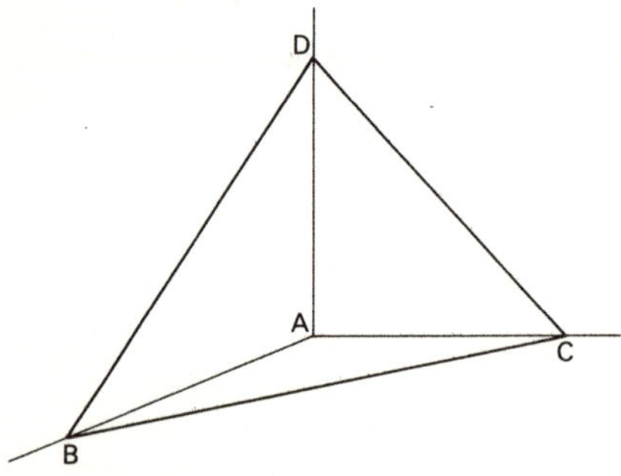

FIGURE 1.4

fixed Cartesian axes Ox, Oy, Oz and the face BCD has its normal in a fixed direction whose direction cosines are $(\ell,m,n) = \hat{\underset{\sim}{n}}$. Then, if Δ_1, Δ_2, Δ_3, Δ are the areas of the faces ACD, ABD, ABC, BCD respectively, we have $\Delta_1 = \ell\Delta$, $\Delta_2 = m\Delta$, $\Delta_3 = n\Delta$. Also, if h is the length of the perpendicular from A to the face BCD, the volume of the tetrahedron is $\frac{1}{6}h\Delta$. (Note that we shall shortly let $h \to 0$, keeping ℓ, m and n unchanged.) Then, if ρ is the density of the fluid at A and $\underset{\sim}{u} = (u_1, u_2, u_3)$ its velocity there, the x-component of the momentum of the fluid is $\frac{1}{6}h\Delta\,\rho\,u_1$ to the first order of the small quantity h and its rate of increase is $\frac{1}{6}h\Delta\,\dfrac{\partial(\rho u_1)}{\partial t}$. The cause of this increase is threefold:

(i) the body force (X,Y,Z) per unit mass has x-component $\frac{1}{6}h\,\Delta\rho X$;

(ii) the pressure forces on the faces ACD and BCD of the tetrahedron (inside the tetrahedron the internal forces are equal and opposite and so cancel out, and the forces on the other two faces have no component parallel to the x-axis) have x-components $p_1\Delta_1$ and $-p\ell\Delta$;

(iii) there is an inflow of x-momentum carried by the fluid across the surface of the tetrahedron, and this is calculated as follows. We consider the fluid which crosses the element of surface δS between the times t and $t + \tau$ (Figure 1.5).

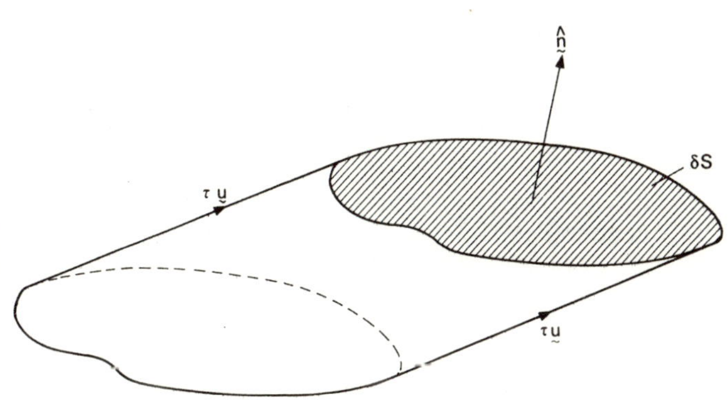

FIGURE 1.5

This is precisely that contained at time t in the cylinder whose base is δS and whose generators are parallel to $\underset{\sim}{u}$. If a quantity f (measured per unit mass) is carried by the fluid across δS, the amount carried between times t and $t + \tau$ is ρf times the volume $\tau\underset{\sim}{u}.\hat{\underset{\sim}{n}}\delta S$ of the fluid in the cylinder, where $\hat{\underset{\sim}{n}}$ is the unit vector normal to δS. Hence the rate of transfer of f across the element of surface is $\rho f\underset{\sim}{u}.\hat{\underset{\sim}{n}}\delta S$.

So the rate of inflow of x-momentum across the surface of the tetrahedron is

$$\iint_{\partial V} \rho u_1 (\underset{\sim}{u}.\hat{\underset{\sim}{n}})dS = \rho u_1^2 \Delta_1 + \rho u_1 u_2 \Delta_2 + \rho u_1 u_3 \Delta_3 -$$

$$- \rho u_1 \underset{\sim}{u}.\hat{\underset{\sim}{n}}\Delta + \text{terms of order } h,$$

and this is of order h, since

$$\underset{\sim}{u}.\hat{\underset{\sim}{n}}\Delta = (u_1 \ell + u_2 m + u_3 n)\Delta = u_1 \Delta_1 + u_2 \Delta_2 + u_3 \Delta_3.$$

Hence we have:

$$\frac{1}{6} h \Delta \, \frac{\partial (\rho u_1)}{\partial t} = \frac{1}{6} h \Delta \rho X + (p_1 - p) \ell \Delta + \text{terms of order } h.$$

It follows that $p_1 - p$ is of order h, and as we can choose h as small as we please, we have $p_1 = p$. Similarly $p_2 = p$ and $p_3 = p$, and this is true whatever the direction $\hat{\underset{\sim}{n}}$ of the normal to the surface BCD.

EXERCISES

1 Only two among the five variables of state (pressure p, density ρ, temperature T, entropy S and internal energy E) are independent, and they are related by the equation

$$T \, dS = dE + p \, d\left(\frac{1}{\rho}\right).$$

Show that

$$\rho^2 \left(\frac{\partial T}{\partial \rho}\right)_S = \left(\frac{\partial p}{\partial S}\right)_\rho, \quad \rho^2 \left(\frac{\partial S}{\partial \rho}\right)_T = - \left(\frac{\partial p}{\partial T}\right)_\rho,$$

$$\rho^2 \left(\frac{\partial T}{\partial p}\right)_S = - \left(\frac{\partial p}{\partial S}\right)_p, \quad \rho^2 \left(\frac{\partial S}{\partial p}\right)_T = \left(\frac{\partial \rho}{\partial T}\right)_p.$$

Show also that E is a function of T alone if and only if p/T is a function of ρ alone.

2 The specific heats at constant pressure and at constant volume are respectively

$$c_p = T\left(\frac{\partial S}{\partial T}\right)_p, \quad c_v = T\left(\frac{\partial S}{\partial T}\right)_\rho$$

and their ratio is $\gamma = c_p/c_v$. Using the results of the previous question, show that

$$\left(\frac{\partial p}{\partial \rho}\right)_S = \left(\frac{\partial p}{\partial \rho}\right)_T.$$

The Dieterici equation of state for a gas is

$$p = \frac{R \rho T}{1 - \beta \rho} \exp\left(- \frac{\alpha \rho}{RT}\right)$$

where α, β, R are all constant. Find an expression for the speed of sound, a, in the gas where

$$a^2 = (\partial p/\partial \rho)_S.$$

3 A heat-insulated rigid container contains a perfect gas whose mass is M and a simple pendulum of length b whose bob has mass m. Initially the bob is held so that the pendulum is at an angle α with the vertical, the temperature of the system is T_1 and the system is at rest. At time t = 0, the bob is released and the pendulum begins to oscillate. The system finally comes to rest with the bob in its equilibrium position and the temperature of the system at T_2. Show that

$$T_2 = T_1 + \frac{mgb(1 - \cos\alpha)}{MJc_v},$$

where J is the mechanical equivalent of heat. Find also the increase in entropy in the system.

4 A heat-insulated rigid container contains a body moving along an arbitrary trajectory with a time-dependent speed U in a perfect gas. There is a drag force D on the body which is a function of U. Initially the body is at rest and the gas is in equilibrium at temperature T_1.

Motion begins at time t = 0 and continues until time $t = t_0$ when the body is again at rest and the gas is in equilibrium at temperature T_2. Show that the increase in entropy is

$$S_2 - S_1 = \int_0^{t_0} \frac{UD}{T} \, dt,$$

where T is the temperature of the gas at time t.

5 A heat-insulated, rigid, vertical cylinder is closed at the top by a piston and contains a perfect gas which is in equilibrium at pressure p_1 and temperature T_1.

A weight is placed on the piston, which begins to perform oscillations. Ultimately a new position of equilibrium is reached and the pressure of the gas is p_2 and its temperature is T_2. Show that

$$T_2 = T_1\big[p_1 + (\gamma - 1)p_2\big]/\gamma p_1.$$

Chapter 2

DESCRIPTION AND VISUALIZATION OF FLOW

2.1 DESCRIPTION OF THE FLOW OF A FLUID

We begin this chapter with a brief account of the way we
describe and visualize the macroscopic flow of a fluid.
Throughout we shall use an Eulerian coordinate system*
in which we take as independent variables the position
vector $\underset{\sim}{r}$ and the time t. As dependent variables we
take the velocity $\underset{\sim}{u}$ and the variables of state p, ρ
and T. Then if an element of fluid which is at P,
whose position vector is $\underset{\sim}{r}$, at time t has velocity
$\underset{\sim}{u}(\underset{\sim}{r}, t)$ and density $\rho(\underset{\sim}{r}, t)$, its position at a small
time τ later will be $\underset{\sim}{r} + \tau\underset{\sim}{u}$ (to the first order in τ);
hence at time $t + \tau$ its velocity is $\underset{\sim}{u}(\underset{\sim}{r} + \tau\underset{\sim}{u}, t + \tau)$
and its density is $\rho(\underset{\sim}{r} + \tau\underset{\sim}{u}, t + \tau)$. Hence the change
in its velocity is

$$\underset{\sim}{u}(\underset{\sim}{r} + \tau\underset{\sim}{u}, t + \tau) - \underset{\sim}{u}(\underset{\sim}{r},t) = (\tau\underset{\sim}{u}.\nabla)\underset{\sim}{u} + \tau\frac{\partial\underset{\sim}{u}}{\partial t}$$

to the first order in τ (using Taylor's theorem), and
the change in its density is

$$\tau\underset{\sim}{u}.\nabla\rho + \tau\frac{\partial p}{\partial t}$$

to the same order. It follows that its acceleration,

---

* The alternative is a Lagrangian system in which $\underset{\sim}{r}$ is
  the position coordinate of a given element of the fluid
  at time t. In the Eulerian system we can think of
  ourselves as standing on the bank of a river and
  watching it flow past; in the Lagrangian system we are
  in a boat drifting with the river.

24

which is the limit as $\tau \to 0$ of the ratio of the change in its velocity to τ, is $(\underset{\sim}{u}.\nabla)\underset{\sim}{u} + \frac{\partial \underset{\sim}{u}}{\partial t}$; and similarly the rate of change of its density is $(\underset{\sim}{u}.\nabla)\rho + \frac{\partial \rho}{\partial t}$. The operator* $\underset{\sim}{u}.\nabla + \partial/\partial t$ is one which occurs often in fluid mechanics and we shall denote it by D/Dt. Thus, by definition,

$$\frac{D\underset{\sim}{u}}{Dt} = (\underset{\sim}{u}.\nabla)\underset{\sim}{u} + \frac{\partial \underset{\sim}{u}}{\partial t} \, , \quad \frac{D\rho}{Dt} = (\underset{\sim}{u}.\nabla)\rho + \frac{\partial \rho}{\partial t} \, . \tag{9}$$

Evidently $D\underset{\sim}{u}/Dt$ is the acceleration of the fluid at P, or the rate of change of the velocity of the element of fluid which, at time t, is at P. We usually call this the rate of change 'following the motion'.[†] Similarly $D\rho/Dt$ is the rate of change of the density 'following the motion'. Although we have discussed the situation only for the rates of change of velocity and density of the element, it is evident that the rate of change of any vector or scalar quantity 'following the motion' is given by this same operator D/Dt.

A flow is said to be *steady* if the dependent variables are functions of $\underset{\sim}{r}$ only and not of the time t. This does not mean, of course, that the velocity or density, etc., of a given element of fluid do not depend on t, because at different times it will (in general) be at different places. Thus, although in steady flow $\partial\underset{\sim}{u}/\partial t = 0$ and $\partial\rho/\partial t = 0$, the rates of change following the motion $(D\underset{\sim}{u}/Dt$ and $D\rho/Dt)$ are not zero in general.

There are certain unsteady flows which, by using a frame of reference moving with a constant uniform velocity $\underset{\sim}{U}$, may be reduced to steady flow. For example, to an observer on the bank of a canal the flow caused by a boat sailing up the river is unsteady: but to an observer in the boat the flow appears quite steady. Whenever we discuss steady flows in future we shall include in the discussion the corresponding unsteady flow, where this is of interest, obtained by using such a moving coordinate system.

---

* It is common nowadays to denote the operator by d/dt, but for students meeting it for the first time it is probably better to use D/Dt, as we have done here, to avoid confusion with an ordinary derivative with respect to time.

† It is sometimes known as the 'convective derivative'.

An alternative way of looking at the process described in the previous paragraph is to think of a velocity $-\underset{\sim}{U}$ superposed on the whole system (the boundaries as well as the fluid). This is similar to the technique which uses the principle of *superposition of velocities*. Here we suppose that $\underset{\sim}{u}_1$, $\underset{\sim}{u}_2$ are velocity fields such that on a specified surface S, the velocity of the fluid relative to S is zero.* Then the velocity field defined by $\underset{\sim}{u}_1 + \underset{\sim}{u}_2$ also satisfies the same boundary condition. As an example of this we consider the flows defined, using a fixed set of Cartesian axes $Oxyz$, by

$$\underset{\sim}{u}_1 = \left\{ U \left[1 - \frac{a^2(x^2 - y^2)}{(x^2 + y^2)^2} \right], \; - \frac{2Ua^2xy}{(x^2+ y^2)^2}, \; 0 \right\}, \qquad (10)$$

$$\underset{\sim}{u}_2 = \left\{ - \frac{\kappa y}{2\pi(x^2 +y^2)}, \; \frac{\kappa x}{2\pi(x^2 +y^2)}, \; 0 \right\}. \qquad (11)$$

It will be shown in the next section that the first flow field represents the inviscid flow past the cylinder $x^2 + y^2 = a^2$ whose velocity at infinity is U parallel to Ox. The second represents an inviscid circulating flow round the same cylinder, whose speed falls off to zero as we move away from the cylinder. Then there is also a flow past the cylinder defined by

$$\underset{\sim}{u} = \left\{ U \left[1 - \frac{a^2(x^2-y^2)}{(x^2 + y^2)^2} \right] - \frac{\kappa y}{2\pi(x^2 +y^2)}, \right.$$
$$\left. - \frac{2Ua^2xy}{(x^2 +y^2)^2} + \frac{\kappa x}{2\pi(x^2 +y^2)}, \; 0 \right\}. \qquad (12)$$

This flow will be discussed in the next section.

A point at which the velocity of the fluid is zero is called a *stagnation point*. Thus, in the flow defined by equation (10), the points $(-a,0)$ and $(a,0)$ are stagnation points, but the flow defined by equation (11) has no stagnation points. For the flow defined by equation (12) we have to distinguish between two separate cases: when $\kappa \geq 4\pi Ua$ the points in the x-plane where

---

* This is true for a real fluid. For an inviscid fluid it is sufficient for both flows to have their normal component of relative velocity zero on the surface S.

$x = 0$ and

$$y = \frac{\kappa}{4\pi U} \pm \left(\frac{\kappa^2}{16\pi^2 U^2} - a^2 \right)^{\frac{1}{2}}$$

are stagnation points, and when $\kappa \leq 4\pi Ua$ the points on the cylinder at which $y = \frac{\kappa}{4\pi U}$ are stagnation points.

2.2 STREAMLINES

The most useful way to visualize the flow of a fluid is to sketch the *streamlines*. These are curves which, at every point, are tangential to the velocity at that point. There is one and only one streamline through a given point unless that point happens to be a stagnation point. We shall show later in this chapter that the spacing between the streamlines is directly related to the magnitude of the velocity: in particular, for incompressible flow, the distance between two streamlines increases (or decreases) as the speed of the fluid between them decreases (or increases). More generally, in subsonic flow the closer the streamlines the faster the flow, and in supersonic flow the closer the stream-lines the slower the flow.

The streamlines give a picture of the instantaneous motion of the fluid. When the motion is steady, this means that a particular element of fluid always travels along the same streamlines, for it certainly moves along the streamline instantaneously and, at the next instant, it moves along the streamline through the neighbouring point - and this is, of course, the same curve since the flow is steady. For unsteady motion, however, this new streamline is *not* the same streamline (in general) and so the element moves along a curve which, at any instant, touches the instantaneous streamline at the point where the element is.

We shall now discuss the streamline pattern of several more or less simple flows which will be defined by their velocity field $\underset{\sim}{u}$ in terms of fixed rectangular Cartesian axes $Oxyz$.

EXAMPLE 2.2.1

$$\underset{\sim}{u} = \left\{ U\left[1 - \frac{a^2(x^2 - y^2)}{(x^2 + y^2)^2} \right], \ - \frac{2Ua^2 xy}{(x^2 + y^2)^2}, \ 0 \right\}, \quad x^2 + y^2 \geq a^2.$$

This is more easily vizualized in terms of
cylindrical polar coordinates (r,θ,z); in this
system, we have

$$u_r = U\left(1 - \frac{a^2}{r^2}\right)\cos\theta, \quad u_\theta = -U\left(1 + \frac{a^2}{r^2}\right)\sin\theta, \quad u_z = 0,$$

$$r \geq a.$$

Hence there is always a transverse component u_θ
of velocity except where $\theta = 0$ or π; this is
anticlockwise for $-\pi < \theta < 0$ and clockwise for
$0 < \theta < \pi$. Further, there is always a radial
component u_r of velocity except where $r = a$ or
$\theta = \pm\frac{\pi}{2}$. Note also that for large values of r,
$\underset{\sim}{u}$ approaches a velocity U parallel to the x-axis
($\theta = 0$). Finally, there is no component of
velocity parallel to the z-axis, and so any
streamline lies in a plane z = constant.
    It follows that $r = a$, $\theta = 0$ and $\theta = \pi$ are
all streamlines, and that the directions of the
velocity in the four quadrants of a plane z =
constant are as shown schematically in Figure 2.1(a).

(a)

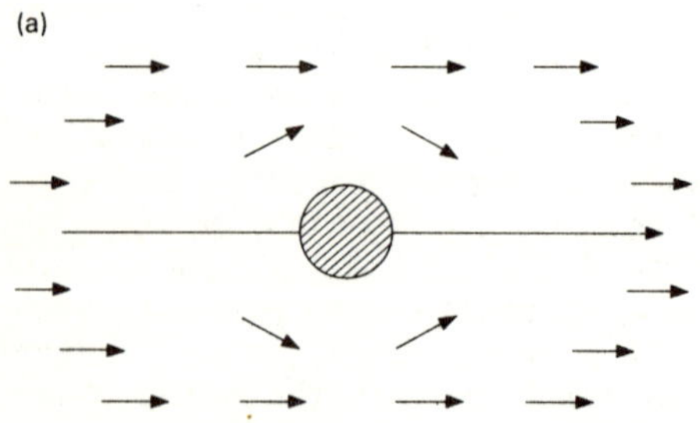

FIGURE 2.1(a)

These obviously connect up to give a system of
streamlines as in Figure 2.1(b).

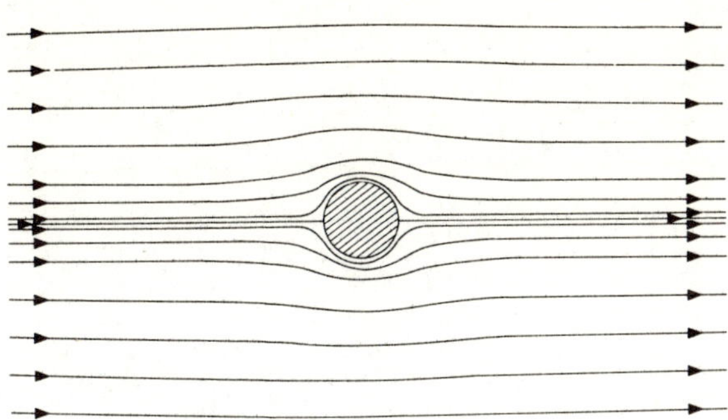

FIGURE 2.1(b)

EXAMPLE 2.2.2

$$\underset{\sim}{u} = \left\{ -\frac{\kappa\, y}{2\pi(x^2+y^2)} \, , \quad \frac{\kappa\, x}{2\pi(x^2+y^2)} \, , \quad 0 \right\} .$$

Again it is most convenient to express this in terms of the cylindrical polar coordinates (r,θ,z). Then

$$u_r = 0, \quad u_\theta = \frac{\kappa}{2\pi r} , \quad u_z = 0 .$$

It is immediately obvious that the streamlines are the circles r = constant, z = constant, that when $\kappa > 0$ the flow is anticlockwise, and that the speed of the fluid falls off as r increases. The streamline pattern is therefore as shown in Figure 2.2.

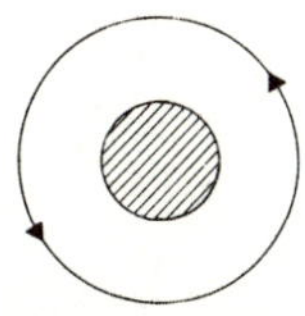

FIGURE 2.2

EXAMPLE 2.2.3

$$\underset{\sim}{u} = \left\{ U\left[1 - \frac{a^2(x^2-y^2)}{(x^2 + y^2)^2} \right] - \frac{\kappa y}{2\pi(x^2+y^2)} \, , \right.$$

$$\left. \frac{\kappa x}{2\pi(x^2+y^2)} - \frac{2Ua^2xy}{(x^2+y^2)^2} \, , \, 0 \right\} \, , \, x^2 + y^2 \geq a^2.$$

This flow can be most easily vizualized as the sum of the flows described in Examples 2.2.1 and 2.2.2. Thus over the upper surface of the cylinder (that is, where $y > 0$) the flow of 2.2.1 is slowed down, and over the lower surface (that is, where $y < 0$) it is speeded up. If κ is sufficiently large, the component of $\underset{\sim}{u}$ parallel to the x-axis is negative in the neighbourhood of the cylinder for $y = a$; otherwise it is positive. The streamlines for a case when $\kappa < 4\pi\rho a$ are shown in Figure 2.3, and those for $\kappa > 4\pi Ua$ in Figure 2.4.

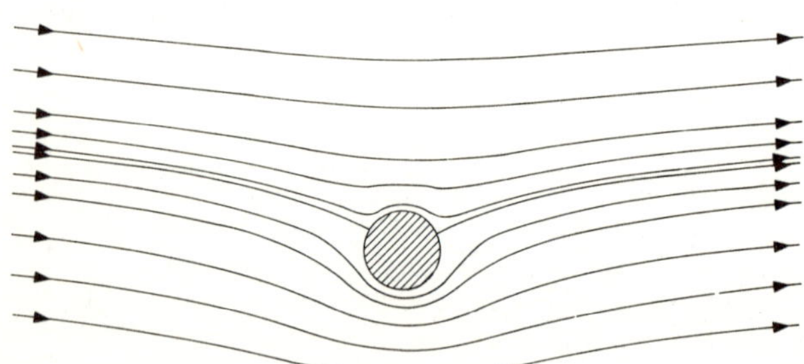

FIGURE 2.3

EXAMPLE 2.2.4

$$\underset{\sim}{u} = \left\{ - \frac{\kappa y}{2\pi(x^2+y^2)} \, , \, \frac{\kappa x}{2\pi(x^2+y^2)} \, , \, - W \right\}.$$

Referred to cylindrical polar coordinates (r,θ,z), this is

$$\underset{\sim}{u} = \left\{ 0 \, , \, \frac{\kappa}{2\pi r} \, , \, - W \right\}.$$

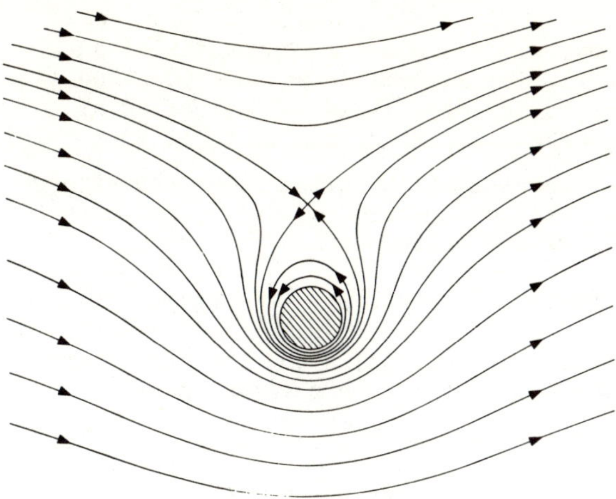

FIGURE 2.4

The streamlines evidently still lie on the cylinders
r = constant, but they no longer lie in planes
z = constant. They are in fact helices on the

cylinders r = constant whose pitch is $4\pi^2 Wr^2/\kappa$.
Two streamlines are shown in Figure 2.5.

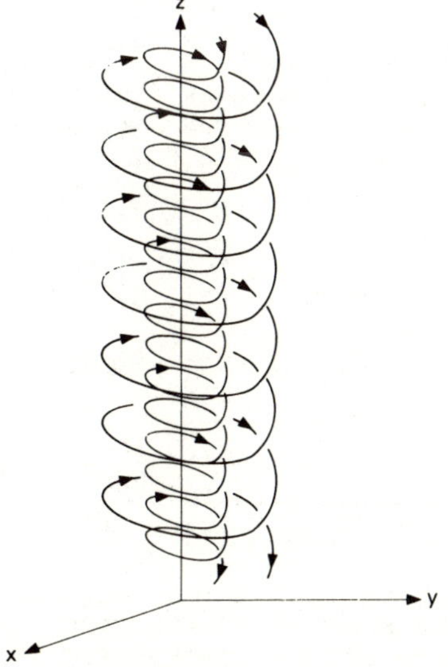

FIGURE 2.5

In the above examples, the motion is steady, so the
elements of fluid actually travel along the streamlines.
If we wish to find the equations of the streamlines we

note that, if dx, dy, dz are the components of an
element of a particular streamline at some point on it,
then the vector (dx, dy, dz) is in the direction of
$\underset{\sim}{u} = (u_1, u_2, u_3)$, the velocity of the fluid at that point.
If s is some parameter which varies along the stream-
line (it may be distance measured along it, but is not
necessarily so), we can write

$$\frac{dx}{ds} = k(s)u_1, \quad \frac{dy}{ds} = k(s)u_2, \quad \frac{dz}{ds} = k(s)u_3,$$

where $k(s)$ varies (in general) along the streamline.*
This is usually written in the more formal manner

$$\frac{dx}{u_1} = \frac{dy}{u_2} = \frac{dz}{u_3} . \tag{13}$$

In Example 2.2.2, for example, we have

$$\frac{\kappa(x^2+y^2)dx}{-2\pi y} = \frac{\kappa(x^2+y^2)dy}{2\pi x} = \frac{dz}{0} ,$$

and so $dz = 0$ always, and $xdx + ydy = 0$. These imply
that $z = $ constant on a streamline and that $x^2 + y^2 = $
constant also, and this is exactly the same as we found
before from first principles.

   If u_1, u_2, u_3 are explicitly functions of t as

well as of x, y and z, then we must treat t as a
constant parameter when we solve the equations (13)
for the streamlines (since the streamlines are defined
for instantaneous t).

   EXAMPLE 2.2.5

$$\underset{\sim}{u} = \left\{ U\, e^{ky}\sin[k(x-ct)], \; - U\, e^{ky}\cos[k(x-ct)], \; 0 \right\}, \tag{14}$$

$$0 < U \ll c, \; k > 0, \; y < 0.$$

The streamlines are given by

$$\frac{dx}{U\, e^{ky}\sin[k(x-ct)]} = \frac{dy}{-U\, e^{ky}\cos[k(x-ct)]} = \frac{dz}{0} .$$

---

* Note that $k(s)$ may also be a function of the time
  t but, for a streamline, there is no variation in
  t and so this may be regarded merely as a parameter.

Hence they lie in the planes given by $dz = 0$ (that is, $z = $ constant) and the cylinders given by

$$\frac{dy}{dx} = - \frac{\cos[k(x-ct)]}{\sin[k(x-ct)]} \; .$$

This can be integrated immediately and we find that the streamlines lie on the cylinders

$$ky = \text{constant} - \log\{|\sin[k(x-ct)]|\}.$$

The instantaneous streamlines when $t = 0$ in a plane $z = $ constant are shown in Figure 2.6, and the pattern moves along parallel to Ox with speed c.

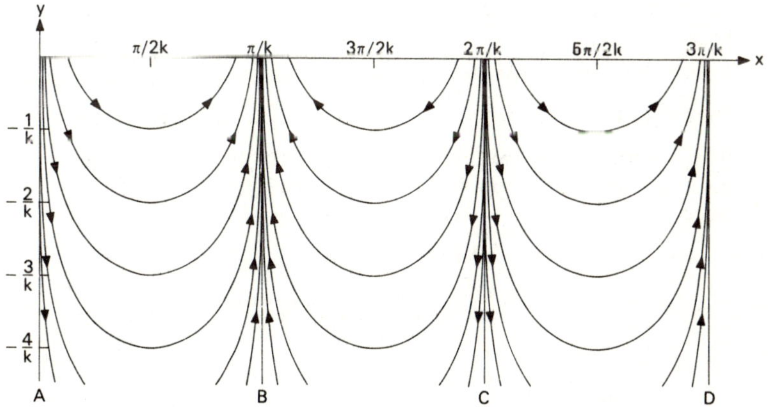

FIGURE 2.6

To obtain the path of an element of fluid, we note that, if the coordinates of this element at time t are (X,Y,Z), then at this instant

$$\frac{dX}{dt} = U\,e^{kY}\sin[k(X-ct)], \quad \frac{dY}{dt} = -\,U\,e^{kY}\cos[k(X-ct)],$$

$$\frac{dZ}{dt} = 0 \; .$$

By differentiating the expression for dX/dt, we obtain

$$\frac{d^2 X}{dt^2} = kc\,\frac{dY}{dt} \; ,$$

and this gives

$$Y = Y_0 + \frac{1}{kc}\frac{dX}{dt} = Y_0 + \frac{U}{kc}e^{kY}\sin[k(X-ct)].$$

To find a corresponding expression for X in the general case is impossible, but it is easy to verify that, if U/c is so small as to be negligible compared with unity, then

$$X = X_0 + \frac{U}{kc}e^{kY}\cos[k(X-ct)]$$

satisfies the given equations to this order and this is a good approximation to the true flow when U/c is small enough. Here X_0 and Y_0 are

constants of integration and serve to identify the particular element of fluid under discussion. Also, of course, Z = constant and each element moves along a curve in a plane perpendicular to the z-axis.

To sketch these paths when $Y \leq 0$ (which is a case of practical interest, as we shall see in Chapter 10), we note that $k(Y - Y_0)$ is of order

U/c, and so is small compared with unity. Hence to the order of approximation already used, we have

$$e^{kY} = e^{kY_0}[1 + k(Y - Y_0)+...]$$

$$= e^{kY_0}[1 + O(U/c)]$$

$$= e^{kY_0}.$$

The paths of the elements of fluid are therefore given approximately by

$$X = X_0 + \frac{U}{kc}e^{kY_0}\cos[k(x-ct)],$$

$$Y = Y_0 + \frac{U}{kc}e^{kY_0}\sin[k(x-ct)],$$

$$z = \text{constant},$$

and the fluid moves in circles whose radii decrease as $-Y_0$ increases (Figure 2.7).

It is possible to get some idea of how the two concepts fit in together by considering the motion of an element of fluid as the set of streamlines shown in Figure 2.6 travels from left to right.

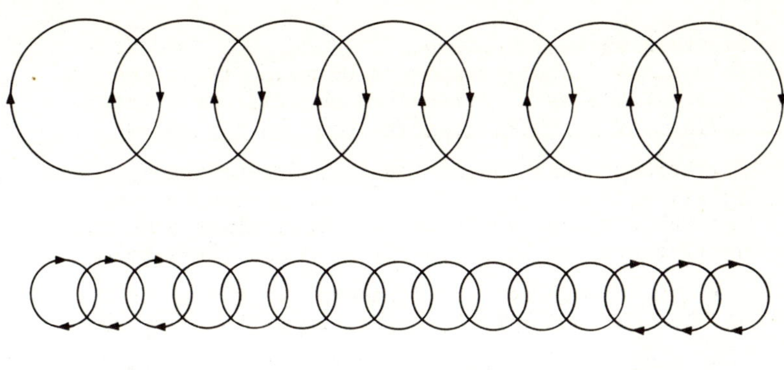

FIGURE 2.7

We suppose that, at time t_0, the element is moving downwards with the streamline CC associated with it, as shown in Figure 2.8: at a later instant t_1, the streamline through its new position is one

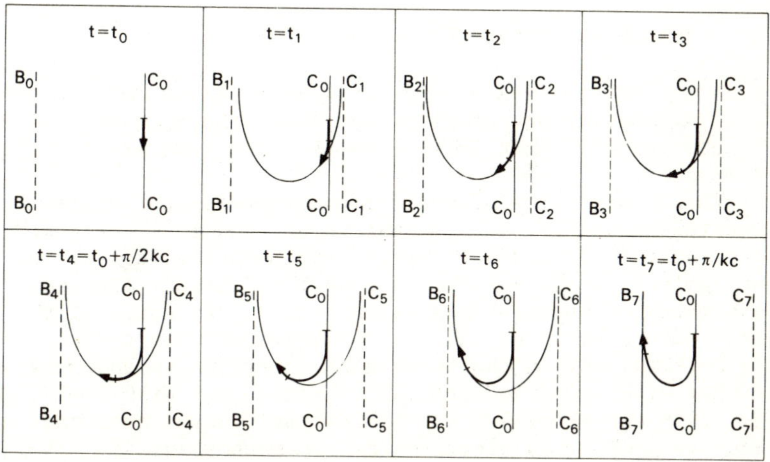

FIGURE 2.8

of those which is between BB and CC, and so
it now has a small component of velocity to the
left as well as a vertical component. At a later
instant t_2, the streamline pattern has moved
further over and the appropriate streamline
(still between BB and CC, but getting nearer
to BB) is more nearly horizontal. This goes on
until the instant t_4, say, when the streamline
through the element is horizontal at the element.
At the next instant t_5, the streamline through
the element is nearer to BB than to CC and
so has an upward component of velocity as well as
one to the left. This component becomes more and
more predominant until the streamline BB has
moved far enough to pass through the element, and
then the element is moving vertically upwards.
After this the streamlines between BB and AA
pass successively through the element, and these
all have a component of velocity to the right.
At first they have an upward component as well,
and then a downward one so that, when AA reaches
the element it is back in the same place as it was
at time t_0 (at least to the order of approximation with which we are dealing). The cycle is
then repeated indefinitely. The first part of
one such cycle is shown in Figure 2.8.

2.3 LAMINAR AND TURBULENT FLOW

When the streamlines are fairly simple curves and change
only slowly (or not at all) with time, the flow is said
to be *laminar*. When this is not so, the flow is
turbulent.

This description of the difference between laminar
and turbulent flow is somewhat vague, and it is extremely
difficult to find a satisfactory mathematical definition
of laminar flow. We often say that in laminar flow we
can think of the fluid as flowing in layers which slip
past each other, and this is the origin of the term.
Turbulent flow, on the other hand, contains an enormous
number of unsteady eddies (whose sizes vary with
position and time) and the fluid itself moves from one
eddy to another as time goes on. Common cases of
turbulent flow are found in the wake behind an obstacle
placed in a stream, in the smoke rising from a cigarette

and in the steam emerging from the spout of a kettle (see
Plate 1). In the last two cases, we have the interesting
phenomenon of the motion being laminar to start with and
becoming turbulent (rather suddenly) later on. This kind
of phenomenon was first noticed by Reynolds in 1883: he
observed it by placing some dye in water flowing along a
tube. At first the dye remained as a 'thread' and was
easily visible; but further downstream, this 'thread'
broke up, dispersed, and the dye could no longer be
distinguished. The existence of eddies is not, however,
a sufficient criterion for distinguishing between laminar
and turbulent flow as it is perfectly possible to find
eddies in laminar flows. For example, we could regard
the flow described by equation (11) as an eddy, and yet
this flow is certainly laminar. In Figure 2.9 we give
two examples of laminar flow in which eddies feature
prominently, and in which they are not quite steady;

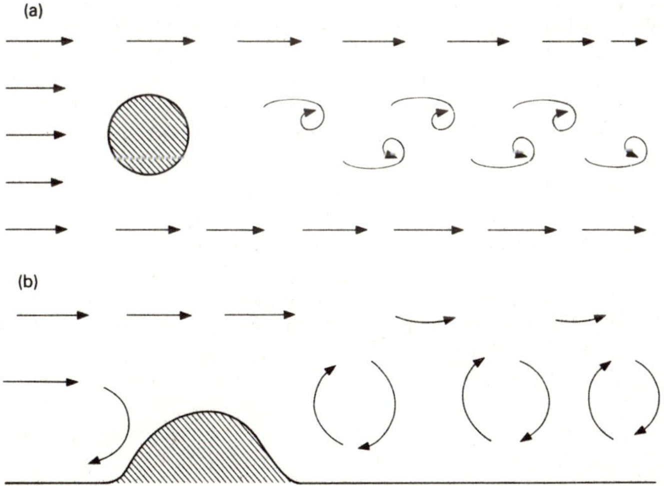

FIGURE 2.9

however, they do form a regular pattern. On the other
hand, irregularity is one of the features of turbulent
flow. It is usually fairly clear in practice which kind
of flow we have to deal with.
    It is worth noting that, in turbulent flow, the fluid
velocity varies considerably from point to point and from
time to time. But if averaged over a fairly short
distance or time interval (but much longer than the
infinitesimal distances and time intervals of Chapter 1)
the mean velocity may vary only slowly. It is usually

this mean velocity we discuss when we consider turbulent
flow, but it is clear that the 'streamlines' corresponding to this mean velocity have no physical significance.

Most of the flows we shall discuss in the rest of
this book are laminar, and it should be assumed that this
is so unless it is stated to the contrary.

2.4 STREAMTUBES

Suppose we have a simple closed curve C in the fluid
and that this curve is not a streamline nor does it pass
through a stagnation point of the flow. Then through
every point of C we can draw a streamline and so
construct a tube through which fluid is flowing - that
is, there is no flow of fluid across its walls since
these consist entirely of streamlines. Such a tube is
called a *streamtube*.

If we are neglecting viscosity, it is clear that in
steady motion the streamtube could be replaced by a fixed
solid tube. For the rest of this chapter, whenever we
discuss motion in a tube, it is possible to think of this
either as inviscid fluid flowing through a fixed pipe or
as fluid (usually, but not necessarily, inviscid) flowing
through a streamtube which is part of the flow. We shall
consider primarily cases in which the area of cross-
section of the tube varies slowly with distance along the
tube, and we shall assume throughout that the variables
of state and the magnitude of the component of velocity
parallel to the axis* are all constant over a cross-
section of the tube. For a small streamtube this is
justifiable, since we can take the cross-section to be
infinitesimally small. For a finite tube the assumption
is less good, but it is a good approximation in many
cases (as we shall see in the next chapter).

2.5 STEADY, LAMINAR, INVISCID MOTION IN A FIXED TUBE OR CHANNEL

We measure distance along the tube by means of the
coordinate x, and suppose that its cross-sectional area
at the station x is A, where A is either constant
or varies only slowly with x. The pressure p, the
density ρ, the temperature T and the component of
velocity u parallel to the axis of the tube are assumed,

---

* The axis is the line joining the centroids of the
  cross-sections of the tube.

like A, to be functions of x only. The component of
velocity normal to the axis is assumed to be negligible
compared with u: this is true only for laminar flow in
a tube with slowly varying A. We shall take two
stations x_1, x_2 on the axis of the tube and designate

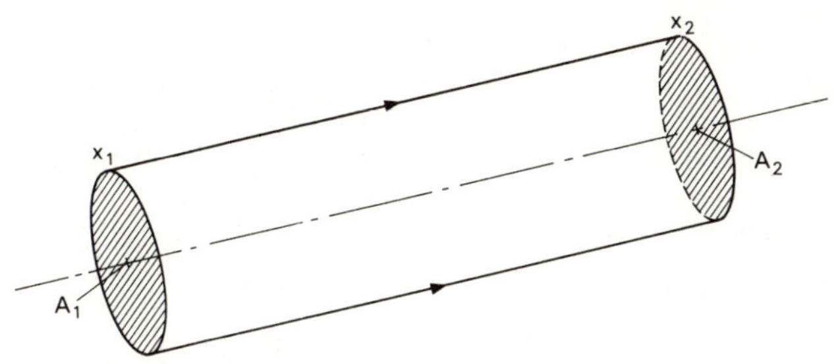

FIGURE 2.10

all the dependent variables at these stations with
suffices 1 and 2 respectively; for simplicity we shall
always have fluid flowing from x_1 to x_2 (that is,

x_2 is always taken to be downstream of x_1).

   As always we now apply the conservation laws of mass
and energy, and, in some cases, of momentum. When the
tube is not of constant cross-section or when it is not
straight, however, the latter equation must contain terms
involving the reactions from the walls of the tube, and
these are unknown; it must therefore be used in this
case to determine these forces and is of no use to
determine the flow itself.
   First, then, we consider the mass of fluid inside the
part of the tube between the stations x_1 and x_2.
Since the motion is steady, this is constant. Further,
there is no flow across the sidewalls of the tube as
these are streamlines. Hence the mass of fluid entering
the tube at x_1 per unit time is equal to that leaving
at x_2. It follows that

$$\rho_1 u_1 A_1 = \rho_2 u_2 A_2 . \tag{15}$$

(This result is true, however A varies with x, even
if it is not slowly.) Alternatively this may be
expressed in the form that the quantity $\rho u A$ is constant
along the tube. Note that if the fluid is incompressible,

so that $\rho_1 = \rho_2$, then uA is a constant along the tube. It follows that, for such a fluid, u is inversely proportional to A for a given tube - a result which was anticipated in Section 2.2.

Next we consider the energy balance. We assume that any external force, $\underset{\sim}{F}$ per unit mass, acting on the fluid is conservative; that is, we have $\underset{\sim}{F} = - \text{grad} \, \Omega$. Thus Ω is the potential energy of the fluid per unit mass. The forces on the fluid due to the walls of the tube are normal to the direction of flow (for an inviscid fluid) and so do no work. There is an inflow of energy into the section of the tube between the stations x_1 and x_2 through the cross-section at x_1 and an outflow through the cross-section at x_2; this energy is made up of kinetic energy, potential energy and internal energy. The pressure forces at the ends of the section of tube also do work, and these must balance the difference between the energy inflow and the energy outflow (since the work done by the pressure forces in the interior of the region is zero). It follows that, if the flow is adiabatic (that is, if there is no heat put into or taken out of the fluid),

$$p_1 u_1 A_1 - p_2 u_2 A_2 \equiv \text{rate of working of pressure forces}$$

$$= \text{rate of outflow of energy} \\ \text{across} \quad x_2 - \text{rate of inflow across } x_1$$

$$\equiv u_2 A_2 \left(\tfrac{1}{2} \rho_2 u_2^2 + \rho_2 \Omega_2 + \rho_2 E_2 \right) - \\ - u_1 A_1 \left(\tfrac{1}{2} \rho_1 u_1^2 + \rho_1 \Omega_1 + \rho_1 E_1 \right)$$

where E is the internal energy per unit mass of the fluid. Using equation (15) in this we obtain

$$\rho_1 u_1 A_1 \left(\frac{p_1}{\rho_1} - \frac{p_2}{\rho_2} \right) = \rho_1 u_1 A_1 \left[\left(\tfrac{1}{2} u_2^2 + \Omega_2 + E_2 \right) - \\ - \left(\tfrac{1}{2} u_1^2 + \Omega_1 + E_1 \right) \right].$$

Hence

$$\frac{p_1}{\rho_1} + E_1 + \tfrac{1}{2} u_1^2 + \Omega_1 = \frac{p_2}{\rho_2} + E_2 + \tfrac{1}{2} u_2^2 + \Omega_2, \tag{16}$$

or, in another form, the quantity $\frac{p}{\rho} + E + \tfrac{1}{2} u^2 + \Omega$ is constant along the tube. This equation is one form of *Bernoulli's equation* and is valid for steady adiabatic flow. We can take the tube to be a stremtube of vanishingly small cross-section and so obtain the result that (for inviscid flow) the quantity $\frac{p}{\rho} + E + \frac{1}{2}u^2 + \Omega$ is constant along a streamline: the constant may vary from one streamline to another.

We can usually simplify Bernoulli's equation to some extent, for we usually have an expression for E. For an incompressible gas or for a liquid, E is essentially constant, and so we have

$$\frac{p}{\rho} + \tfrac{1}{2}u^2 + \Omega = \text{constant} \tag{17}$$

along each streamline for the steady incompressible flow of an inviscid fluid. For a perfect compressible gas we have $E = c_v T$ and $p = R\rho T$. Hence, using equation (5),

$$\frac{p}{\rho} + E = RT + c_v t = c_p T. \tag{18}$$

It follows that

$$c_p T = \tfrac{1}{2}u^2 + \Omega = \text{constant} \tag{19a}$$

along each streamline for the steady adiabatic compressible flow of an ideal* gas.

It usually happens that, in motions of a gas in which compressibility is important, any changes in Ω (usually a gravitational potential) are negligibly small compared with changes in $\tfrac{1}{2}u^2$; for example, a change in speed from zero to 200 m/s involves a change in $\tfrac{1}{2}u^2$ of 2×10^4 m^2/s^2. To achieve such a change in the gravitational potential we need a change of height of 2 km, and in most compressible problems such large changes of height do not occur. So equation (19a) reduces to

$$c_p T + \tfrac{1}{2}u^2 = \text{constant} \tag{19b}$$

along each streamline for the steady adiabatic compressible flow of an ideal gas. In meteorological problems, when we do have to consider such vertical distances, the motions are very much slower and equation (17) holds. Evidently for such problems, and also for a liquid (whose density is much greater than that of a gas) it is necessary to keep the potential term in. Sometimes it is useful, however, to define the *hydrostatic pressure* p_s by means of the relation

$$p_s + \rho\Omega = \text{constant} \tag{20}$$

---

* An *ideal* gas is one which is both perfect and inviscid.

throughout the fluid; thus p_s is the pressure at a point due to the (conservative) external forces when there is no motion in the fluid. The value of the constant on the right-hand side of equation (20) is usually not important, and depends in any case on the value of Ω which is always defined only within an additive arbitrary constant; it is often convenient to arrange for it to be zero. We can now define the quantity $p_d = p - p_s$, and call it the *dynamic pressure*. It is evident that equation (17) now reduces to

$$\frac{p_d}{\rho} + \tfrac{1}{2}u^2 = \text{constant} \tag{21}$$

along each streamline for the steady incompressible flow of an inviscid fluid.

If it happens that all the fluid originates from a region in which conditions are uniform, the constants on the right-hand sides of equations (17), (19) and (21) are naturally independent of the particular streamline chosen. But in general they vary from one streamline to another.

It is useful to have a conventional notation for these constants, and there are several possibilities, some useful in one problem and some in another. First we define p_0 as the value which p would take if the fluid were reduced to rest at some point of the stream-line. This is called the *stagnation pressure*. The corresponding temperature T_0 and density ρ_0 are called the *stagnation temperature* and *stagnation density* respectively. For an ideal gas, of course, $p_0 = R\rho_0 T_0$. Then Bernoulli's equation in the forms (19b) and (21) becomes

$$c_p T + \tfrac{1}{2}u^2 = c_p T_0 \tag{22}$$

along each streamline for the steady adiabatic flow of an ideal compressible gas, and

$$\frac{p_d}{\rho} + \tfrac{1}{2}u^2 = \frac{p_0}{\rho} \tag{23}$$

along each streamline for the steady flow of an incompressible, inviscid fluid of density ρ. In practice, of course, the fluid cannot actually be reduced to rest (otherwise there could be no flow along a streamtube); the relations (22), (23) are therefore,

strictly speaking, definitions of the stagnation
temperature and pressure.

For the compressible case, it is often more convenient
to use as a dependent variable the speed, a, at which
small disturbances are propagated in the fluid. This is
the speed of sound and we shall show later that, for a
gas in which p is a function of ρ alone, $a^2 = dp/d\rho$.
In particular, if the motion is isentropic (as it often
is) we have $p = K\rho^\gamma$ and so $a^2 = \gamma p/\rho = \gamma RT$. If a_0
is the value of a when $T = T_0$ (that is, if a_0 is
the speed of sound in the fluid at rest) we can use
equation (6) to write equation (22) in the form

$$\frac{a^2}{\gamma - 1} + \tfrac{1}{2}u^2 = \frac{a_0^2}{\gamma - 1} \tag{24}$$

along each streamline for the isentropic steady motion
of an ideal compressible gas.

This equation shows that as the fluid speed increases
along a streamline the speed of sound decreases; evidently
it could happen that the two become equal. If this does
happen we denote the point where it does so as the *sonic
point* and the speed there as the *sonic speed*. We
designate variables at this point by means of an asterisk.
Thus, at the sonic point, $u \equiv u^* = a^* = (\gamma RT^*)^{\frac{1}{2}}$, and
the right-hand side of equation (24) can then be expressed
as

$$\frac{a^{*2}}{\gamma - 1} + \tfrac{1}{2}a^{*2} = \tfrac{1}{2}\frac{\gamma + 1}{\gamma - 1}a^{*2}.$$

So an alternative form of the equation is

$$\frac{a^2}{\gamma - 1} + \tfrac{1}{2}u^2 = \tfrac{1}{2}\frac{\gamma + 1}{\gamma - 1}a^{*2} \tag{25}$$

along each streamline for the isentropic steady motion of
an ideal compressible gas. Note that the sonic speed is
defined by equation (25) even when there is no sonic
point on the streamline.

It may be useful to summarize below the most usual
forms of Bernoulli's equation for steady flow.

For the steady adiabatic incompressible flow of an
inviscid fluid of density ρ we have, along each
streamline,

$$\frac{p}{\rho} + \tfrac{1}{2}u^2 + \Omega = \text{constant}, \quad \frac{p_d}{\rho} + \tfrac{1}{2}u^2 = \frac{p_0}{\rho}.$$

For the steady adiabatic flow of an ideal gas we have, along each streamline,

$$c_p T + \tfrac{1}{2}u^2 = c_p T_0 ;$$

when the flow is also isentropic we have, along each streamline

$$\frac{a^2}{\gamma - 1} + \tfrac{1}{2}u^2 = \frac{a_0^2}{\gamma - 1} = \tfrac{1}{2}\frac{\gamma + 1}{\gamma - 1} a^{*2} .$$

When we have to consider the momentum of the fluid in the tube, we usually restrict ourselves to the cases in which the cross-sectional area A is constant and the tube is straight. Otherwise, as we have already seen, an unknown force must be included in the equations. To determine the force, we need to evaluate an integral of the form $\iint p\ell\, dS$ where ℓ is the cosine of the angle which the normal to the surface of the tube makes with the x-direction. If, however, A is constant and the tube is straight, then ℓ is zero and there is no effect on the x-component of momentum due to the side walls. In this case, the only forces which have components in the x-direction are the pressure forces at the ends and the body forces. These have the effect of creating momentum in the fluid as it passes from x_1 to x_2 and we have, when p does not vary across the tube,

$$p_1 A_1 - p_2 A_2 - \int_{x_1}^{x_2} \rho A \frac{\partial \Omega}{\partial x}\, dx \equiv \text{x-component of force on fluid}$$

$$= \text{outflow of momentum through } x_2 - \text{inflow through } x_1$$

$$\equiv \rho_2 u_2^2 A_2 - \rho_1 u_1^2 A_1 ;$$

since A is constant, this may be written in the form

$$p_1 + \rho_1 u_1^2 + \int_{x_0}^{x_1} \rho \frac{\partial \Omega}{\partial x}\, dx = p_2 + \rho_2 u_2^2 + \int_{x_0}^{x_2} \rho \frac{\partial \Omega}{\partial x}\, dx ,$$

where x_0 is some standard reference point on the tube.

Hence

$$p + \rho u^2 + \int_{x_0}^{x} \rho \frac{\partial \Omega}{\partial x} \, dx = \text{constant} \tag{26}$$

along the tube. When Ω is zero (or more precisely, when $\partial\Omega/\partial x$ is zero) this becomes

$$p + \rho u^2 = \text{constant} \tag{27}$$

for the flow of a fluid under the action of no external forces along a straight tube of constant cross-section.

In the next chapter we shall discuss the use of these equations in some real problems.

EXERCISES

1 At time t the velocity of the fluid at the point whose vector position is $\underset{\sim}{r}$ is $\underset{\sim}{u}(\underset{\sim}{r},t) = f(t)\underset{\sim}{U}(\underset{\sim}{r})$.
Show that the elements of fluid move along the stream-lines. Give an example to illustrate this.

2 Referred to Cartesian axes, the velocity of the fluid in a steady flow is $(2x^2y, -2xy^2, 0)$. Sketch the streamlines of the flow.

Determine the position at time t of the element of fluid which is at $(1,1,0)$ initially, sketch its path and indicate (by arrows) its velocity at times $t = 0.1$, 0.2 and 0.3.

3 The velocity components of a fluid at the point whose vector position is $\underset{\sim}{r} = (x,y,z)$ are (u,v,w). Find the equations of the streamlines and sketch them in the following cases
  (a) $u = x$, $v = y$, $w = z$;
  (b) $u = y - z$, $v = z - x$, $w = x - y$;
  (c) $u = -y$, $v = x$, $w = 3$;
  (d) $u = y^2$, $v = y$, $w = -z$;
  (e) $u = 2yz$, $v = xz$, $w = -3xy$;
  (f) $u = -y/(x^2 + y^2)$, $v = x/(x^2 + y^2)$, $w = 0$.
Which of these velocity distributions corresponds to the flow of an incompressible fluid?

4 Bernoulli's equation for the steady, one-dimensional adiabatic flow of an ideal gas (under the action of no body forces) may be written in the form $c_p T + \frac{1}{2}u^2 = c_p T_0$.

The local speed of sound, a, is given by the local value of $(dp/d\rho)^{\frac{1}{2}}$ and $M = u/a$ is the local Mach number of the flow. Show that the local values of the temperature T, the density ρ, and the pressure p, can be expressed in terms of M as follows:

$$\frac{T_0}{T} = 1 + \tfrac{1}{2}(\gamma - 1)M^2, \quad \frac{\rho_0}{\rho} = \left[1 + \tfrac{1}{2}(\gamma - 1)M^2\right]^{1/(\gamma-1)},$$

$$\frac{p_0}{p} = \left[1 + \tfrac{1}{2}(\gamma - 1)M^2\right]^{\gamma/(\gamma-1)},$$

where the suffix zero indicates the stagnation value of a quantity.

5   If the incompressible form $\dfrac{p}{\rho} + \tfrac{1}{2}u^2 = \dfrac{p_0}{\rho}$ of Bernoulli's equation is used in error for the steady, one-dimensional adiabatic flow of a compressible gas (under the acation of no body forces), show that

$$\frac{p_0}{p} = 1 + \tfrac{1}{2}\gamma M^2.$$

This formula is used (instead of the correct one found in the previous question) to calculate the local Mach number, using measured values of p_0 and p. Show that, for small values of M, the fractional error in the calculated value of M is $M^2/8$.

6   Using Bernoulli's equation for the adiabatic flow of an ideal gas as given in question 4, and using the notation of that question, show that

$$M = \frac{u}{a_0}\left[1 - \tfrac{1}{2}(\gamma - 1)\left(\frac{u}{a_0}\right)^2\right]^{-\frac{1}{2}}.$$

Find expressions for T_0/T, ρ_0/ρ and p_0/p, in terms of u/a_0.

7   The Venturi tube illustrated in Figure 2.11 is a device for measuring flow velocities in a pipe along which is flowing an incompressible inviscid fluid of density ρ. It may be assumed that the speeds u_1, u_2 are uniform across the pipe at each of the two sections

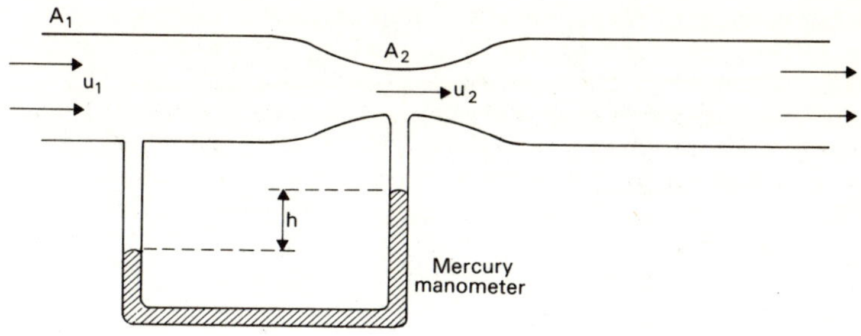

FIGURE 2.11

shown. If ρ' is the density of mercury, show that

$$u_1 = \left[\frac{2\rho' ghA_2^2}{\rho(A_1^2 - A_2^2)}\right]^{\frac{1}{2}} .$$

8 The manometer in the Venturi tube of the previous question is connected, as shown in Figure 2.12, to a pitot tube (instead of to a static pressure hole) in

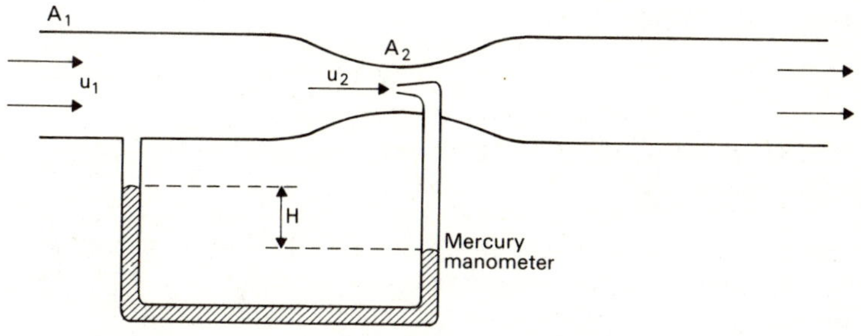

FIGURE 2.12

the contraction. It may be assumed that the disturbance to the flow due to the presence of the pitot tube is negligible. Show that

$$\frac{H}{h} = \frac{A_2^2}{A_1^2 - A_2^2} .$$

9 A vertical chimney of height h contains an incompressible gas of density ρ and is surrounded by an incompressible atmosphere of density ρ', where $\rho' > \rho$. The pressure of the gas inside the chimney is the same as that in the atmosphere outside, both at the top and the bottom of the chimney. The atmosphere outside the chimney is at rest everywhere and the speed of the gas at the bottom of the chimney is negligible. Show that the gas emerges from the top of the chimney with speed w where

$$w = \left[2gh\left(\frac{\rho'}{\rho} - 1\right)\right]^{\frac{1}{2}}.$$

10 A liquid whose density is ρ flows along a horizontal pipe whose cross-sectional area changes from A_1 to A_2 (with $A_2 > A_1$) downstream. In the smaller bore the pressure is p_1 and the speed of the liquid is u_1. When the change in bore is abrupt, it is found that the pressure in the larger pipe at the junction is everywhere p_1 and, far enough downstream for the flow to have uniform speed right across the pipe, the pressure is p_2 and the speed u_2. Find an expression for $p_1 - p_2$ in terms of ρ, u_1 and u_2.

   If the junction had been smooth so that there was no energy loss there, the pressure far downstream would have been p_2'. Show that

$$p_2' - p_2 = \tfrac{1}{2}\rho(u_1 - u_2)^2.$$

What is the relationship of this expression to the loss of energy at the abrupt junction?

11 An ideal gas flows out from a reservoir; show that, as long as the speed in the reservoir is negligible, and the flow is adiabatic, it can never reach a speed greater than u_m, where $u_m^2 = 2a_0^2/(\gamma - 1)$.

   What are the corresponding limiting values of T and M ? Explain these values physically.

12 A two-dimensional jet of liquid has uniform speed U, uniform pressure p, and width b. As shown in Figure 2.13, it strikes a plane wall whose normal is at an angle α with the jet, and it then splits into two

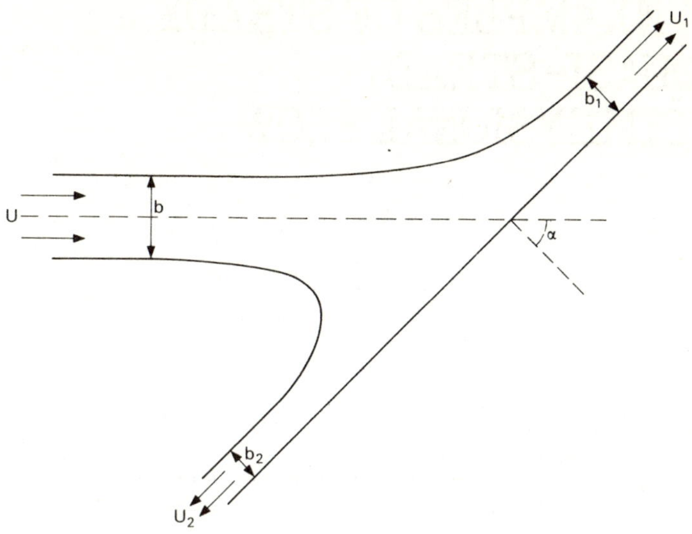

FIGURE 2.13

wall-jets having speeds U_1, U_2 and widths b_1, b_2 respectively. There are no body forces on the liquid. Show
  (a) that the pressure in each of the wall-jets is independent of distance from the wall once the flow in the wall-jet has become parallel to the wall;
  (b) that the speeds U_1, U_2 do not depend on the distance from the wall, and hence that $U_1 = U_2 = U$;
  (c) that $b_1 = \frac{1}{2}b(1 + \sin\alpha)$ and $b_2 = \frac{1}{2}b(1 - \sin\alpha)$.

13 An inviscid liquid is in steady, laminar motion and has a free surface in contact with a gas at constant pressure. Show that:
  (a) the free surface is a stream surface, and
  (b) if surface tension effects are neglected, and there are no body forces on the liquid, the speed of each element of fluid in the surface is independent of time.

Chapter 3

SOME EXAMPLES OF STEADY AND QUASI-STEADY ONE-DIMENSIONAL FLOW

3.1 SOME QUALITATIVE RESULTS

In this chapter we shall deal with some particular cases of one-dimensional flow which are of importance both because they illustrate a number of important physical principles and also because they have important applications either in the laboratory or in nature. In each case we shall begin by discussing a steady flow and then, in two cases, go on to consider the motion referred to a moving frame of reference. The flows, referred to the moving frame of reference, are no longer steady, but because they can so readily be reduced to steady-flow problems we call them quasi-steady and it is appropriate to discuss them in this chapter.

First, however, we shall describe briefly some simple experiments most of which can be carried out by the reader, and which illustrate the way in which the equations deduced in the last chapter can be used qualitatively. We shall be concerned with incompressible flow only and so, using equation (15), we expect the fluid speed to vary inversely with the cross-sectional area of the region through which the fluid flows. Also we expect equation (23) to hold and, since gravitational effects will be negligible for the fluid concerned (which will be air), the pressure itself is actually equal to the dynamic pressure p_d which occurs in this equation; it follows that, if the velocity increases, then the pressure decreases - and conversely. The net result is that the pressure and the cross-sectional area increase or decrease together.

The easiest experiment to perform is to take a rectangular sheet of paper ABCD (half a piece of A4 file paper is excellent), hold it by one edge AB held

horizontally so that the opposite edge CD falls freely
(Figure 3.1(a)). Then blow across the top of the paper

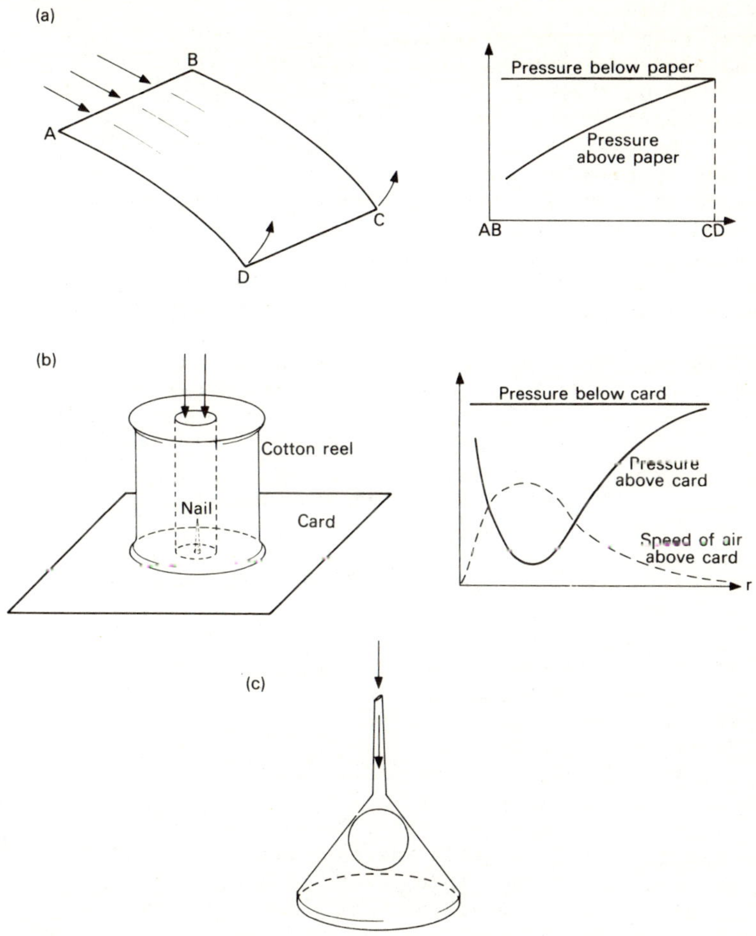

FIGURE 3.1

at right angles to AB. At first sight we would expect
this to have no effect on the paper, but a little thought
shows that this will not happen: if we consider the
forces on the paper due to the air surrounding it, we see
that those below the paper are due to the pressure of air
more or less at rest, while that above the paper is due
to air which has a speed greater than zero. It follows
that the forces due to the air pressure on the paper has
an upward component and this should tend to counteract

the force due to gravity on the paper and the edge CD
should rise. It is easy to verify experimentally that
this does happen.

   Next take a piece of card (half a postcard will do)
and hold it horizontally with a light nail sticking
vertically up through its centre; place one flat end of
a cotton reel on the card so that the nail is in the
hole in the reel (Figure 3.1(b)). Now blow down the
hole and release the card (a sustained flow of air is
required and experimenters with only weak lung power
may need to use a compressed air supply; most people
can sustain the flow for a few seconds). It is found
that, instead of the card with its nail falling away, it
remains almost touching the reel. This is because, as
indicated in the associated pressure graph, the air
speed falls off* with distance r from the axis of
symmetry, since it has to pass through a larger area,
and so the pressure rises as r increases. But at the
edge of the card the pressure at the top is the same as
that everywhere underneath the card: hence the pressure
is, nearly everywhere, less at the top than at the
bottom of the card and there is a resulting upward force.
As long as the air speed is large enough, this upward
force will overcome the weight of the card and nail and
these remain in place.

   For the next experiment, slightly more elaborate
equipment is required: a funnel, a ping-pong ball and
preferably (though not necessarily) a compressed air
supply. The funnel is held inverted and the ball held
inside it; air is then blown down the funnel and it is
found that the ball stays in position without further
support (Figure 3.1(c)). It is left as an exercise to
the reader to explain qualitatively why this might be
expected to happen: exactly the same kind of argument
can be used as for the other cases. (If a continuous
stream of air is not available for this experiment, a
stream of water will work just as well and the experiment
can be carried out in the kitchen sink.)

3.2 THE LAVAL NOZZLE

We consider the reversible adiabatic flow of an ideal
compressible gas through a tube having a contraction as
shown in Figure 3.2. The equation (15) may be written

---

* This does not apply near the axis itself, where the
  flow is not one-dimensional anyway.

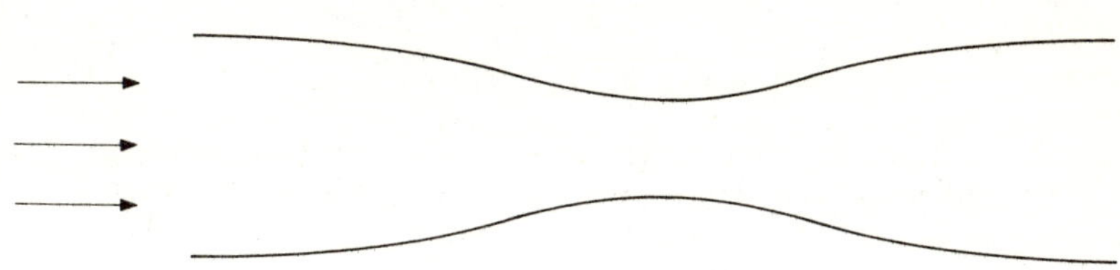

FIGURE 3.2

$$\rho u A = \text{constant} = \rho * a * A * \qquad (28)$$

where $a*$ is the sonic speed, $\rho*$ is the density corresponding to this speed and $A*$ is a constant defined by the equation: if the speed in the tube ever becomes sonic, then $A*$ is clearly the value of A when this happens. Also equation (25) can be written as

$$\frac{2a^2}{\gamma - 1} + u^2 = \frac{\gamma + 1}{\gamma - 1} a*^2 \qquad (29)$$

and, since the motion is reversible and adiabatic, it is also isentropic, and

$$a^2 = \gamma RT \quad \text{and} \quad \rho^\gamma \propto \rho T.$$

So we have

$$\left(\frac{a}{a*}\right)^2 = \frac{T}{T*} = \left(\frac{\rho}{\rho*}\right)^{\gamma-1} = \left(\frac{a*A*}{uA}\right)^{\gamma-1},$$

using equation (28). Hence, writing $u = Ma$, where M is the *Mach number*, we have

$$\frac{A}{A*} = \frac{1}{M}\left(\frac{a*}{a}\right)^{1+\frac{2}{\gamma-1}} = \frac{1}{M}\left(\frac{a*}{a}\right)^{\frac{\gamma+1}{\gamma-1}}.$$

Finally, using equation (29), this gives

$$\frac{A}{A*} = \frac{1}{M}\left[\frac{2}{\gamma+1}\left(1 + \frac{\gamma-1}{2}M^2\right)\right]^{\frac{1}{2}\frac{\gamma+1}{\gamma-1}}. \qquad (30)$$

To see how the function of M on the right-hand side of this equation varies with M, we note that, since

$\gamma > 1$, $\frac{1}{2}\frac{\gamma+1}{\gamma-1} > \frac{1}{2}$ and so $A/A* \to \infty$ as $M \to \infty$. Also

$A/A^* \to \infty$ as $M \to 0$. In addition we find that

$$\frac{2}{A}\frac{dA}{dM} = \frac{2(M^2 - 1)}{M[1 + (\gamma - 1)M^2]},\tag{31}$$

and it follows at once that dA/dM is positive when $M > 1$ and negative when $M < 1$. Also $A/A^* = 1$ when $M = 1$, as we would expect from the definitions of A^* and M. The form of the relation defined in equation (30) is therefore as shown in Figure 3.3. Note that,

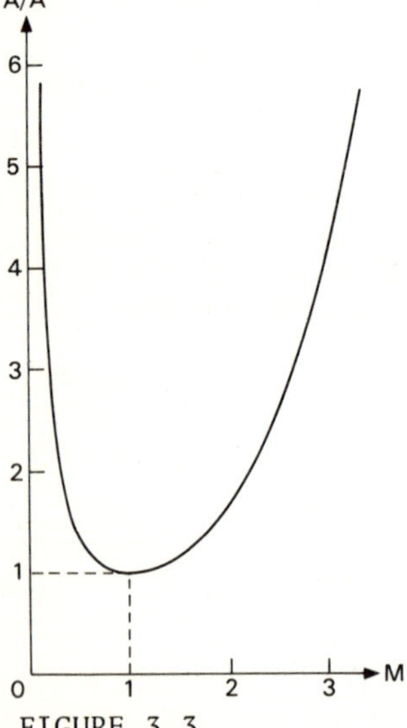

FIGURE 3.3

if the flow is ever sonic, it can only be so when A is a minimum: that is, it is either sonic at the throat of the contraction or not at all.

We note also that equation (29) may be rewritten in the form

$$\left(\frac{u}{a^*}\right)^2 = \frac{(\gamma + 1)M^2}{2 + (\gamma - 1)M^2}.\tag{32}$$

So u increases steadily from zero to $(\gamma + 1)/(\gamma - 1)$ as M increases from zero to infinity, as indicated in Figure 3.4.

If the flow is subsonic (that is, if $M < 1$) upstream, then as the fluid gets nearer to the throat so that A decreases, we see from Figure 3.3 that M increases and so, using equation (32) the flow gets faster. If the flow is still subsonic when it reaches the throat, then it can never become supersonic since it cannot cross

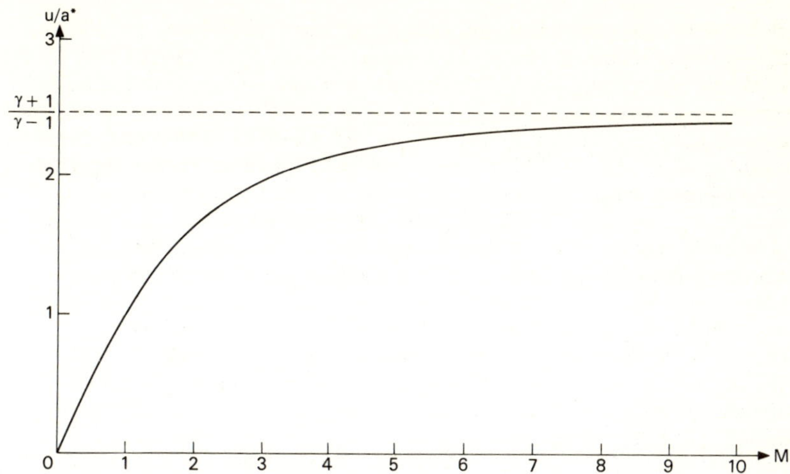

FIGURE 3.4

the minimum of Figure 3.3; in this case it slows down
again as the fluid enters the expanding region. It is
conceivable, however, that the minimum of the graph is
reached and in this case the fluid can become supersonic
(that is, M > 1) in the expanding region: in this case
the Mach number increases as A increases, and so does
the speed u. This is the result stated in Section 2.2
that, in subsonic flow, the fluid speed is faster when
the streamlines are closer together and in supersonic
flow, it is slower.
    The results discussed above are used in the const-
ruction of wind tunnels in the laboratory to produce
supersonic flow. One of these is shown in diagrammatic
form in Figure 3.5. The gas is stored in the large

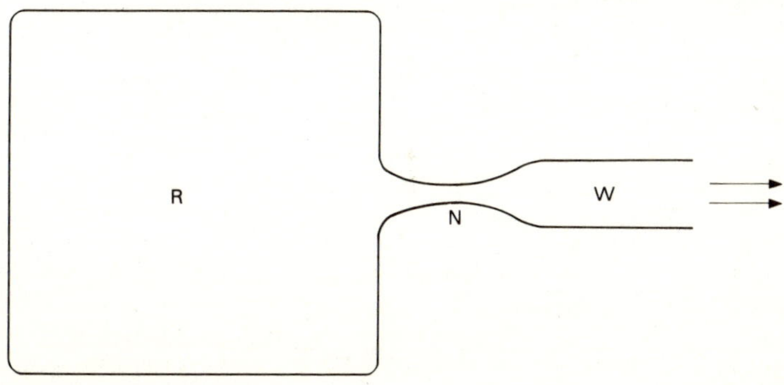

FIGURE 3.5

reservoir R at high pressure; a tap is opened to allow the gas to flow through the Laval nozzle N into the working section W which is at a much lower pressure. Under suitable conditions the flow in the working section W will be supersonic and it is worth noticing that the Mach number of the flow depends only on the ratio of the cross-sectional areas of the working section and of the throat.* Since the flow, once supersonic, has to become subsonic again when the gas reaches the atmosphere, some mechanism must be supplied to enable this to happen. Sometimes this is done by putting a second nozzle (the exit nozzle) downstream of the working section so that the stream can become sonic again at the second throat (Figure 3.6(a)). Sometimes the exit nozzle is formed automatically by the thickening boundary layers on the walls (Figure 3.6(b)), as described in Chapter 6. In yet another method, the flow is allowed to become sub-sonic by passing through a stationary normal shock wave at the end of the working section (Figure 3.6(c)); this mechanism is described in Section 3.3 below. Other

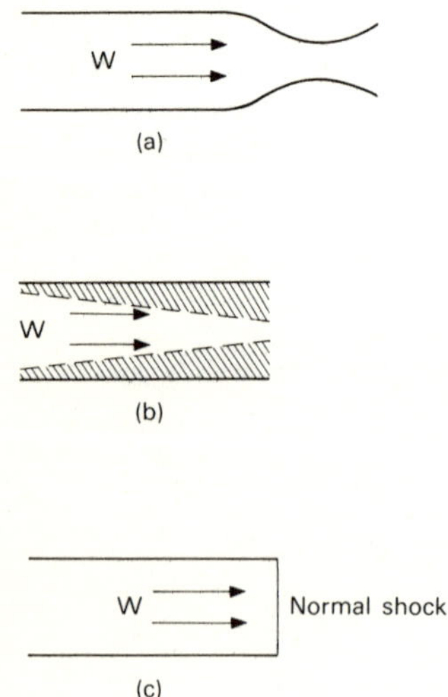

FIGURE 3.6

* In practice a small allowance has to be made for the effect of tangential stresses in the neighbourhood of the boundaries.

mechanisms (such as an oblique shock system) may also occur.

To discuss under what conditions the flow can become supersonic is beyond the scope of this book. We merely state that for supersonic flow to occur it is necessary both for the initial (that is, before the tap is turned on to start the flow) ratio of the pressures in the reservoir R and the working section W to be greater than a certain critical value (this critical ratio increases as the Mach number of the flow in the working section increases), and for a sufficiently large pressure ratio to be maintained.

3.3 THE NORMAL PLANE SHOCK

We now discuss the adiabatic compressible flow of an ideal gas along a tube of uniform cross-sectional area. We look to see if there is a steady flow which might have a discontinuity somewhere in the tube - that is, somewhere between the stations x_1 and x_2 (see Figure 3.7).

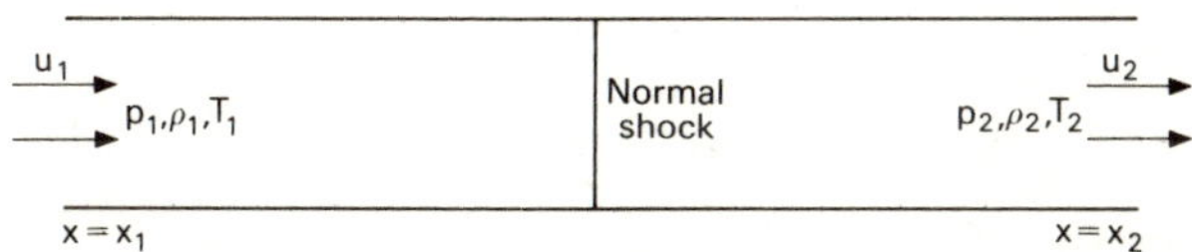

FIGURE 3.7

There will be conservation of mass, so equation (15) holds and we have

$$\rho_1 u_1 = \rho_2 u_2. \tag{33}$$

There are no external forces on the body of the fluid (the variations in gravitational force are negligible even if the tube is vertical - usually it is horizontal, and then the component parallel to the tube is in fact zero), so equation (27) holds and we have

$$p_1 + \rho_1 u_1^2 = p_2 + \rho_2 u_2^2. \tag{34}$$

Finally, since there are no external forces, Bernoulli's equation (19b) may be written, using (1) and (6), as

$$\frac{\gamma}{\gamma - 1}\frac{p_1}{\rho_1} + \tfrac{1}{2}u_1^2 = \frac{\gamma}{\gamma - 1}\frac{p_2}{\rho_2} + \tfrac{1}{2}u_2^2. \tag{35}$$

In addition to these equations, we need to use the inequality of the Second Law of Thermodynamics stated in Section 1.5. This may, using equation (7), be written in the form

$$p_1\rho_1^{-\gamma} \le p_2\rho_2^{-\gamma}. \tag{36}$$

The relations (33), (34), (35) and (36) are sometimes called the Rankine-Hugoniot equations; they are consistent and sufficient to determine the flow at the station x_2 when that at x_1 is known.

We note first that they have one obvious, but trivial, solution given by $u_2 = u_1$, $p_2 = p_1$, $\rho_2 = \rho_1$, $T_2 = T_1$: we now look to see if there is another solution. Further, we note that if any one of the four variables is unchanged, then so are all the others: for example, if $u_2 = u_1$, then equation (33) implies that $\rho_2 = \rho_1$ and (34) that $p_2 = p_1$. Now we can combine equations (33) and (34) to give

$$p_2 - p_1 = \rho_1 u_1 (u_1 - u_2). \tag{37}$$

It is convenient to introduce a new variable α which is the ratio of the pressure downstream to that upstream: thus

$$\alpha = p_2/p_1. \tag{38}$$

Then equation (37) may be written in the form

$$u_1 - u_2 = \frac{p_1}{\rho_1 u_1}(\alpha - 1). \tag{39}$$

Also equation (35) becomes

$$\tfrac{1}{2}(u_1^2 - u_2^2) = \frac{\gamma}{\gamma - 1}\left(\frac{p_2}{\rho_2} - \frac{p_1}{\rho_1}\right),$$

and so, using equations (33) and (38), we have

$$\tfrac{1}{2}(u_1 - u_2)(u_1 + u_2) = \frac{\gamma}{\gamma - 1}\frac{p_1}{\rho_1}\left(\frac{\alpha u_2}{u_1} - 1\right).$$

Using equation (39) in this expression we obtain

$$\frac{1}{2}\frac{p_1}{\rho_1 u_1}(\alpha - 1)(u_1 + u_2) = \frac{\gamma}{\gamma - 1}\frac{p_1}{\rho_1 u_1}(\alpha u_2 - u_1),$$

which gives

$$\frac{u_2}{u_1} = \frac{\rho_1}{\rho_2} = \frac{\alpha(\gamma - 1) + \gamma + 1}{\alpha(\gamma + 1) + \gamma - 1} . \tag{40}$$

Substituting this in equation (39) we have

$$\frac{2(\alpha - 1)}{\alpha(\gamma + 1) + \gamma - 1} = \frac{p_1}{\rho_1 u_1^2}(\alpha - 1).$$

But $\alpha \neq 1$ and $a_1^2 = \gamma p_1/\rho_1$; also

$$M_1^2 \equiv \frac{u_1^2}{a_1^2} = \frac{\rho_1 u_1^2}{\gamma p_1}, \quad M^2 \equiv \frac{u_2^2}{a_2^2} = \frac{\rho_2 u_2^2}{\gamma p_2} .$$

It follows that

$$2\gamma M_1^2 = \alpha(\gamma + 1) + \gamma - 1, \tag{41}$$

and so

$$\frac{p_2}{p_1} \equiv \alpha = \frac{2\gamma M_1^2 - \gamma + 1}{\gamma + 1} . \tag{42}$$

We also have

$$M_2^2 \equiv \frac{\rho_2 u_2^2}{\gamma p_2} = \frac{\rho_1 u_1^2}{\gamma \alpha p_1}\frac{u_2}{u_1} = \frac{M_1^2}{\alpha}\frac{\alpha(\gamma - 1) + \gamma + 1}{\alpha(\gamma + 1) + \gamma - 1} ,$$

and so

$$M_2^2 = \frac{\alpha(\gamma - 1) + \gamma + 1}{2\alpha\gamma} \tag{43}$$

This may be combined with equation (42) to give

$$M_2^2 = \frac{(\gamma - 1)M_1^2 + 2}{2\gamma M_1^2 - \gamma + 1} \tag{44}$$

So we have unique expressions for u_2/u_1, ρ_2/ρ_1, p_2/p_1 and M_2^2 in terms of M_1, the known condition upstream.

So far in the discussion there is nothing to limit the value of M_1 for which flow of this kind is possible, but we shall now show that it cannot happen unless

$M_1 > 1$. To do this we consider equation (36): this may be written in the form

$$1 \leq \frac{p_2^{1/\gamma}}{p_2} \frac{\rho_1}{p_1^{1/\gamma}} = f(\alpha),$$

where

$$f(\alpha) = \alpha^{1/\gamma} \frac{\alpha(\gamma - 1) + \gamma + 1}{\alpha(\gamma + 1) + \gamma - 1}. \tag{45}$$

Differentiating logarithmically we have

$$\frac{f'(\alpha)}{f(\alpha)} = \frac{(\alpha - 1)^2(\gamma^2 - 1)}{\alpha\gamma[\alpha(\gamma - 1) + \gamma + 1][\alpha(\gamma + 1) + \gamma - 1]}.$$

Since $\gamma > 1$ and $\alpha > 0$ always, we see that $f(\alpha)$ is always positive and that $f'(\alpha)$ is never negative; hence the graph of $f(\alpha)$ is of the form shown in Figure 3.8.

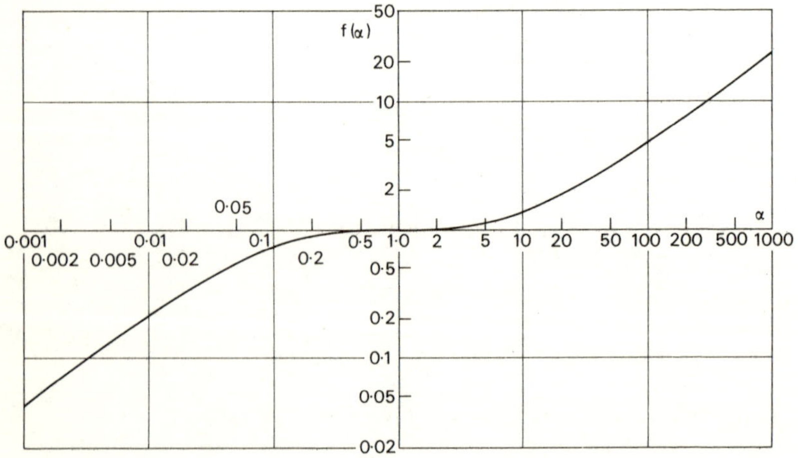

FIGURE 3.8

It is easily verified that $f(\alpha) = 1$ when $\alpha = 1$. It follows at once that, since we must have $f(\alpha) \geq 1$ in order to satisfy the Second Law of Thermodynamics, the motion is possible only if $\alpha \geq 1$. But we have already seen that the case $\alpha = 1$ is trivial, so the type of motion in which we are interested can occur only if $\alpha > 1$. This means that the flow must be from a low-

pressure region to a high-pressure region, from a supersonic region to a subsonic region (see equations (41) and
(43)), from a low-density region to a high-density region
(see equation (40)) - these are all different ways of
saying the same thing. Thus we cannot have a discontinuity in a steady flow from a subsonic region to a
supersonic region. But we can have a supersonic flow
which turns subsonic through a stationary discontinuity -
and this, as stated in the previous section, is a
mechanism which can be used to turn the flow subsonic
again downstream of the working section in a supersonic
wind tunnel.
    The 'discontinuity' to which we have referred is, of
course, on a macroscopic scale: on a molecular scale it
turns out that it has a thickness of about three or four
mean free paths. The energy of the molecules on the
upstream side consists mainly of translational energy,
the random part of their motion being comparatively
small. As they meet a region where the molecules are
much closer together (the more dense region) they are
unable to maintain their forward motion; the flow is
adiabatic, so their energy remains constant and after a
few collisions with the molecules of the denser air,
most of it is random and a comparatively small part is
translational. The irreversibility is a typical example
of the fact that any change of state tends to become
more random rather than less, which is a loose
formulation of the Second Law of Thermodynamics.
    If we refer the motion to a frame of reference moving
with the uniform supersonic velocity of the upstream
flow (or, if we superpose a velocity u_1 in the opposite

direction to the flow), we have the situation of a plane
shock wave travelling with speed u_1 into a region of

gas at rest as shown in Figure 3.9. The flow behind the

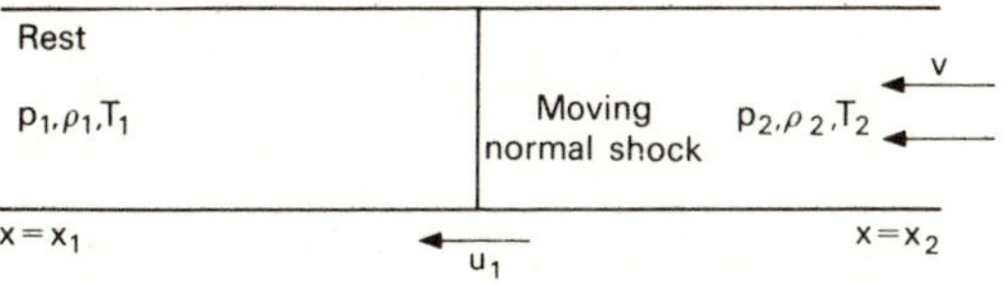

FIGURE 3.9

shock wave has a speed $v = u_1 - u_2$ and Mach number M
where $M^2 = \rho_2 v^2 / \gamma p_2$. (Care must be taken because, with
this notation, the suffix 2 refers to upstream conditions,

not to downstream ones as in the steady case.) Using equation (40) we have

$$\frac{v}{u_1} = 1 - \frac{u_2}{u_1} = \frac{2(\alpha - 1)}{\alpha(\gamma + 1) + \gamma - 1} \qquad (46)$$

and, using (39) and (40),

$$M^2 = \frac{\rho_1 u_1}{\gamma \alpha p_1} \frac{(u_1 - u_2)^2}{u_2} = \frac{(\alpha - 1)(u_1 - u_2)}{\alpha \gamma u_2} .$$

Hence the upstream Mach number M is given by

$$M^2 = \frac{2(\alpha - 1)^2}{\alpha \gamma [\alpha(\gamma - 1) + \gamma + 1]} . \qquad (47)$$

This may be either greater or less than unity so, for a moving shock, the upstream flow may be either supersonic or subsonic. Differentiating equation (47) logarithmically we obtain

$$\frac{2}{M} \frac{dM}{d\alpha} = \frac{\alpha(3\gamma - 1) + \gamma + 1}{\alpha(\alpha - 1)[\alpha(\gamma - 1) + \gamma + 1]}$$

and this is positive for $\alpha > 1$ (since $\gamma > 1$). So, as α increases from unity to infinity, M^2 increases steadily from zero to the value $2/\gamma(\gamma - 1)$. Thus, there is a maximum Mach number which can be achieved behind the shock wave, and this depends on the value of γ for the gas concerned. Since $\gamma < 2$ in practice, this maximum Mach number is always greater than unity. For a monatomic gas, $\gamma = 5/3$ and so the maximum Mach number is 1.34 ; for a diatomic gas, $\gamma = 7/5$ and the maximum Mach number is 1.89 .

These results for the moving shock are used in the *shock tube*. Here a long uniform tube is divided into two parts by a diaphragm. One part (usually the shorter) is filled with gas at high pressure and the other is either partially evacuated or left at atmospheric pressure. At some instant $(t = 0)$ the diaphragm is burst, so that there is a discontinuity of pressure and a shock wave will travel along the tube. This is conveniently represented in an x-t diagram, as shown in Figure 3.10; here, for simplicity, the position of the diaphragm is shown to be in the middle of the tube. At the time $t = t_1 > 0$

shown in the diagram, the shock wave has reached the position B and the contact surface (that is, the surface dividing the fluid which was originally in the low-

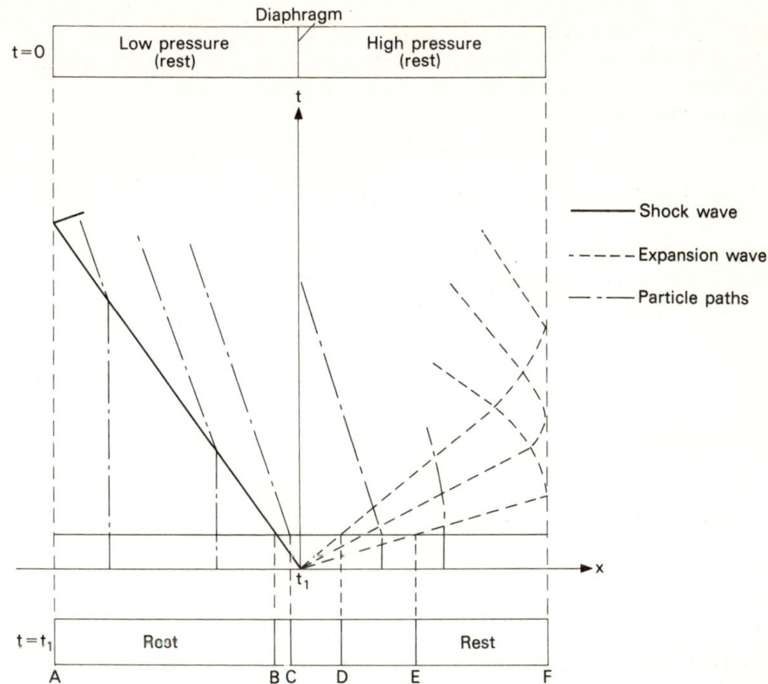

FIGURE 3.10

pressure region from that which was originally in the
high-pressure region) has reached C. Thus the fluid
is at rest in AB, and in the notation of this section,
B moves towards A with speed u_1 and the fluid

between B and C (including C itself) moves towards
A with speed v . An element of fluid originally in
the low-pressure region remains at rest until B reaches
it and then suddenly starts moving with speed v . All
these are clearly shown in the x-t diagram.

On the other side of the contact surface is an
expansion wave and we have seen that this cannot be a
sharp discontinuity. It is beyond the scope of the
present book to discuss such a wave in detail and we
merely describe briefly what happens in the region CF
without explanation. In the region CD the velocity
and pressure of the fluid, but not its temperature and
density, are the same as in region BC ; in the region
DE (which gets larger as time goes on) there is a
continuous variation of all the flow quantities; and in
EF the fluid is still unaffected by the burst of the

diaphragm. Once the shock wave and the expansion wave reach the end of the tube, complicated reflections occur: these can be included in the diagram, but we do not discuss them here.

3.4 THE RIVER BORE WITH NO HORIZONTAL FORCES

Here we consider the flow of a liquid in a straight channel of uniform width b whose horizontal base is the surface $z = 0$ (see Figure 3.11). As in the case

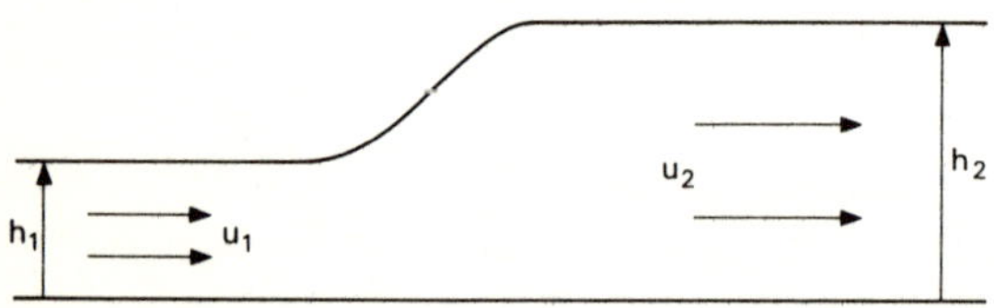

FIGURE 3.11

of the plane shock wave, we look to see if there can be any sharp discontinuities in a flow which is uniform everywhere except in the neighbourhood of the discont- inuities. We suppose that h is the height of the fluid at the station x so that $A = bh$, and we shall also take ρ as constant throughout this section. Then equation (15), which expresses the principle of conservation of mass, becomes

$$h_1 u_1 = h_2 u_2 . \tag{48}$$

The momentum equation has to be modified slightly from the form obtained in equation (26) because the pressure is mainly hydrostatic and is not constant over the cross-section. By considering the equilibrium of a vertical cylinder of fluid between the level z and the surface $(z = h)$ it is easy to see that, since the vertical component of velocity is negligible, $p = \Pi + g\rho(h - z)$ where Π is the pressure of the atmosphere above the surface of the liquid. Hence the horizontal component of the force due to pressure on a vertical cross-section of height h is

$$\int_0^h pb \, dz = b\int_0^h [\Pi + g\rho(h - z)]dz = \Pi bh + \tfrac{1}{2}g\rho bh^2.$$

Between the stations x_1 and x_2 there will be a region where h is not constant: in this region, of course, there will be a horizontal component of force from the pressure on the surface. It is easy to show (and is left as an exercise to the reader) that the magnitude of this component is

$$\int_{h_1}^{h_2} \Pi b \ dz = \Pi b (h_2 - h_1).$$

Hence there is a resultant force whose horizontal component has magnitude

$$(\Pi b h_1 + \tfrac{1}{2} g \rho b h_1^2) - (\Pi b h_2 + \tfrac{1}{2} g \rho b h_2^2) + \Pi b (h_2 - h_1)$$
$$= \tfrac{1}{2} g \rho b (h_1^2 - h_2^2).$$

This must be balanced by a change in the momentum flux through the ends of the region, and this has magnitude $\rho u_2^2 b h_2 - \rho u_1^2 b h_1$. Hence we have

$$\tfrac{1}{2} g \rho b (h_1^2 - h_2^2) = \rho u_2^2 b h_2 - \rho u_1^2 b h_1,$$

and this may be written, after dividing by ρb, in the form

$$\tfrac{1}{2} g (h_1^2 - h_2^2) = u_2^2 h_2 - u_1^2 h_1. \tag{49}$$

Now equations (48) and (49) may easily be combined to give

$$\tfrac{1}{2} g (h_1^2 - h_2^2) = \frac{u_1^2 h_1}{h_2}(h_1 - h_2)$$

and so, since $h_2 \neq h_1$, we have

$$\frac{g h_2 (h_1 + h_2)}{2 h_1} = u_1^2, \qquad \frac{g h_1 (h_1 + h_2)}{2 h_2} = u_2^2. \tag{50}$$

So far we have not discussed what happens to the energy. Once again this is a somewhat more complicated problem than the one we solved in the last chapter. The rate of work done by the pressure forces at the cross-section at the station x is

$$\int_0^h pub \ dz = \int_0^h [\Pi + g\rho(h - z)]ub \ dz = ubh(\Pi + \tfrac{1}{2}g\rho h).$$

So the rate of work done on the fluid between x_1 and x_2 by the pressure forces is, using equation (48),

$$\tfrac{1}{2}\rho gbh_1^2 u_1 - \tfrac{1}{2}\rho gbh_2^2 u_2 = \tfrac{1}{2}\rho gbh_1 u_1 (h_1 - h_2). \tag{51}$$

Also the rate of inflow of kinetic energy through the cross-section at x_1 is $\tfrac{1}{2}\rho u_1^3 bh_1$, and that of potential energy is

$$\int_0^{h_1} g\rho bu_1 z \; dz = \tfrac{1}{2}\rho gbh_1^2 u_1,$$

and there are corresponding expressions for the outflow at x_2. There is, therefore, a net increase of kinetic and potential energy in the region at the rate

$$- (\tfrac{1}{2}\rho u_2^3 bh_2 + \tfrac{1}{2}\rho gbh_2^2 u_2) + (\tfrac{1}{2}\rho u_1^3 bh_1 + \tfrac{1}{2}\rho gbh_1^2 u_1)$$

$$= \tfrac{1}{2}\rho bh_1 u_1 \left[1 - \frac{(h_1 + h_2)^2}{2h_1 h_2} \right] g(h_1 - h_2), \tag{52}$$

using equation (50). In addition work is done on the fluid by the pressure forces and this is given by equation (51). Hence there is a net increase in energy in the region at a rate given by the sum of the expressions given in equations (51) and (52), that is of magnitude

$$g\rho bu_1 (h_2 - h_1)^3 / 4h_2.$$

If $h_1 > h_2$ this expression is negative, and so there must be a source of energy somewhere; but this is physically impossible unless there is an artificial source, a situation which we do not consider. On the other hand, if $h_2 > h_1$, there is a positive gain in energy throughout the region, and for the flow to be possible there must be a mechanism whereby this can be lost. Although such a mechanism cannot exist in the idealized flow we have defined here, it is not difficult to imagine one occurring in the real flow of a real fluid which does not affect the mass and momentum equations used in the discussion. Usually this energy loss occurs through a turbulent region where there is a sudden jump in the depth of the water, or by a series of waves on the surface downstream of the jump; in

either case the disturbances occur somewhere between the
stations x_1 and x_2, and do not effect the mass and
momentum flow at the ends of the region, and it is these
which were used in deriving the equations. The sudden
jump is called a *bore* or a *hydraulic jump*.

It is interesting to note the parallel with the
theory of the gas shock discussed in the previous section.
Both allow discontinuities (or, at least, near discontinuities) and both are irreversible - but the reason for
the irreversibility is different in each case. Further
there is no energy loss in the flow of a gas through a
shock, but there is in the flow of a liquid through a
bore.

In order to demonstrate another similarity between
the two types of flow, we anticipate something we shall
show in Chapter 4 - that the speed with which small
changes in u are propagated in a channel of height

h is $(gh)^{\frac{1}{2}}$. Then we define the non-dimensional ratio

$$F = u/(gh)^{\frac{1}{2}}, \tag{53}$$

and call it the *Froude number*. This corresponds to the
Mach number in the flow of a gas, as this is also the
ratio of the speed of the flow to the speed at which
small disturbances are propagated. When $F = 1$ the
flow is said to be *critical*; otherwise it is *subcritical*
if $F < 1$ and *supercritical* if $F > 1$.

In the present problem we denote the upstream and
downstream Froude numbers by F_1 and F_2 respectively.

Then, using equation (50), we see that

$$F_1^2 = \frac{(h_1 + h_2)h_2}{2h_1^2}, \quad F_2^2 = \frac{(h_1 + h_2)h_1}{2h_2^2}. \tag{54}$$

It follows that, since $h_2 > h_1$ the upstream Froude

number is greater than unity and the downstream one is
less than unity. So the only physically possible type
of flow is from a supercritical region to a subcritical
region - and this corresponds again to the flow through
a stationary shock, in which the gas could flow only
from a supersonic region to a subsonic one.

A very well-known example of a stationary bore was
pointed out by Lighthill in the Ramsden Memorial Lecture
on 'Shock Waves' to the Manchester Literary and
Philosophical Society in 1958. This is not the one-
dimensional bore of the present section, but is a
circular one: it is present in the kitchen sink when

water from the tap is allowed to run into it. A
stationary one-dimensional bore with no momentum loss
also occurs in nature downstream of a bore with no energy
loss, as discussed in the next section.

By using a frame of reference moving with the fluid
downstream of the stationary bore, we have immediately
the case of the moving bore (Figure 3.12). This moves

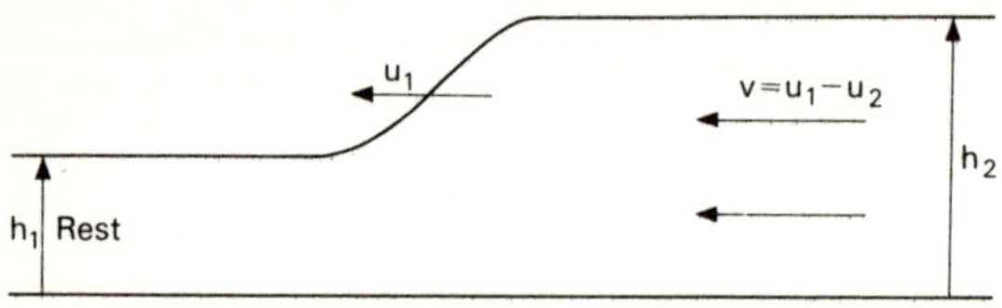

FIGURE 3.12

with speed u_1 into a fluid whose height is h_1 and
which is at rest. The fluid behind the bore has a height
h_2 and moves with speed

$$v = u_1 - u_2 = (h_2 - h_1)\left[\frac{g(h_1 + h_2)}{2h_1 h_2}\right]^{\frac{1}{2}}, \tag{55}$$

using equation (50). Such a phenomenon sometimes occurs
in rivers: an outstanding example is that which occurs in
the Tsien Tang river, and this is sometimes 3 metres high.
The most famous one in Great Britain is that which occurs
in the River Severn: here a sudden rise of nearly a metre
is often observed (see Plate 2).

3.5 THE BORE WITH NO ENERGY LOSS

Two possible causes of this type of bore are suggested
in Figure 3.13: in each case an obstacle is placed in
the way of a uniform flow of fluid whose height is h_1

and whose speed is u_1. There will be a force on the

fluid due to the presence of the obstacle, but we shall
assume that there is no loss of energy. The mass flow
is as in the previous section so that equation (48) still
holds. The momentum balance is no longer given by
equation (49), but this time the net rate of gain of
energy (which is equal to the sum of the expressions
in equations (51) and (52)) is zero. Thus we have

$$u_1 h_1 = u_2 h_2,$$

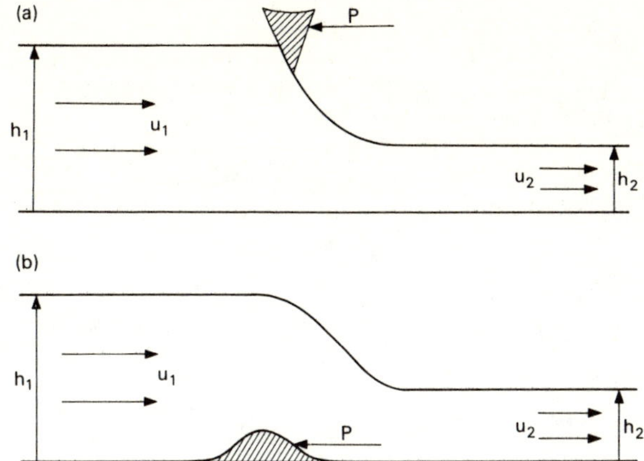

FIGURE 3.13

using equation (48), and

$$\tfrac{1}{2}u_1^2 + gh_1 = \tfrac{1}{2}u_2^2 + gh_2. \tag{56}$$

Note that equation (56) is in fact Bernoulli's equation for the surface streamline: in the previous section we could not use it in this form because there could be a break in the streamline where the flow was turbulent at the discontinuity. Substituting from equation (48) into equation (56), we have (after some manipulation)

$$u_1^2 = \frac{2gh_2^2}{h_1 + h_2} \,,\, u^2 = \frac{2gh_1^2}{h_1 + h_2}.$$

Defining the appropriate Froude numbers F_1, F_2 as in the previous section, we obtain

$$F_1^2 = \frac{2h_2^2}{h_1(h_1 + h_2)} \,,\quad F^2 = \frac{2h_1^2}{h_2(h_1 + h_2)} \,, \tag{57}$$

and so

$$\frac{F_1^2}{F_2^2} = \frac{h_2^3}{h_1^3}. \tag{58}$$

Also, equation (56) gives directly

$$(\tfrac{1}{2}F_1^2 + 1)gh_1 = (\tfrac{1}{2}F_2^2 + 1)gh_2$$

and this, using equation (58), gives

$$\left(\frac{F_1^2 + 2}{F_2^2 + 2}\right)^3 = \frac{h_2^3}{h_1^3} = \frac{F_1^2}{F_2^2} \tag{59}$$

This is essentially a cubic in F_2^2 which can be **factorized into the form**

$$F_1^2(F_2^2 - F_1^2)\left[F_2^4 + (6 + F_1^2)F_2^2 - 8/F_1^2\right] = 0.$$

Now $F_1 \neq 0$ and $F_1 \neq F_2$ since $h_1 \neq h_2$. So the square bracket must be zero and we have

$$F_2^2 = \tfrac{1}{2}\left\{\left[(6 + F_1^2)^2 + 32/F_1^2\right]^{\frac{1}{2}} - (6 + F_1^2)\right\}, \tag{60}$$

where the positive sign of the square root is taken since $F_2^2 \geq 0$. It is easy to verify from equation (60) that, since $F_1^2 > 0$, F_2^2 decreases monotonically as F_1^2 increases, that $F_2^2 \to \infty$ when $F_1^2 \to 0$, that $F_2^2 \to 0$ when $F_1^2 \to \infty$ and that $F_2^2 = 1$ when $F_1^2 = 1$. Hence the relation between F_2^2 and F_1^2 is as shown in Figure 3.14.

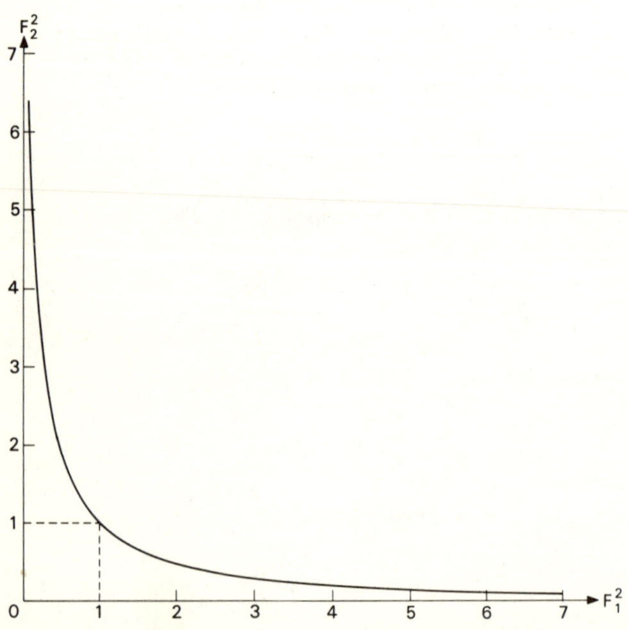

FIGURE 3.14

So, if $F_1 < 1$, then $F_2 > 1$ and if $F_1 > 1$, then $F_2 < 1$. Thus if the flow upstream is subcritical, then the flow downstream is supercritical - and will probably become subcritical again through a hydraulic jump as described in the previous section. Such a system often occurs in nature when a shallow mountain stream flows over a large stone: downstream of the stone the water is shallower than upstream and the flow is supercritical. Very often such a flow becomes subcritical again a few feet downstream of the disturbance.

There is nothing in the theory to suggest that this type of bore cannot occur when the upstream flow is subcritical (that is, when $h_1 < h_2$) but in practice this does not occur, and it is plausible that it should not do so, as we shall see.

To calculate the force P exerted by the stream on the obstacle per unit width (that is, the force on the obstacle, which is assumed to extend right across the channel, is Pb) we consider the momentum balance. Using some of the results we obtained in the derivation of equation (49), we have

$$P = \frac{1}{b}(\text{pressure forces on cross-sections at } x_1 \text{ and } x_2$$
$$+ \text{ flow of momentum across these cross-sections})$$
$$= \frac{1}{b}(\tfrac{1}{2}g\rho bh_1^2 - \tfrac{1}{2}g\rho bh_2^2 - \rho u_2^2 bh_2 + \rho u_1^2 bh_1)$$
$$= \rho g[h_1(\tfrac{1}{2} + F_1^2) - h_2(\tfrac{1}{2} + F_2^2)].$$

Using equation (57) we obtain, after some manipulation,

$$P = \frac{g\rho(h_1 - h_2)^3}{2(h_1 + h_2)}. \tag{61}$$

This implies that, as long as $h_1 > h_2$, there is a loss of momentum as the stream flows past the obstacle, and this is what we would expect. But there is no basic physical principle which says that this must be so, since P is a force applied from outside the system.

EXERCISES

1 An ideal gas flows adiabatically along a horizontal pipe whose slowly varying cross-section is A. The motion is steady. Show that the density ρ and the area A increase or decrease together when the flow is

subsonic, but that the opposite is true when the flow is supersonic.

When the Mach number M of the flow is very much less than unity, show that

$$\frac{A}{\rho}\frac{d\rho}{dA} \simeq M^2.$$

2 An ideal gas, whose ratio of specific heats is γ, flows adiabatically through a Laval nozzle, the motion being steady. The rate of mass flow is m, the stagnation pressure is p_0, the stagnation density is ρ_0 and the stagnation speed of sound is a_0. The area of cross-section at the throat is α, the Mach number of the flow at the throat is M_T and the pressure at the nozzle exit (far downstream from the throat) is p_1. Show that

(a) $m = \rho_0 a_0 \alpha M_T \left[1 + \tfrac{1}{2}(\gamma - 1)M_T^2\right]^{-(\gamma+1)/2(\gamma-1)}$;

(b) m cannot exceed the value $\rho_0 a_0 \alpha \left[\dfrac{2}{\gamma + 1}\right]^{(\gamma+1)/2(\gamma-1)}$;

(c) when the flow downstream of the nozzle is supersonic (so that $M_T = 1$), then $p_1 < p_0 \left(\dfrac{2}{\gamma + 1}\right)^{\gamma/(\gamma-1)}$.

3 Using the Rankine-Hugoniot equations, show that

$$u_1 u_2 = a^{*2},$$

where u_1 is the velocity of the gas upstream of a normal shock, u_2 is that downstream, and a^* is the sonic speed for the flow.

4 A shock moves with speed U along a tube of uniform cross-section into a gas which is initially at rest; the speed of the gas immediately behind the shock is V, its Mach number is $M_2 = V/a_2$ (where $a_2^2 = \gamma p_2/\rho_2$), the ratio of its pressure to the pressure in front of the shock is α, and the ratio of specific heats is everywhere γ. Use the results for a stationary normal shock to show that

(a) $\dfrac{U - V}{U} = \dfrac{\gamma + 1 + \alpha(\gamma - 1)}{\gamma - 1 + \alpha(\gamma + 1)}$;

(b) $M_2^2 = \dfrac{2(\alpha - 1)^2}{\gamma\alpha[\alpha(\gamma - 1) + \gamma + 1]}$;

(c) M_2 increases monotonically with α and can never exceed the value $[2/\gamma(\gamma - 1)]^{\frac{1}{2}}$.

5   A horizontal semi-infinite pipe of uniform cross-section contains an ideal gas which is initially at rest. Suddenly a piston begins to move along the pipe with constant speed V.   Determine the flow in the gas into which the piston is moving.

6   An ideal gas, whose ratio of specific heats is γ, flows steadily and adiabatically along a pipe of uniform cross-section from region 1 of the pipe through a stationary normal shock into region 2 of the pipe.   The ratio of the pressure jump across the shock to the upstream pressure is $(p_2 - p_1)/p_1 = \varepsilon$, where $\varepsilon \ll 1$. Show that

$$\frac{\rho_2 - \rho_1}{\rho_1} = - \frac{u_2 - u_1}{u_1} \simeq \frac{\varepsilon}{\gamma} ,$$

$$M_1^2 = 1 + \frac{\gamma + 1}{2\gamma}\varepsilon, \quad M_2^2 \simeq 1 - \frac{\gamma + 1}{2\gamma}\varepsilon,$$

and that the change in entropy is

$$S_2 - S_1 \simeq \frac{\gamma^2 - 1}{12\gamma^2} c_v \varepsilon^3 .$$

7   An ideal gas, whose ratio of specific heats is γ, passes through a stationary plane shock which makes an angle ϕ with the direction of $\underset{\sim}{q}_1$, the velocity of the gas upstream of the shock.   Downstream of the shock, the gas moves with velocity $\underset{\sim}{q}_2$ and the angle between $\underset{\sim}{q}_1$ and $\underset{\sim}{q}_2$ is θ.   Write down the equations for:

  (a) the continuity of the tangential velocity,
  (b) the conservation of mass,
  (c) the momentum balance, and
  (d) the conservation of energy

across the shock.   If M_1 is the Mach number of the flow in the region upstream of the shock, show that

$$\cot\theta = \tan\phi\left[\frac{(\gamma + 1)M_1^2}{2(M_1^2\sin^2\phi - 1)} - 1\right].$$

Show also that, if M_2 is the Mach number downstream of the shock, then

$$M_2^2 = \frac{(\gamma - 1)M_1^2 \sin^2\phi + 2}{2\gamma M_1^2 \sin^2\phi - \gamma + 1} \operatorname{cosec}^2(\phi - \theta).$$

8 An ideal gas flows along a tube of uniform cross-section. At a certain point in the tube, a quantity of heat is added to the gas (that is, its stagnation temperature is increased). Far enough away from the place where the heat is added, the flow may be assumed to be uniform. When the oncoming flow is supersonic, show that when the heat has been added (a) the Mach number and the fluid speed decrease, and (b) the temperature increases. When the oncoming flow is subsonic, show that the Mach number and the fluid speed increase. Can you say anything about the temperature in the second case?

What is the downstream Mach number when the stagnation temperature is increased by $T_1(1 - M_1^2)^2/2(\gamma + 1)M_1^2$, where the suffix 1 denotes upstream values?

9 The plane shock of question 4 meets and is reflected from a rigid plane wall parallel to itself. If the pressure behind the incident shock is p_2 and that behind the reflected shock is p_3, show that

$$\frac{p_3}{p_2} = \frac{(3\gamma - 1)\alpha - \gamma + 1}{(\gamma - 1)\alpha + \gamma + 1},$$

where α and γ are as before.

If W is the speed of the reflected shock, find an expression for W/U in terms of α and γ, and show that W is less than U.

10 A long horizontal channel has slowly varying breadth. Water flows along the channel, and it may be assumed that its velocity is uniform over any vertical section of the channel. If the depth of the fluid is H and its velocity is $2\sqrt{(gH)}$ in magnitude at a point where the breadth is B, show that steady flow can occur only if the breadth of the channel is nowhere less than $B/\sqrt{2}$.

11 The bottom of a long channel of uniform breadth is in the surface $z = f(x)$, where the z-axis points vertically upwards and the x-axis is parallel to the sides of the channel. Water flows steadily along the

channel in the positive direction of the x-axis, the
velocity being uniform over each vertical section of
the channel. In a certain region of the channel $f(x)$
is slowly increasing. If the speed of the fluid is
u when its depth is h , show that u^2 > or < gh
according as the free surface rises or falls there.

12 Water flows steadily through an open channel of
rectangular cross-section whose breadth is b ; the
depth of the water is h and the volume flow per unit
time is q .   The water originates from a large deep
lake, and the depth of the bottom of the channel below
the free surface of the lake is d .   The quantities
b, d and h all vary slowly along the channel, and it
may be assumed that the velocity is uniform across the
channel.   Show that, when the speed of the water in the
lake is negligible compared with that in the channel,

$$\frac{h^2}{d^2}\left(1 - \frac{h}{d}\right) = \frac{q^2}{2gb^2d^3} \,.$$

Show that such a steady flow can occur only if b^2d^3
is not less than a certain value. Find this value in
terms of q and show that, if there is a point where
b^2d^3 takes this value, the depth of the water in the
channel there is twice the difference in heights of the
surfaces there and in the lake.

13 A hydraulic jump (or bore) moves along a channel of
uniform width into water which is initially at rest and
of depth h .   The water behind the jump has depth αh .
The jump is subsequently reflected from a rigid plane
parallel to the jump. If the depth of water behind the
reflected jump is βh , show that

$$\beta = \tfrac{1}{2}\left\{\alpha - 1 + \left[(\alpha - 1)^2 + 4\alpha^3\right]^{\frac{1}{2}}\right\}.$$

(It may be assumed that there are no momentum losses
except at the reflection.)

Chapter 4

SOME EXAMPLES OF UNSTEADY ONE–DIMENSIONAL FLOW

4.1 THE GENERAL CASE

In this chapter we shall still restrict ourselves to the study of the flow of an inviscid fluid in one dimension, but the flow will no longer be assumed to be steady. We shall consider the flow along a straight tube and measure distance along the tube by the space coordinate x , exactly as for the steady flows we have already discussed; but this time we suppose tbat all the variables depend on x and also on the time t , but not on any other variables. The tube cannot now be regarded as a streamtube, in general, because such a streamtube would not always contain the same fluid. The walls of the tube may, however, be either fixed (as in an organ pipe) or free surfaces (as in shallow water waves).

We consider the fluid in a region of the tube between the fixed stations x and $x + \xi$, where we shall take ξ to be small (see Figure 4.1). The cross-section at

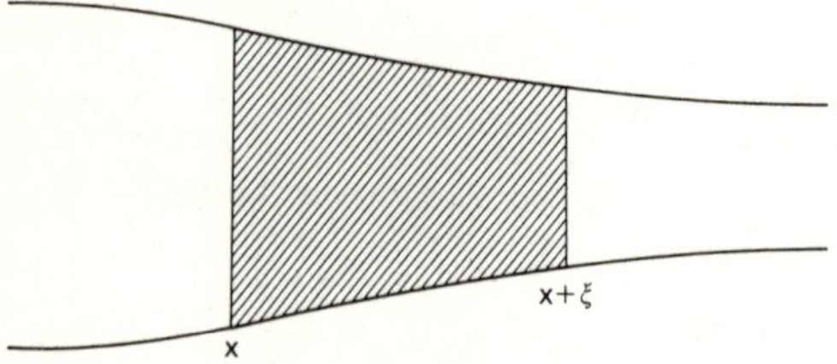

FIGURE 4.1

the station x has area A , exactly as in the steady case except that we now have $A = A(x,t)$. The mass of fluid in this region is, to the first order in ξ,

$\rho A \xi$ and its rate of increase is $\xi \dfrac{\partial}{\partial t}(\rho A)$. This rate of increase of mass can be caused only by inflow through the cross-section at x and outflow through that at $x + \xi$; hence, using suffices to denote where quantities are to be evaluated, we have

$$\xi \frac{\partial}{\partial t}(\rho A) = (\rho u A)_x - (\rho u A)_{x+\xi}$$

$$= - \xi \frac{\partial}{\partial x}(\rho u A)$$

to the first order in ξ. Hence the equation expressing conservation of mass is

$$\frac{\partial}{\partial t}(\rho A) + \frac{\partial}{\partial x}(\rho u A) = 0 . \tag{62}$$

Note that, if ρ and A are independent of time, this equation reduces to $\frac{\partial}{\partial x}(\rho u A) = 0$ and this integrates to give equation (15), as in the steady case.

We next consider the momentum of the fluid in the region. The rate of increase of this is $\xi \frac{\partial}{\partial t}(\rho u A)$, and this increase is caused partly by inflow and outflow of momentum through the ends of the region as it is carried by the fluid, and partly by the pressure forces on the fluid. (We shall consider only problems in which there is no component of external body force on the fluid in the x-direction.) The rate at which the momentum is increased by being carried by the fluid across the ends of the region is

$$(\rho u^2 A)_x - (\rho u^2 A)_{x+\xi} = - \xi \frac{\partial}{\partial x}(\rho u^2 A)$$

to the first order in ξ. If we assume that the pressure is constant over any given cross-section,* the forces due to the pressure are made up of three parts:

(i) $(pA)_x$, the force on the cross-section at x,

(ii) $-(pA)_{x+\xi}$, that on the cross-section at $x + \xi$,

and

(iii) $\iint_S p\ell\,dS$, that on the curved part S of the

surface containing the region under consideration, where

---

* If this is not so, as in Section 4.4, we must divide the tube into a large number of tubes of cross-section sufficiently small for the variation to be negligible. The rest of the argument is then true for each of the elementary tubes.

ℓ is the cosine of the angle made with the x-axis by the outward pointing normal $\hat{\tilde{n}}$ of the element δS of the surface, as shown in Figure 4.2. Now we have

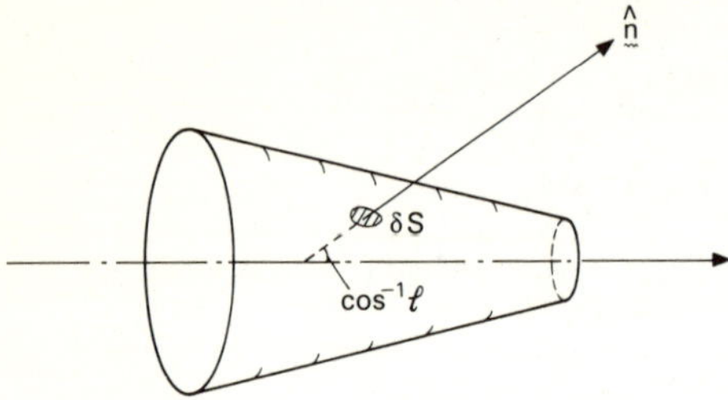

FIGURE 4.2

$$\iint_S p\ell\, dS = p \iint_S \ell\, dS$$

to the first order in ξ, and as can be seen from the diagram,

$$\iint_S \ell\, dS = A_{x+\xi} - A_x = \xi\frac{\partial A}{\partial x}$$

to the same order. Hence, to this order, the pressure forces on the fluid are given by

$$(pA)_x - (pA)_{x+\xi} + p\xi\frac{\partial A}{\partial x} = -\xi\frac{\partial}{\partial x}(pA) + p\xi\frac{\partial A}{\partial x} = -\xi A\frac{\partial p}{\partial x}.$$

Putting all this together, we have

$$\xi\frac{\partial}{\partial t}(\rho uA) = -\xi\frac{\partial}{\partial x}(\rho u^2 A) - \xi A\frac{\partial p}{\partial x}$$

to the first order in ξ. Cancelling the factor ξ and rearranging we have

$$u\left[\frac{\partial(\rho A)}{\partial t} + \frac{\partial(\rho uA)}{\partial x}\right] + \rho A\left(\frac{\partial u}{\partial t} + u\frac{\partial u}{\partial x}\right) = -A\frac{\partial p}{\partial x}.$$

This may be combined with equation (62) to give

$$\rho\frac{\partial u}{\partial t} + \rho u\frac{\partial u}{\partial x} = -\frac{\partial p}{\partial x}. \tag{63}$$

Note that the operator D/Dt defined by equation (9) reduces in this case to $\frac{\partial}{\partial t} + u\frac{\partial}{\partial x}$. So an alternative form for equation (63) is

$$\rho \frac{Du}{Dt} = - \frac{\partial p}{\partial x} . \tag{64}$$

For the purposes of this chapter we shall not need the energy equation, and so shall leave its derivation to Chapter 7, where we discuss the more general case.

4.2 SOUND WAVES IN A PIPE

Here we shall consider what happens when a gas contained in a tube of constant cross-section suffers a small disturbance from equilibrium when its density ρ_0 is a constant. We shall consider the case when the disturbance is in the form of longitudinal vibrations and we shall assume that, throughout the motion, the pressure p is a function of the density ρ only (and not explicitly of temperature). Such a gas is said to be *barotropic*. Then we shall write $a^2 = dp/d\rho$ and a_0^2 for the value of a^2 when $\rho = \rho_0$. Under these circumstances, equation (63) can be written

$$\rho \frac{\partial u}{\partial t} + \rho u \frac{\partial u}{\partial x} = - a^2 \frac{\partial \rho}{\partial x} . \tag{65}$$

Now we assume that $\rho - \rho_0$ and u are small quantities of the first order, and that the same is true of their derivatives. Then, if we neglect small quantitites of the second order, equation (62) becomes

$$\frac{\partial \rho}{\partial t} + \rho_0 \frac{\partial u}{\partial x} = 0 , \tag{66}$$

since A is constant. Also equation (65) is, to the same order,

$$\rho_0 \frac{\partial u}{\partial t} = - a_0^2 \frac{\partial \rho}{\partial x}, \tag{67}$$

where we have used the fact that $a^2 - a_0^2$ is small since $\rho - \rho_0$ is small. Equations (66) and (67) combine to give

$$\frac{\partial^2 \rho}{\partial t^2} = a_0^2 \frac{\partial^2 \rho}{\partial x^2}, \quad \frac{\partial^2 u}{\partial t^2} = a_0^2 \frac{\partial^2 u}{\partial x^2}. \tag{68}$$

These are examples of the one-dimensional wave equation.*
Hence both the density disturbance and the velocity are
propagated along the tube with speed a_0. Since this
type of motion is that which occurs when sound is
propagated, the quantity a_0 is called the 'speed of
sound in a medium at rest'. It should be noticed that,
although the speed u of the fluid is small, the speed
a_0 of propagation is usually large. The speed of
propagation can be thought of as the speed at which the
molecules of the fluid can transmit information (about
ρ or about u); because of the high speed of the
molecules, and the many collisions which take place,
this transmission of information can occur very rapidly.
The speed of the fluid, however, is the magnitude of
the mean velocity of the molecules and this is small.
To find out what the value of a_0 is, we need to
know the form of the relation between p and ρ.
Newton thought that the temperature would remain constant
throughout the motion: if this were the case, we would
have, by differentiating equation (1),

$$a^2 = (\partial p/\partial \rho)_T = RT .$$

For air at normal temperatures this gives a value for
a_0 which is about 20 per cent too small. What seems
to happen in fact is that the motion is isentropic -
this implies that the motions are so fast that there
is not time for a significant exchange of heat between
elements of the fluid during one period of the
oscillation. In this case we have, using equation (8),

$$a^2 = (\partial p/\partial \rho)_{isentropic} = K\gamma\rho^{\gamma-1} = \gamma p/\rho = \gamma RT. \qquad (69)$$

---

* It is easily verified that the solution of the equation
$\partial^2 y/\partial t^2 = c^2 \, \partial y/\partial x^2$ is $y = y_1(x + ct) + y_2(x - ct)$,
where y_1 and y_2 are arbitrary functions. The first
term in this solution can be thought of as a function
of x which travels with speed c in the negative
direction of the x-axis without change of shape, and
the second as a function of x which travels with
speed c in the position direction of the x-axis
without change of shape. The speed c is often
referred to as the *speed of propagation* or the *wave
speed*.

This gives a value for a_0 of 340 m/s for air at a temperature 280°K, and this agrees with experiment. We shall therefore use equation (69) as the appropriate equation for a^2; in fact we did this in Section 2.5.

When we solve the equation (68) for the speed u, we use the boundary conditions that $u = 0$ at a fixed end of the pipe and that at an end of the pipe open to the atmosphere, the pressure is independent of time. This second condition means that ρ is independent of time also (since p is a function of ρ) and so $\partial\rho/\partial t = 0$. Hence, using equation (66), the condition at an open end is $\partial u/\partial x = 0$. (When we solve the equation (68) for ρ, it is easy to verify that the corresponding boundary condition at a closed end is $\partial\rho/\partial x = 0$, and we have already seen that at an open end $\partial\rho/\partial t = 0$.)

Suppose now that we have a pipe (for example, an organ pipe) of length b; when the air vibrates in the pipe its speed satisfies the second of equations (68) which has the general solution

$$u = \sum_k (A_k \cos kx + B_k \sin kx)\cos(ka_0 t + \alpha_k). \quad (70)$$

We shall consider such a pipe with three different sets of boundary conditions.

CASE 1   A pipe closed at both ends: $u = 0$ when $x = 0$ and when $x = b$.

Since $u = 0$ when $x = 0$, we have

$$\sum_k A_k \cos(ka_0 t + \alpha_k) = 0$$

for all values of t. This is possible only if $A_k = 0$ for all k.

Since $u = 0$ when $x = b$, we have

$$\sum_k B_k \sin kb \, \cos(ka_0 t + \alpha_k) = 0$$

for all values of t, and this is possible only if $\sin kb = 0$ (the solution in which $B_k = 0$ for all k is trivial). It follows that $kb = n\pi$ where n is an integer. Hence, rewriting B_k, α_k as B_n, α_n respectively, we have

$$u = \sum_{n=1}^{\infty} B_n \sin \frac{n\pi x}{b} \cos\left(\frac{n\pi a_0 t}{b} + \alpha_n\right).$$

The constants α_n, B_n can be determined from the initial conditions if these are known. We see that the only frequencies which can be present in the disturbance are (in radians per second) $n\pi a_0/b$. This means that the lowest frequency present is $\pi a_0/b$, and this is called the

fundamental frequency. Integer multiples of this frequency are called *harmonics*, the frequency $n\pi a_0/b$ being known as the (n-1)th *overtone*.

Thus, for a pipe closed at both ends, it is possible for all the harmonics to be present. In practice it may happen that some of them are absent (that is, if some of the B_n are zero), but this would be an accident of the initial conditions rather than a restriction inherent in the tube itself.

CASE 2   A pipe open at both ends: $\partial u/\partial x = 0$ when $x = 0$ and $x = b$.
   Differentiating equation (70), we obtain

$$\frac{\partial u}{\partial x} = \sum_k (-kA_k \sin kx + kB_k \cos kx)\cos(ka_0 t + \alpha_k).$$

Since $\partial u/\partial x$ when $x = 0$, we have

$$\sum_k kB_k \cos(ka_0 t + \alpha_k) = 0$$

for all values of t.   Hence $B_k = 0$ for all values of k.
   Since $\partial u/\partial x = 0$ when $x = b$, we have

$$\sum_k kA_k \sin kb \cos(ka_0 t + \alpha_k) = 0$$

for all values of t, and (as before) this is possible only if $\sin kb = 0$. Hence, as in case 1, $kb = n\pi$. The pipe has therefore exactly the same fundamental frequency $\pi a_0/b$ and all the harmonics may again be present.   The expression for u may in this case be written

$$u = \sum_{n=0}^{\infty} A_n \cos\frac{n\pi x}{b} \cos\left(\frac{n\pi a_0 t}{b} + \alpha_n\right).$$

TABLE 4.1   Sound waves in a pipe of constant width

Case 1: closed at both ends	Case 2: open at both ends	Case 3: closed at one end, open at the other
Fundamental $\omega = \dfrac{\pi a_0}{b}$, $\lambda = 2b$	Fundamental $\omega = \dfrac{\pi a_0}{b}$, $\lambda = 2b$	Fundamental $\omega = \dfrac{\pi a_0}{b}$, $\lambda = 4b$
First harmonic $\omega = \dfrac{2\pi a_0}{b}$, $\lambda = b$	First harmonic $\omega = \dfrac{2\pi a_0}{b}$, $\lambda = b$	First harmonic $\omega = \dfrac{3\pi a_0}{2b}$, $\lambda = \dfrac{4b}{3}$
Second harmonic $\omega = \dfrac{3\pi a_0}{b}$, $\lambda = \dfrac{2b}{3}$	Second harmonic $\omega = \dfrac{3\pi a_0}{b}$, $\lambda = \dfrac{2b}{3}$	Second harmonic $\omega = \dfrac{5\pi a_0}{2b}$, $\lambda = \dfrac{4b}{5}$

CASE 3   A pipe closed at one end and open at the other: $u = 0$ when $x = 0$ and $\partial u/\partial x = 0$ when $x = b$.

As in case 1 the condition that $u = 0$ when $x = 0$ gives the result that $A_k = 0$ for all values of k. Applying the other boundary condition, we have

$$\sum_k kB_k \cos kb \, \cos(ka_0 t + \alpha_k) = 0$$

for all values of t, and this is possible only if $\cos kb = 0$. We must therefore have $kb = (2n + 1)\pi/2$ where n is an integer. The expression for u is therefore

$$u = \sum_{n=0}^{\infty} B_n \sin\left[\frac{(2n + 1)\pi x}{2b} \cos \frac{(2n + 1)\pi a_0 t}{2b} + \alpha_n\right],$$

and the possible frequencies (in radians per second) are $(2n + 1)\pi a_0/2b$. In this case, therefore, the fundamental frequency is $\pi a_0/2b$ (which is half that of a pipe of the same length closed at both ends or open at both ends) and only the odd harmonics are present.

In Table 4.1 are listed the first three possible frequencies, ω, of oscillation in the three different cases discussed above, together with the corresponding wavelengths λ, which are given by the relation $\omega\lambda = 2\pi a_0$. Together with each pair (of frequency and wavelength) is a diagram which essentially incorporates the graphs of u and $-u$ for the appropriate part of the solution at a fixed value of t. As t varies, the only thing that happens to these diagrams is that their amplitude varies. It is easy to see from the diagrams how it comes about that in cases 1 and 2 all the harmonics are permissible, but only the odd ones in case 3.

4.3 THE PROPAGATION OF FINITE DISTURBANCES IN AN IDEAL GAS

Here we consider reversible adiabatic motions in a pipe of constant cross-section as in the previous section, but we no longer assume that u and $\rho - \rho_0$ are small. In this case, equation (62) becomes

$$\frac{\partial \rho}{\partial t} + u \frac{\partial \rho}{\partial x} + \rho \frac{\partial u}{\partial x} = 0 \tag{71}$$

and equation (65) is still

$$\rho \frac{\partial u}{\partial t} + \rho u \frac{\partial u}{\partial x} + a^2 \frac{\partial \rho}{\partial x} = 0.$$

Since the gas is an ideal gas we have equation (69), and this gives $a^2 = K\gamma\rho^{\gamma-1}$. Differentiating this logarithmically, we obtain

$$\frac{2}{a} = (\gamma - 1) \frac{1}{\rho} \frac{d\rho}{da} .$$

So equations (71) and (65) may be written

$$\left(\frac{2}{\gamma - 1}\right)\frac{\partial a}{\partial t} + \left(\frac{2u}{\gamma - 1}\right)\frac{\partial a}{\partial x} + a \frac{\partial u}{\partial x} = 0, \qquad (72a)$$

$$\frac{\partial u}{\partial t} + u \frac{\partial u}{\partial x} + \left(\frac{2a}{\gamma - 1}\right)\frac{\partial a}{\partial x} = 0. \qquad (72b)$$

Adding the two equations (72) we have

$$\frac{\partial}{\partial t}\left[u + \left(\frac{2a}{\gamma - 1}\right)\right] + (u + a) \frac{\partial}{\partial x}\left[u + \left(\frac{2a}{\gamma - 1}\right)\right] = 0, \qquad (73)$$

and subtracting them, we have

$$\frac{\partial}{\partial t}\left[u - \left(\frac{2a}{\gamma - 1}\right)\right] + (u - a) \frac{\partial}{\partial x}\left[u - \left(\frac{2a}{\gamma - 1}\right)\right] = 0. \qquad (74)$$

Suppose we now consider one of the curves in the x-t plane (compare Figure 3.10 in Section 3.3) which is defined by $dx/dt = u + a$. If we measure s along the curve and consider the variation of the quantity $f = u + \frac{2a}{\gamma - 1}$ along the curve, we have

$$\frac{\partial f}{\partial s} = \frac{\partial f}{\partial t}\frac{dt}{ds} + \frac{\partial f}{\partial x}\frac{dx}{ds}$$

$$= \frac{dt}{ds}\left[\frac{\partial f}{\partial t} + \frac{dx}{dt}\frac{\partial f}{\partial x}\right]$$

$$= \frac{dt}{ds}\left[\frac{\partial f}{\partial t} + (u + a) \frac{\partial f}{\partial x}\right],$$

since on this particular curve $dx/dt = u + a$, by definition. But, using equation (73), we see that the square bracket in the expression for $\partial f/\partial s$ vanishes. Hence $f = u + \frac{2a}{\gamma - 1}$ does not vary along the curve. Another way of saying this is that any disturbance (whether small or large) to the quantity $u + \frac{2a}{\gamma - 1}$ at some instant is propagated in the x-t plane along a curve given by $dx/dt = u + a$; that is, it is propagated locally with speed $u + a$. This, in turn, is another way of saying that the disturbance is propagated with speed a, relative to the fluid in the same direction as the fluid is moving.

Similarly, by consideration of equation (74), we can show that the quantity $u - \dfrac{2a}{\gamma - 1}$ is constant along curves on which $dx/dt = u - a$. This means that a disturbance to this quantity at some instant is propagated locally with speed u - a, that is with speed a relative to the fluid, but in the opposite direction to that in which the fluid is moving.

The families of curves in the x-t plane defined by $\dfrac{dx}{dt} = u \pm a$ are called *characteristics:* usually the curves of the family corresponding to the plus sign (along which forward propagation occurs) are called positive characteristics, and those corresponding to the negative sign (along which backward propagation occurs) are called negative characteristics.

It is instructive to see what happens to a disturbance to the quantity $f = u + \dfrac{2a}{\gamma - 1}$, which is propagated along a positive characteristic, by studying the x-t diagram (see Figure 4.3). The form of the function when t = 0 is shown in the upper part of the figure: it has a hump in the centre (between B and E) and a region in the middle of the hump (between C and D) in which the value of the function is constant. We shall assume that an increase in f implies an increase in u + a and that both are positive everywhere. The lines in the x-t diagram are the lines whose slopes are u + a, and so disturbances are propagated along these lines. (The exact slope of these lines depends on the values of u + a when t = 0.) It is easy to see that the part of the disturbance which is at the rear of the hump in the graph of f against x gradually spreads out as time increases, whereas that in the front gets narrower. Ultimately the characteristics in the x-t diagram from the region CD intersect those from EF, thus giving two different values for f at a given point at a given time. This is physically impossible and so a discontinuity forms - a shock wave.

If it happens, as it sometimes does (for example in a shock tube), that the quantity $u - \dfrac{2a}{\gamma - 1}$ remains constant in some region of the flow, knowledge of the variation of $u + \dfrac{2a}{\gamma - 1}$ is sufficient to determine the

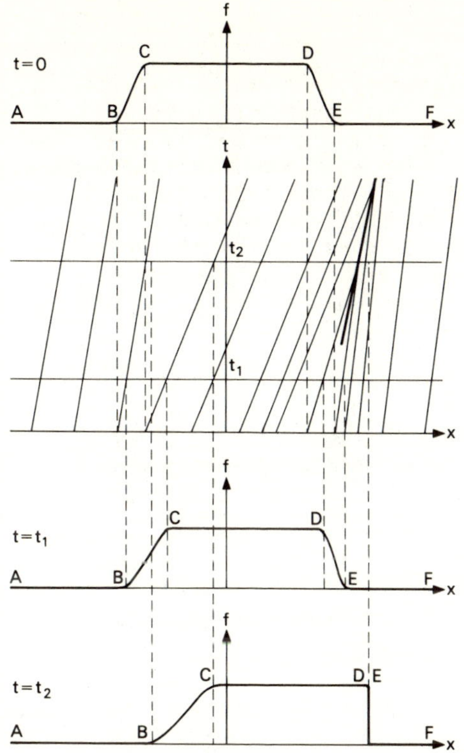

FIGURE 4.3

flow. In particular, if there is a discontinuity in $u + \dfrac{2a}{\gamma - 1}$, then clearly u and a are discontinuous as well. In fact such discontinuities are to be expected in general, although the argument is more complicated when neither $u - \dfrac{2a}{\gamma - 1}$ nor $u + \dfrac{2a}{\gamma - 1}$ is constant: in such a case we have to consider the propagation of both quantities along their own characteristics.

The fact that the quantity $a = (dp/d\rho)^{\frac{1}{2}}$ is the characteristic speed of propagation of disturbances relative to the fluid, and that these disturbances are audible to the human ear as sound, is the justification for the term 'speed of sound' used in Section 2.5.

4.4 SHALLOW WATER WAVES

Here we consider a liquid (usually, but not necessarily, water) flowing along a horizontal channel of constant width b, whose height h is a function of x, the distance along the channel. The shallowness of the liquid is a necessary condition for the velocity u to be independent of z (and we shall see in Chapter 10 that the required condition for this is that the depth of the fluid is very much less than the wavelength of the wave motion under consideration).
Equation (62) now becomes

$$\frac{\partial h}{\partial t} + \frac{\partial}{\partial x} (hu) = 0 ; \tag{75}$$

equation (63) was derived for the special case in which $\partial p / \partial z = 0$, and so it cannot be applied directly here. As in Section 3.4 we find that, since we are neglecting vertical motions,

$$p = \Pi + g\rho (h - z),$$

and so the x-components of the pressure forces on the fluid between the stations x and $x + \xi$ are

(i) $\left(\displaystyle\int_0^h pb \, dz \right)_x = (\Pi hb + \tfrac{1}{2}\rho gh^2 b)_x$ on the cross-

section at x,

(ii) $\left(\displaystyle\int_0^h pb \, dz \right)_{x+\xi} = (\Pi hb + \tfrac{1}{2}\rho gh^2 b)_{x+\xi}$ on that at

$x + \xi$, and

(iii) $\displaystyle\iint_S p\ell \, dS$, where S is now the free surface of

the liquid,

since the forces on the side walls and the base of the canal have no component in the x-direction. These add up to give, to the first order in ξ,

$$- \xi \frac{\partial}{\partial x} (\Pi hb + \tfrac{1}{2}\rho gh^2 b) + \Pi b [h(x + \xi) - h(x)]$$

$$= - gb\rho \xi h \frac{\partial h}{\partial x} .$$

The equation for momentum balance is therefore

$$\xi \frac{\partial}{\partial t} (\rho ubh) = - \xi \frac{\partial}{\partial x} (\rho u^2 bh) - \rho gb\xi h \frac{\partial h}{\partial x},$$

and this may be combined with equation (75) to give

$$\frac{\partial u}{\partial t} + u\frac{\partial u}{\partial x} + g\frac{\partial h}{\partial x} = 0 . \tag{76}$$

We consider first the case in which we have a small disturbance to a liquid at rest which has an undisturbed depth h_0. Then equations (75) and (76) reduce to

$$\frac{\partial h}{\partial t} + h_0\frac{\partial u}{\partial x} = 0 , \quad \frac{\partial u}{\partial t} + g\frac{\partial h}{\partial x} = 0 \tag{77}$$

to the first order in the small quantities u and $h - h_0$ and their derivatives. These equations combine to give

$$\frac{\partial^2 h}{\partial t^2} = gh_0\frac{\partial^2 h}{\partial x^2} , \quad \frac{\partial^2 u}{\partial t^2} = gh_0\frac{\partial^2 u}{\partial x^2} ; \tag{78}$$

so disturbances are propagated with speed $(gh_0)^{\frac{1}{2}}$, as we anticipated in Section 3.4.

When u and $h - h_0$ are not small, we cannot approximate to equations (75) and (76), but we can manipulate them into a more convenient form. If we multiply equation (75) by $(g/h)^{\frac{1}{2}}$, we obtain

$$\left(\frac{g}{h}\right)^{\frac{1}{2}}\frac{\partial h}{\partial t} + \left(\frac{g}{h}\right)^{\frac{1}{2}} u\frac{\partial h}{\partial x} + (gh)^{\frac{1}{2}}\frac{\partial u}{\partial x} = 0,$$

which may be written in the form

$$\frac{\partial}{\partial t}\left(2(gh)^{\frac{1}{2}}\right) + u\frac{\partial}{\partial x}\left(2(gh)^{\frac{1}{2}}\right) + (gh)^{\frac{1}{2}}\frac{\partial u}{\partial x} = 0. \tag{79}$$

Adding equations (76) and (79) we have

$$\frac{\partial}{\partial t}\left[u + 2(gh)^{\frac{1}{2}}\right] + \left[u - (gh)^{\frac{1}{2}}\right]\frac{\partial}{\partial x}\left[u + 2(gh)^{\frac{1}{2}}\right] = 0, \tag{80}$$

and, subtracting them, we have

$$\frac{\partial}{\partial t}\left[u - 2(gh)^{\frac{1}{2}}\right] + \left[u - (gh)^{\frac{1}{2}}\right]\frac{\partial}{\partial x}\left[u - 2(gh)^{\frac{1}{2}}\right] = 0. \tag{81}$$

It follows from equation (80) that, in general, any disturbance to the quantity $u + 2(gh)^{\frac{1}{2}}$ propagates with speed $(gh)^{\frac{1}{2}}$ relative to the liquid in the same direction as the fluid is flowing. Similarly, we see, from equation (81), that any disturbance to the quantity

$u - 2(gh)^{\frac{1}{2}}$ propagates with speed $(gh)^{\frac{1}{2}}$ relative to the liquid but in the opposite direction to that in which the fluid is flowing.

It follows that, if we have a surface whose shape is that shown between D and E of Figure 4.4, the part of

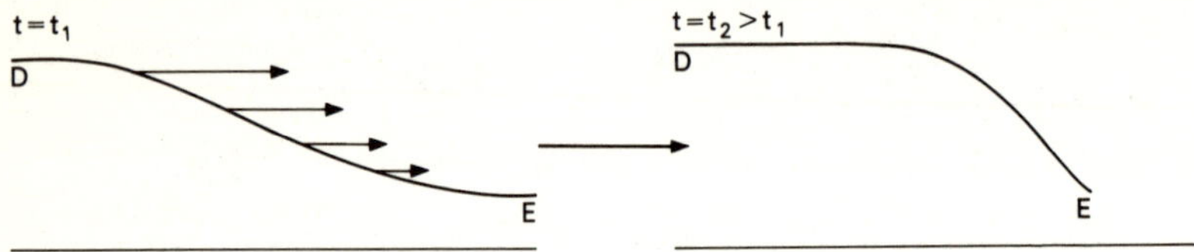

FIGURE 4.4

the disturbance near the top moves faster than that at the bottom and so the 'front' gradually steepens. The theory suggests that this goes on happening until the surface forms a Z-shape; that is, there is a part of this surface which has a downward pointing normal This configuration is physically impossible, and usually breaking occurs at the top (thus causing the loss in energy referred to in Section 3.4). An alternative way of looking at this is to use an x-t diagram, as described in the previous section.

On the other hand, if the shape of the wave is in the opposite sense as shown between B and C in Figure 4.5, it will gradually flatten out for the same

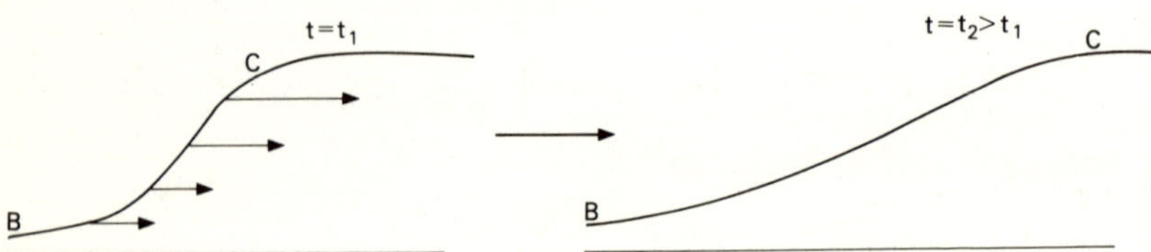

FIGURE 4.5

reason. This confirms the result that a discontinuity can be formed by a disturbance travelling in one direction, but not in another.

EXERCISES

1 An organ pipe of length b is closed at one end and
open at the other. The pressure at the open end is
forced to vary so that it differs from its mean value
by $p_0 \sin \omega t$. Find the motion in the pipe when $2\omega b/\pi a_0$
is not an odd integer, where a_0 is the speed of sound
in the air in the pipe.

2 The region $x < 0$ is occupied by a gas in which the
speed of sound is a_1, and the region $x > 0$ by one in
which it is a_2. A plane sound wave, whose frequency is
ω, is incident normally on the plane $x = 0$ from the
region $x < 0$. Find the amplitude and phase of the wave
transmitted into the region $x > 0$, and those of the
wave reflected back into the region $x < 0$.

3 A straight tube of uniform cross-section A and of
length b is closed at one end and open at the other.
It contains an airtight piston of mass M which can
slide freely along the tube and which, in equilibrium,
is midway along the tube. Show that the periods of the
normal modes of small oscillations of the piston are
$2\pi/\omega$ where

$$\frac{\omega b}{a_0} \tan \frac{\omega b}{a_0} = \frac{2\rho_0 Ab}{M} ,$$

the undisturbed density of the air in the tube being ρ_0
and the corresponding speed of sound being a_0.

4 A straight pipe of uniform cross section is open at
one end and at the other it is closed by an airtight disc
whose distance from the open end at time t is
$b + c \sin \omega t$ where b, c and ω are all given constants
and $c \ll b$. If M is the mean mass of the air contained
in the pipe and a is the corresponding speed of sound,
show that the kinetic energy of the air in the pipe is

$$\frac{1}{4} Mc^2\omega^2 \left[\sec^2\left(\frac{\omega b}{a_0}\right) + \frac{a_0}{\omega b} \tan\left(\frac{\omega b}{a_0}\right) \right] \cos^2\omega t.$$

5 A straight canal of uniform width contains two layers
of liquid, each of uniform thickness h_0 in the
undisturbed state. The density of the lower layer is ρ_1
and that of the upper is $\rho_2 (< \rho_1)$. Show that there are
two possible velocities of shallow waves (for small

disturbances), given by

$$c^2 = gh_0\left[1 \pm \sqrt{(\rho_2/\rho_1)}\right].$$

When ρ_1 and ρ_2 are nearly equal, show that the ratio of the disturbance of the upper free surface to that of the interface is approximately 2 for the faster of the two waves and is very small for the slower one.

6 At time t, the height of the liquid in a horizontal channel of uniform width is $h(x,t)$ where x is a horizontal coordinate measured along the tube, and its velocity is $u(x,t)$ along the channel. At time $t = 0$, it is known that $u = u_0$ and $h = h_0$ in $x_1 \leq x \leq x_2$, where u_0 and h_0 are constants. Show that $u = u_0$ and $h = h_0$ everywhere in

$$x_1 + \left[u_0 + \sqrt{(gh)}\right]t \leq x \leq x_2 + \left[u_0 - \sqrt{(gh)}\right]t,$$

as long as $0 \leq t \leq (x_2 - x_1)/2\sqrt{(gh)}$.
(N.B. It should not be assumed that u and $h - h_0$ are necessarily small.)

Chapter 5

SOME EXAMPLES OF VISCOUS FLOW

5.1 VISCOSITY

All the flows that we have discussed so far have been
adequately described by neglecting viscous forces: that
is, we have assumed that the internal forces within the
fluid are normal to the surfaces on which they act. As
we saw in Chapter 1, it is impossible to satisfy the full
boundary conditions with such an assumption; in all the
cases discussed in Chapters 2, 3 and 4, however, the
boundaries of the flow were so far from the region of
interest that the necessary modification of the boundary
conditions had a negligible effect on the flow. We now
go on to discuss what happens when viscous forces must
be included.

We shall consider in this chapter mostly rectilinear
motions of a fluid in which the velocity is a function
of one variable only. More precisely, we shall consider
flows parallel to the x-axis in which the speed u is a
function only of either z (where Oxyz are rectangular
Cartesian axes) or r (where r is distance measured
from the x-axis). First of all we consider the case in
which u is a function of z only.

The internal stress in this case* on a surface on
which z is constant consists of two parts: a normal
component which is the pressure p and a tangential
component which we shall call τ , and which is
proportional to $\partial u/\partial z$. That is, we have

$$\tau = \mu\frac{\partial u}{\partial z}, \tag{82}$$

where the constant of proportionality, μ , is called the
viscosity of the fluid and depends only on the nature of

---

* The general case is discussed in Appendix 1.

the fluid, not on the particular flow involved. Typical values of the viscosity for some fluids are shown in Table 5.1. It is found that μ is a function of the temperature of the fluid, but not of its density. The ratio $\nu = \mu/\rho$ is one that often occurs in fluid mechanics and is called the *kinematic viscosity*. Typical values of ν are also shown in Table 5.1.

TABLE 5.1 Values of the viscosity μ, and the kinematic viscosity ν, for fluids at various temperatures.

Substance	T (°C)	$10^6 \mu$ (kg/m s)	$10^6 \nu$ (m^2/s)
Air	0	17.1	13.2
"	20	18.1	15.0
Carbon dioxide	20	14.6	7.4
Carbon monoxide	20	17.7	14.2
Helium	20	19.4	109
Hydrogen	20	8.8	97.9
Nitrous oxide	20	14.6	7.4
Oxygen	20	20.0	14.0
Blood plasma of man	20	4,100 - 5,100	3.9 - 4.8
Carbon tetrachloride	0	1,348	0.83
Olive oil	20	8.4×10^4	93
Pitch	15	1.3×10^{15}	1.18×10^{12}
Water	0	1,786.5	1.79
"	20	1,001.9	1.00
"	100	282.9	0.295

On a surface of constant x, the component of stress normal to the surface is the pressure, p; there is also a tangential viscous force, but it turns out that we do not need to know the magnitude of this.

5.2 PLANE COUETTE FLOW

First we consider flow which is caused by the relative motion of two parallel plates bounding the fluid when no pressure gradient is applied. The plate $z = 0$ (which we shall refer to as the lower plate) moves parallel to the x-axis with constant speed U_1, and the plate $z = h$ (the upper plate) moves parallel to the x-axis with constant speed U_2. There is no restriction on the signs

of U_1 and U_2, but we use the convention that either
is positive if it is in the direction of x increasing.
We shall assume that there are no external forces acting
on the fluid.* The system is illustrated in Figure 5.1.

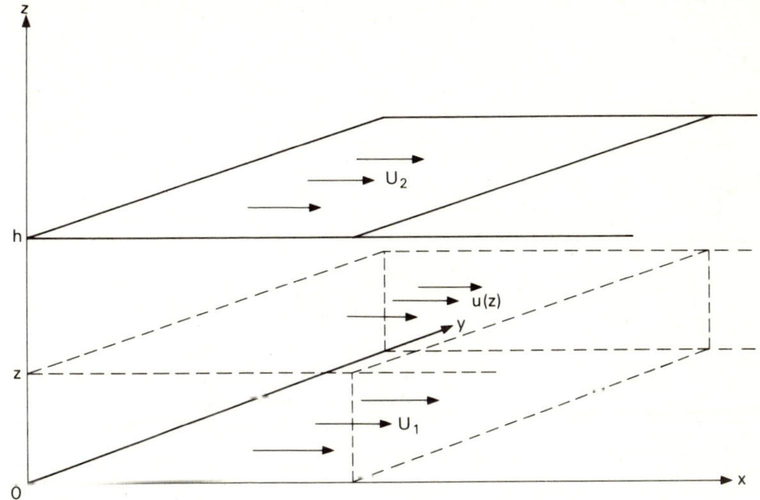

FIGURE 5.1

We would expect the flow of the fluid to be parallel
to Ox and its speed u and pressure p to be
functions of z only, and this is what happens in
practice in many cases. We consider the fluid which
lies between the lower plate and a plane at distance z
(z < h) above it. By consideration of mass conservation,
we see that ρu is independent of x, and hence ρ is
independent of x.
We next consider the momentum of the fluid. It has
none in the z-direction and the tangential components of
stress in the z-direction cancel out; so the pressure
must be independent of z (as we have suggested above);
if it were not, the normal component of force per unit
area on the lower plate would depend on the thickness of
the layer of fluid we are considering, and this would be
ridiculous. If the region we are considering is of width
b and of length L, then the forces on the fluid in it

---

* Such a flow is possible even in the earth's
   gravitational field if both Ox and Oz are
   horizontal. It is left as an exercise to the reader
   to show that the flow is unaltered if Oz is vertical.

are shown in Figure 5.2, where F_0 is the tangential component of the force per unit area exerted on the fluid by the lower plate. Since the motion is steady and

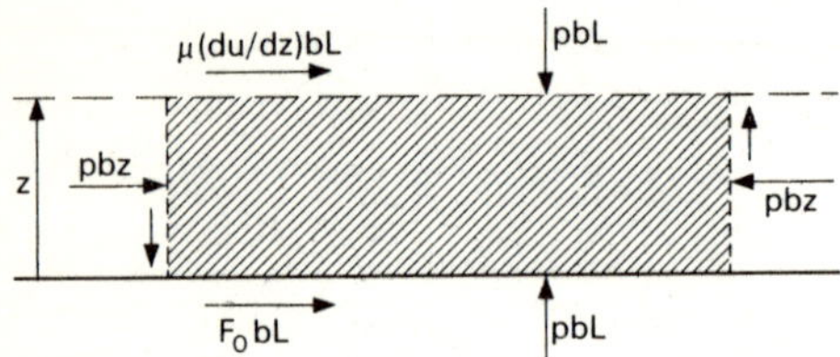

FIGURE 5.2

ρu^2 is independent of x (so that the amount of momentum carried by the fluid in at one end of the region is exactly equal to that carried out at the other), these forces must balance exactly and we have

$$F_0 bL + \mu \frac{du}{dz} bL = 0 ,$$

and this may be written

$$\frac{du}{dz} = - \frac{F_0}{\mu} ,$$

which can be solved to give

$$u = A - \frac{F_0}{\mu} z ,$$

where A is an arbitrary constant of integration. We have two boundary conditions: that $u = U_1$ when $z = 0$ and $u = U_2$ when $z = h$. It appears at first sight, therefore, that the problem is overdetermined since we have only one constant of integration and two boundary conditions. However, a moment's thought reminds us that F_0 is an unknown quantity which has yet to be determined - so we do, after all, need two boundary conditions. Applying them we find that $A = U_1$ and

$$F_0 = \frac{\mu(U_1 - U_2)}{h} , \tag{83}$$

and so

$$u = U_1 + (U_2 - U_1) \frac{z}{h} \, .\tag{84}$$

It is convenient to introduce here the idea of the *velocity profile*. This is nothing more than a graph of u against z, but it is drawn superimposed on a cross-sectional diagram of the system as in Figure 5.3. For

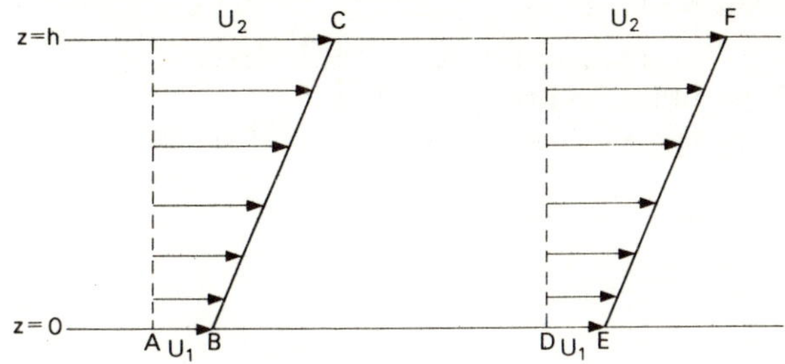

FIGURE 5.3

the present problem the graph is the same for all x, and each is a straight line, so we say that the profile is linear. Note that the graph represents the magnitude of the velocity as a function of z for a given value of x, that value of x being the base line of the graph. Thus, in the diagram, BC is the graph of u as a function of z at the station A - and EF that at the station D .
   Note that F_0, the tangential stress exerted by the lower plate on the fluid is equal to $\mu du/dz$, the value of the tangential stress that would be exerted if there were fluid in the region z < 0 moving in such a way that the motion in the region 0 < z < h had the same velocity profile.

5.3 PLANE POISEUILLE FLOW

As a second example we consider an incompressible fluid between the same two plates as in the previous case, but this time with the plates at rest. The motion is caused by a pressure difference in the fluid: at x = 0 we have $p = p_1$ and at x = L we have $p = p_2$. We shall consider the case in which $p_2 < p_1$.

Since the fluid is incompressible, ρ is constant, and so the principle of conservation of mass tells us that u is independent of x.

If the motion is parallel to the x-axis, we have no momentum parallel to Oz and so, as for plane Couette flow, p is independent of z. This is really an assumption and depends on the way in which the pressures are applied at $x = 0$ and at $x = L$. As long as p_1 and p_2 are independent of z, then the assumption is reasonable, and we shall make it here. The forces on the fluid in the region of width b, height z and length L are as shown in Figure 5.4, and these must

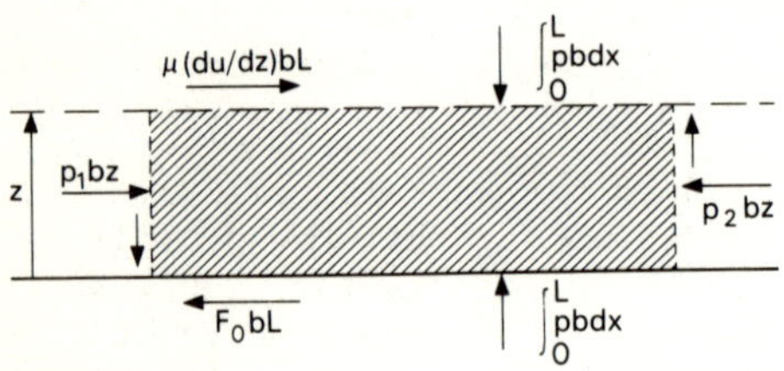

FIGURE 5.4

balance for steady flow. Here F_0 is the force per unit area on the fluid exerted by the lower plate: we have put it in the negative direction of Ox because we would expect the plate to have a dragging effect on the fluid. We therefore have

$$F_0bL - \mu \frac{du}{dz} bL + p_2bz - p_1bz = 0 \, ,$$

and this may be written

$$\frac{du}{dz} = \frac{F_0}{\mu} - \left(\frac{p_1 - p_2}{\mu L}\right)z \, .$$

Integrating this we obtain

$$u = B + \left(\frac{F_0}{\mu}\right)z - \left(\frac{p_1 - p_2}{2\mu L}\right)z^2 \, ,$$

where B is an arbitrary constant of integration. As for Couette flow, we have a second unknown constant in this expression as F_0 has yet to be determined. Hence we can apply the two boundary conditions that $u = 0$ when $z = 0$ and when $z = h$. It follows that $B = 0$, and

$$F_0 = \frac{h}{2L} (p_1 - p_2), \qquad (85)$$

and so

$$u = \frac{p_1 - p_2}{2\mu L} z(h - z). \qquad (86)$$

The velocity profile this time is parabolic (that is, the graph of u as a function of z is a parabola), and the maximum velocity is

$$u_{max} = \frac{(p_1 - p_2)h^2}{8\mu L}, \qquad (87)$$

and occurs in midstream where $z = \frac{1}{2}h$. The velocity profiles are as shown in Figure 5.5 and, once again, are independent of x.

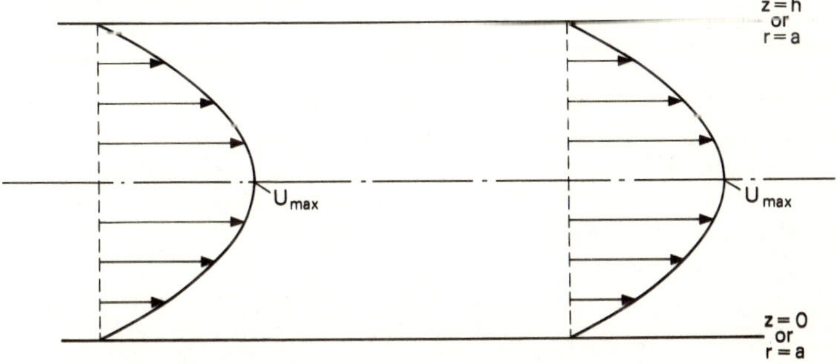

FIGURE 5.5

It is sometimes useful to calculate the rate of mass flow across a station x. In this case it is (per unit width)

$$M = \int_0^h \rho u \, dz = \frac{\rho(p_1 - p_2)}{2\mu L} \int_0^h (hz - z^2) \, dz$$

$$= \frac{\rho(p_1 - p_2)h^3}{12\mu L}$$

$$= \frac{2}{3} \rho h u_{max}, \qquad (88)$$

and this gives a mean speed equal to two-thirds of the maximum speed.

5.4 POISEUILLE FLOW

We now go on to consider the more common flow of an
incompressible fluid in a closed pipe instead of between
two plates which extend (in theory) to infinity. We
shall confine ourselves to the case of a right circular
tube whose radius is a, and whose axis is Ox. The
tube is stationary and flow is induced by applying
pressure p_1 at x = 0 and pressure p_2 at x = L,
this pressure being uniformly distributed across the tube
in each case.

We consider the fluid within a right circular cylinder
of radius r (r < a) whose axis is Ox. We expect the
flow to be parallel to Ox and a function of r only.
The stresses on this fluid are:

 (i) a radial pressure on the curved surface of the
   cylinder,

 (ii) a tangential force on the curved surface of the
   cylinder which is parallel to Ox and of
   magnitude $\mu du/dr$ per unit area,

 (iii) a normal force $\pi p_1 r^2$ parallel to Ox on the
   face x = 0, and

 (iv) a normal force $-\pi p_2 r^2$ parallel to Ox on the
   face x = L.

Here we have assumed, as before, that p is a function
of r only. Thus the forces are as shown in Figure 5.6,

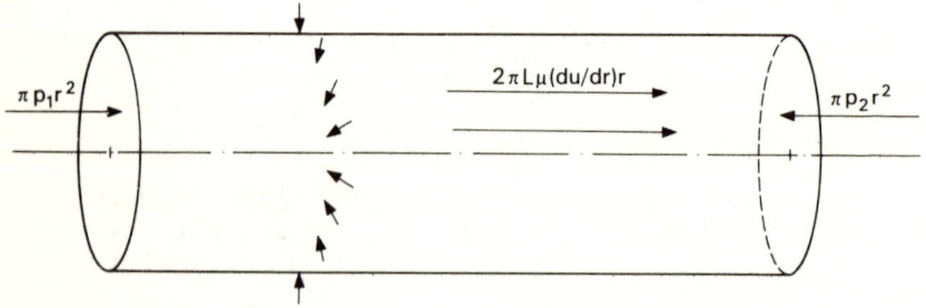

FIGURE 5.6

and must (as before) balance out. So we have

$$2\pi L \mu \frac{du}{dr} r + \pi p_1 r^2 - \pi p_2 r^2 = 0 \, .$$

Hence

$$\frac{du}{dr} = - \frac{(p_1 - p_2)}{2\mu L} r,$$

and this integrates to give

$$u = C - \frac{(p_1 - p_2)}{4\mu L} r^2,$$

where C is an arbitrary constant of integration. Here we have only one unknown, and only one boundary condition - that $u = 0$ when $r = a$. Hence

$$C = \frac{(p_1 - p_2)a^2}{4\mu L},$$

and so

$$u = \frac{(p_1 - p_2)}{4\mu L}(a^2 - r^2). \tag{89}$$

The velocity profile is again parabolic as shown in Figure 5.5, and the maximum speed is

$$u_{max} = \frac{(p_1 - p_2)a^2}{4\mu L}, \tag{90}$$

and occurs on the axis $(r = 0)$ of the tube.

The rate of mass flow across a cross-section is

$$M = \int_0^a 2\pi\rho u r \, dr = \frac{2\pi(p_1 - p_2)\rho}{4\mu L} \int_0^a (a^2 r - r^3) dr$$

$$= \frac{\pi(p_1 - p_2)\rho a^4}{8\mu L}$$

$$= \tfrac{1}{2}\pi a^2 \rho u_{max}. \tag{91}$$

This gives a mean speed equal to one-half of the maximum speed.

The force per unit length exerted by the pipe on the fluid is

$$F_1 = 2\pi a \mu \left(\frac{du}{dr}\right)_{r=a} = - \frac{\pi(p_1 - p_2)a^2}{L}, \tag{92}$$

the negative sign occurring since the pipe has a dragging effect on the fluid.

5.5   FLOW PAST A SPHERE

Here we shall discuss briefly a different type of flow
altogether. This is the 'very viscous' fluid flow past
a sphere: we shall define what we mean by the term 'very
viscous' in the next section. This is not a rectilinear
flow as the previous examples have been, but it is one
in which viscosity is important throughout a large
region of the flow as it was in those examples. We
shall not go into great detail, but merely use a
dimensional argument.

We suppose that we have a fixed sphere of radius a
and that a stream of incompressible fluid of density
and viscosity μ flows past it so that, as shown in
Figure 5.7, a long way from the sphere the velocity of

FIGURE 5.7

the fluid is uniform having magnitude U. We assume that
the flow is axially symmetric. Suppose that the force
required to keep the sphere in place is D : this (from
symmetry) is parallel to Ox and we call it the *drag* on
the sphere. We would expect D to be a function of
a, μ, ρ and U. At first sight it might appear that D
would also be a function of the pressure p_∞ a long way

from the sphere, but we note that if p_∞ is increased

by an amount P, then the pressure everywhere is merely
changed by an amount P. This can have no effect on the
flow (which depends on pressure gradients rather than on
the actual value of the pressure itself) - nor on the
magnitude of the drag.

We can use a dimensional argument and suppose that

$$D \propto a^\alpha \mu^\beta \rho^\gamma U^\delta . \tag{93}$$

Now, using square brackets to denote dimensions, we have

$$[D] = MLT^{-2}, \quad [a] = L, \quad [\mu] = ML^{-1}T^{-1},$$

$$[\rho] = ML^{-3}, \quad [U] = LT^{-1}.$$

where M is mass, L is length and T is time. Substituting in equation (93), we obtain

$$MLT^{-2} = (L^{\alpha})(M^{\beta}L^{-\beta}T^{-\beta})(M^{\gamma}L^{-3\gamma})(L^{\delta}T^{-\delta}).$$

Hence we must have $1 = \beta + \gamma$, $1 = \alpha - \beta - 3\gamma + \delta$, $-2 = -\beta - \delta$, and we have three equations for four unknowns. It is convenient to express everything in terms of β: then

$$\gamma = 1 - \beta, \quad \delta = 2 - \beta, \quad \alpha = 1 + \beta + 3\gamma - \delta = 2 - \beta.$$

Then equation (93) can be written

$$D \propto \rho U^2 a^2 \left(\frac{\mu}{\rho Ua}\right)^{\beta}.$$

This can be generalized to give

$$D = \rho U^2 a^2 f(\mu/\rho Ua), \tag{94}$$

where f is an arbitrary function, and we can get no further by a dimensional argument.

If the viscosity is sufficiently large for viscous forces to be the most important in producing drag, we would expect D to be directly proportional to the viscosity μ since the tangential stress itself is proportional to μ. If this is so we must have $\beta = 1$, and then

$$D \propto \rho U^2 a^2 \frac{\mu}{\rho Ua} = \mu Ua.$$

A more detailed investigation of this problem (with the same assumption that viscous forces are the important ones) gives the constant of proportionality to be 6π. Hence we have

$$D = 6\pi\mu Ua. \tag{95}$$

This formula is due to Stokes and the corresponding flow is usually referred to as Stokes's flow.

In practice it turns out that this is a good formula for the drag as long as $\mu > \rho Ua$, but for values of μ much smaller than this, the drag is greater than that given by such a relation. This is because for much smaller values of the viscosity there is a wake behind

the sphere, and this gives the effect of a much larger
body (sphere plus wake instead of sphere alone) in the
stream.

5.6   THE REYNOLDS NUMBER

We conclude this chapter with a definition.   In equation
(94), the quantity

$$R = \frac{\rho Ua}{\mu} = \frac{Ua}{\nu} \tag{96}$$

occurs as an important parameter.   The equation (94) may
be written

$$D = \tfrac{1}{2}\pi\rho U^2 a^2 f(R), \tag{97}$$

where   f(R)   is an arbitrary function of   R,   and we
have introduced the π because πa^2 is the maximum
cross-sectional area of the sphere and the $\tfrac{1}{2}$ because
it is conventional to do so.
   More generally we consider a flow of fluid whose
density (at some reference point) is ρ,   whose viscosity
is μ and whose speed (at some reference point, not
necessarily the same as for the density) is   U;   then,
if   L   is a typical length of the flow, we define the
Reynolds number of the flow as

$$R = \frac{\rho UL}{\mu}. \tag{98}$$

   The physical significance of the Reynolds number is
easy to see: as in Chapter 1, we can consider the
pressure force to have magnitude of order ρU^2 per unit
area, and (assuming that the spatial derivative of the
velocity is of order   U/L) the viscous force is of order
μU/L   per unit area.   Hence the Reynolds number is a
measure of the ratio of the magnitude of the pressure
forces to that of the viscous forces.   If   R   is very
large, therefore, we might hope to be able to neglect
viscous forces in the flow: in fact this is not quite
possible, and we shall discuss this situation in the
next chapter.   If, however,   R   is small or of the order
of unity, it is obviously necessary to include viscous
as well as pressure forces in any discussion of the flow.
   Sometimes, as we shall find in the next chapter,
there is more than one typical length.   In this case
we have more than one Reynolds number, each of which may
be important for different aspects of the flow.   For the
moment we shall think of   L   as being the diameter or

the length of a body immersed in the flow, or as the width of a channel or as the diameter of a pipe.

In the examples of this chapter, therefore, we might define R as follows:

(i) Plane Couette flow: $R = \dfrac{\rho(U_1 - U_2)h}{\mu}$, and so

$$F_0 = \frac{\rho(U_1 - U_2)^2}{R},$$

(ii) Plane Poiseuille flow: $R = \dfrac{\rho u_{max} h}{\mu}$, and so

$$F_0 = \frac{4\rho u_{max}^2}{R},$$

(iii) Poiseuille flow: $R = \dfrac{\rho u_{max} a}{\mu}$, and so

$$F_1 = -\frac{4\pi\rho u_{max}^2 a}{R}, \quad \text{and}$$

(iv) Stokes's flow: $R = \dfrac{\rho U a}{\mu}$, and so $D = \dfrac{6\pi\rho U^2 a^2}{R}.$

Approximate values of the Reynolds number for several different flows are given in Table 5.2.

TABLE 5.2 Typical Reynolds numbers for various flows

	Typical speed (m/s)	Typical length (m)	R
Aircraft at sea level	100	4	10^7
Wind-tunnel model	50	1	10^6
Large liner at maximum speed	15	300	10^9
Wind past house	10	20	10^7
Wind past man	10	0.3	10^5
Wind past telegraph pole	10	0.15	10^5
Wind past telegraph wire	10	0.001	10^3
Whale swimming in water at 20°C	20	27	10^9
Man swimming in water at 20°C	4	2	10^7
Steel ball falling in olive oil	0.1	0.002	2
Air bubble rising in pitch	7 mm/yr	0.01	10^{-17}

Not only is the Reynolds number an important parameter in the expression for the forces in a fluid, but it is also important (for reasons which are not yet completely understood, and which will not be entered into here) as

a critical parameter for the onset of turbulent flow.
For example, in the flow through a circular pipe we find
that the flow of Section 5.4 is appropriate for
$R = \rho U a/\mu$ less than about 2×10^3. If the Reynolds
number is greater than this, then the flow is turbulent
and the assumptions of that section no longer apply.

Students often find difficulty when they first meet
the concept of the Reynolds number because it is not
clearly defined (that is, there is a certain arbitrar-
iness in the choice of the typical speed and the typical
length). It may be helpful to make the following comments.
When the exact value of the Reynolds number is required,
we have to state precisely which Reynolds number we are
talking about. Thus, the value of R in the expression
$D = 6\pi\rho U^2 a^2/R$ for the drag on a sphere in Stokes's flow
is based on the speed of the fluid at infinity and on
the radius of the sphere. But it often happens that we
need discuss only orders of magnitude. For example, if
we wish to make approximations which are valid only for
$R \gg 1$, we are not at all fussy as to whether $R = 10^6$
or $R = 10^7$, say: so it doesn't matter whether we choose
the radius or the diameter for the typical length L in
the expression (98).

EXERCISES

1 A liquid whose kinematic viscosity is ν is contained
between two fixed, parallel flat plates which are
distance d apart and which make an angle α with the
horizontal. The resulting flow is steady and is everywhere parallel to the lines of greatest slope of the
plates; the rate of volume flow is Q per unit width
of the plates. Show that, if there is no pressure
gradient in the direction of the flow, then

$$\sin \alpha = 12\nu Q/gd^3.$$

2 A layer of water of constant thickness h is bounded
above by air and below by a fixed plane inclined at an
angle α to the horizontal. Assuming that the viscosity
of air is negligible compared with that of water, determine the steady laminar flow due to gravity, and
calculate the volume flux per unit width of the layer.

3 Two fixed horizontal plates are a distance $2b$ apart.
Between them are two layers of a liquid, each of
thickness b, the upper having density ρ_1 and

viscosity μ_1 and the lower having density $\rho_2 \, (> \rho_1)$ and viscosity μ_2. The liquids are in steady flow parallel to the plates. Show that the pressure gradient in the direction of the flow is constant and determine the velocity distribution everywhere in terms of this pressure gradient.

4 At time t a liquid, whose density is ρ and whose viscosity is μ, has velocity components $(u,0,0)$ referred to fixed Cartesian axes Oxyz. It is known that $\partial u/\partial y$ is zero and that, for large values of z, the pressure is independent of x and of y. The external forces on the body of the liquid are parallel to the z-axis, and depend only on z. Show that
  (a) $\partial u/\partial x$ is zero, and
  (b) $\dfrac{\partial u}{\partial t} = \nu \dfrac{\partial^2 u}{\partial z^2}$, where $\nu = \mu/\rho$ is the kinematic
    viscosity.

5 The liquid of question 4 is confined between the horizontal flat plates $z = 0$ and $z = h$ and is at rest. At time $t = 0$, the plate $z = h$ is set in motion in its own plane with velocity U and continues to move with this velocity for $t > 0$; the plate $z = 0$ is held fixed. Show that, at time $t \geq 0$,

$$u = \frac{Uz}{h} + \frac{2U}{\pi} \sum_{n=1}^{\infty} \frac{(-1)^n}{n} \sin \frac{n\pi z}{h} \exp\left(-\frac{n^2\pi^2\nu t}{h^2}\right).$$

6 The liquid of question 4 is at rest in the domain $z > 0$, being bounded below by the horizontal flat plate $z = 0$. At time $t = 0$, the plate begins to move in its own horizontal plane in such a way that the stress it exerts on the liquid is independent of time. Find its velocity as a function of time. (Hint: find the Fourier cosine transform with respect to z of the liquid velocity, and invert this for $z = 0$.)

7 A straight, horizontal pipe of length b and circular cross-section of radius a is open at one end, and the other is attached to a large cylindrical vessel with vertical walls whose cross-sectional area is A. The vessel and pipe contain liquid whose kinematic viscosity is ν and, at time t, the height of the free surface in the vessel above the pipe is h. Show that, when

$A \gg \pi a^2$, the height h is given to a good approximation by

$$h = h_0 \exp(- \pi a^4 gt/8\nu bA),$$

where h_0 is the value of h when $t = 0$.

8 A long, straight tube, whose cross-section is circular and of radius a, is inclined at an angle α to the horizontal. A liquid whose kinematic viscosity is ν flows steadily along the tube, its velocity being everywhere parallel to the tube. The only external force is that due to gravity and there is no applied pressure gradient along the tube. Find the rate of volume flow along the tube.

9 A wire whose cross-section is circular and of radius a is inside a long, horizontal, straight tube whose cross-section is circular and of radius b, the wire and the tube being coaxial. The annular region between the wire and the tube is filled with oil whose kinematic viscosity is ν. The wire is pulled along the tube with constant velocity V and there is no pressure gradient along the tube. Assuming that the velocity of the oil is everywhere parallel to the wire, find the rate of volume flow of oil along the tube.

10 Two coaxial circular cylinders, whose radii are a, b $(b < a)$, have their axes vertical and the space between them is filled with a liquid whose kinematic viscosity is ν. The outer cylinder is fixed and the inner moves downwards with constant speed W. The velocity of the liquid is everywhere vertical and the pressure is uniform throughout the liquid. Show that, if g is the acceleration due to gravity, the speed w of the liquid at a distance r from the axis of symmetry satisfies the equation

$$\frac{d}{dr}\left(r\,\frac{dw}{dr}\right) + \frac{g}{\nu}\,r = 0.$$

Find the force per unit length on the inner cylinder due to the liquid and find the value of the terminal speed of the inner cylinder if its density is the same as that of the liquid and if its motion is due to the action of gravity.

11 Compute the Reynolds number of the flow in each of the
following cases:
  (a) an aeroplane wing with chord length 10 ft flying at
      600 mph at an altitude of 7,000 ft (0°C);
  (b) a man's blood plasma flowing with speed 2 mm/sec at
      20°C in a capillary vessel whose diameter is
      10 microns (1 micron = 10^{-6} m);
  (c) a water droplet of diameter 5 microns moving at a
      speed 2 mm/sec in a cloud chamber whose temperature
      is 0°C;
  (d) a submarine periscope of diameter 16 in moving at
      15 knots in water whose temperature is 20°C
      (1 knot = 6,080 ft/hr);
  (e) a chimney, whose diameter is 20 ft, in a wind of
      30 mph at 20°C.

Chapter 6

FLOW AT HIGH REYNOLDS NUMBERS

6.1 CONDITIONS NEAR THE BOUNDARIES

From Table 5.2 we see that there is a number of
important flows in which the Reynolds number (as defined
in the last chapter) is very large compared with unity:
that means that the viscous forces appear to be very
small compared with the pressure ones (that is, those
normal to the surface on which they act). We would like
to be able to say that under these circumstances the
viscous forces can be neglected and that by considering
the normal ones only we would get a good approximation
to the real flow. In fact we have already done this
with some success in Chapters 3 and 4. We saw in
Chapter 1, however, that we are unable to satisfy the
true boundary conditions if the viscous forces are
neglected. Since it is the presence of the boundaries
which requires the consideration of tangential stresses,
we might hope that it is only near the boundaries that
we need to do this. This idea was first suggested by
Prandtl in 1902.

To get some idea of whether it is possible to do this
or not, we consider the flow of a uniform stream past a
flat plate along $z = 0$, $x > 0$. The uniform stream has
a velocity U parallel to Ox far from the plate.
We shall make the simplifying assumption that when $x < 0$,
$u = U$ everywhere and that when $x > 0$, $u = U$ when
$z > h(x)$ where $h(x)$ is a function of x which has to
be determined; the region $x > 0$, $0 < z < h(x)$ will be
called the *boundary layer* and in it the speed has to
increase monotonically from zero at the plate to U at
$z = h(x)$. This surface, $z = h(x)$, we shall call the
'outer edge', or sometimes simply the 'edge' of the
boundary layer. In this section we shall make the
especially simplifying assumption that u increases

linearly with z in the boundary layer. So we have

$$u = \begin{cases} \dfrac{Uz}{h} & \text{when } x > 0,\ 0 \le z \le h(x) \\[2mm] U & \text{when } x > 0,\ z \ge h(x). \end{cases} \tag{99}$$

To permit the necessary conservation of mass, there must be a component of velocity, w, normal to the plate, but since $w = 0$ both at the plate itself and in the oncoming stream, we can hope that $w \ll u$ everywhere. We consider first the mass flux in the region between $x = 0$ and $x = X$ and between $z = 0$ and $z = h'$ where h' is a constant such that $h(X) < h'$, as shown in Figure 6.1. The rate of inflow of mass (per unit width) across $x = 0$ is $\rho U h'$, the rate of outflow

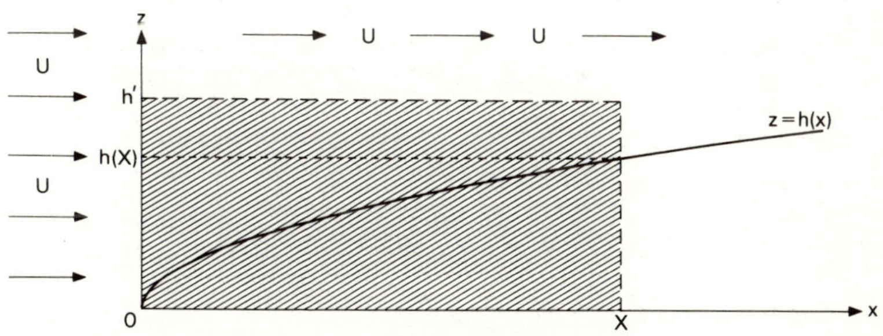

FIGURE 6.1

across $x = X$ is $\rho \displaystyle\int_0^h \dfrac{Uz}{h(X)}\, dz + \rho U[h' - h(X)] =$

$\rho U[h' - \tfrac{1}{2} h(X)]$ and the rate of outflow across $z = h'$

is $\rho \displaystyle\int_0^X w'\, dx$, where w' is the value of w when

$z = h'$; there is no other source (or sink) of mass in the region unless we have suction or blowing at the boundary, and we do not consider this situation here.

Hence these qualities must balance and we have

$$\rho U h' = \rho U [h' - \tfrac{1}{2} h(X)] + \rho \int_0^X w' dx,$$

and this gives the relation

$$\int_0^X w' dx = \tfrac{1}{2} U h(X).$$

Next we consider the momentum balance of the fluid in the same region. As long as the normal component of velocity, w, is sufficiently small the pressure is, to a good approximation, independent of z (this is to be compared with Couette flow in which w is exactly zero, and so is $\partial p / \partial z$); further, it is independent of x outside the boundary layer. It follows that p is constant throughout the fluid, and that the components of force due to the pressure on the faces $x = 0$ and $x = X$ cancel out. Since u is independent of x and z in the neighbourhood of $z = h'$, there is no tangential force on this surface. Hence the resultant component of force parallel to Ox on the fluid (neglecting any body forces which may be present: if the plate is horizontal, gravitational effects are irrelevant, of course) is that due to the plate $z = 0$. This force is (per unit width)

$$\int_0^X \mu \left(\frac{\partial u}{\partial z} \right)_{z=0} dx = \int_0^X \frac{\mu U}{h(x)} dx,$$

where we have used the relation (99) to evaluate $(\partial u / \partial z)_{z=0}$. This force acts in the negative direction of x (since the effect of the plate is to retard the fluid) and is balanced by the rate of loss of the x-component of momentum through the ends $x = 0$ and $x = X$ and through the upper surface $z = h'$ of the region. The rate at which the x-component of momentum enters the region through the end $x = 0$ is $\int_0^{h'} (\rho u^2)_{x=0} dz$, that at which it leaves through the end $x = X$ is $\int_0^{h'} (\rho u^2)_{x=X} dz$ and that at which it leaves through the upper surface $z = h'$ is $\int_0^X \rho (uw)_{z=h'} dx$. But we have chosen h' to be so large that $u = U$ when $z = h'$ and we have already defined w' to be the value of w at $z = h'$; it follows that the last integral is $\int_0^X \rho U w' dx$, and we have

$$\mu U \int_0^X \frac{dx}{h(x)} = \int_0^{h'} \left[(\rho u^2)_{x=0} - (\rho u^2)_{x=X} \right] dz - \int_0^X \rho U w' dx$$

$$= \int_0^{h'} \rho U^2 dz - \int_0^h \rho \left(\frac{Uz}{h} \right)^2 dz - \int_h^{h'} \rho U^2 dz - \rho U \int_0^X w' dx$$

$$= \rho U^2 h' - \tfrac{1}{3} \rho U^2 h(X) - \rho U^2 [h' - h(X)] - \tfrac{1}{2} \rho U^2 h(X)$$

$$= \tfrac{1}{6} \rho U^2 h(X).$$

This is an integral equation for $h(x)$ which is most easily solved by differentiation with respect to X. We obtain

$$\frac{\mu}{h(X)} = \tfrac{1}{6} \rho U \frac{dh}{dX},$$

or, replacing X by the current variable x, we have

$$2h \frac{dh}{dx} = \frac{12\mu}{\rho U}.$$

This may be integrated and, using the obvious boundary condition that $h = 0$ when $x = 0$, we have

$$h^2 = \frac{12\mu x}{\rho U},$$

and so

$$h \simeq 3.5 \left(\frac{\mu x}{\rho U} \right)^{\frac{1}{2}}. \tag{100}$$

This means that the edge of the boundary layer is a parabola, as shown in Figure 6.2; the velocity profiles we have assumed are also shown.

It is clear that h increases fairly slowly with x. We can express this fact more precisely by defining a Reynolds number R_x which depends on the distance downstream from the leading edge $(x = 0)$ of the plate.

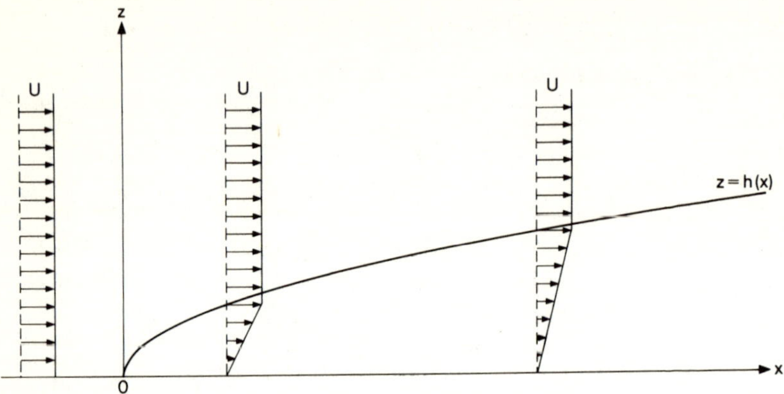

FIGURE 6.2

This is given by

$$R_x = \frac{\rho U x}{\mu} \, , \tag{101}$$

and it is easy to see from equation (100) that

$$\frac{h}{x} = \left(\frac{12}{R_x}\right)^{\frac{1}{2}} . \tag{102}$$

When R_x is of the order 10^6 or greater, as for the flight of an aeroplane or for the flow past models in a wind tunnel in the laboratory, it is evident that h/x is very small. For example, on an aeroplane wing h is rarely more than about 10 cm (which is very small compared with the dimensions of the wing) and on a 25 cm model in a wind tunnel the thickness of the boundary layer is usually less than 2 or 3 mm, which is again very small.

It therefore seems reasonable to think of the flow as divided into two regions: one is a thin layer near the boundary in which viscous forces are important, and the other is the rest of the fluid where viscous forces are negligible. This is supported by the fact that there are two or three flows which can be determined exactly for all Reynolds numbers and for these it is found that, when the Reynolds number of the flow is large, viscous forces are negligible except very near the boundaries.

Also the experimental evidence supports the idea. We
shall, therefore, accept this as a reasonable assumption
on which to base the rest of this chapter.
   The approximations discussed here as being appropriate
to fluid in the neighbourhood of boundaries in the fluid
are also appropriate to certain other flows: whenever
there is a very large velocity gradient in a fluid,
there is a strong probability that the boundary-layer
approximations are valid. Such layers of fluid are
usually referred to as *shear layers,* and often occur in
the wake behind an obstacle.

6.2   THE STEADY BOUNDARY-LAYER EQUATIONS

To derive the equations which are valid in the boundary
layer we should start with the full equations of motion
(derived in Appendix 2) and make appropriate approximations to them (as shown in Appendix 3). It is,
however, possible to derive them in a more simple-minded
fashion which, although not mathematically rigorous,
gives some physical insight into the problem.
   We consider a layer of fluid near to the rigid
surface $z = 0$ such that the flow is nearly parallel to
$z = 0$ (see Figure 6.3) and is independent of y.

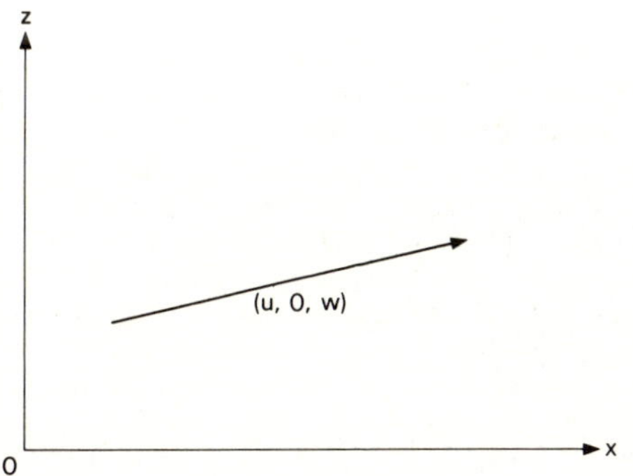

FIGURE 6.3

Thus the components of velocity are $(u,0,w)$ with
$w \ll u$. We assume further that u changes much more
rapidly with z than with x. This boundary layer lies
between the boundary $z = 0$ and an oncoming flow which,
if we were to neglect viscosity, would have a velocity
$(U(x,z),0,0)$ and we shall write $U(x,0) = U(x)$. Since
the boundary layer is so thin, the value of u at its

'outer edge' (that is the value of u in the limit as
z → ∞) is not very different from U(x), and we use
this approximation as a second boundary condition on u.
That it is a good approximation is demonstrated in
Figure 6.4: the value of u as z → ∞ in the boundary

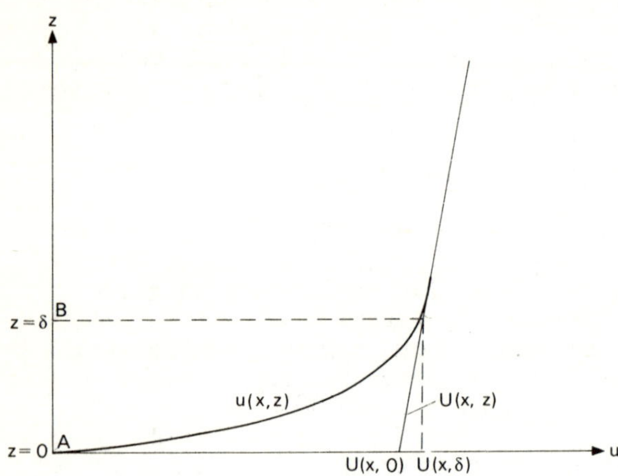

FIGURE 6.4

layer differs little from U(x,δ), the value at the
point B whose coordinates are (x,δ) of the velocity
u obtained using the inviscid theory. But this, in
turn, differs little from U(x,0), the value at the
point A whose coordinates are (x,0) of the velocity
u obtained using the inviscid theory. The whole
argument rests on the idea that δ is large in the
boundary-layer theory (that is $\delta \gg \nu/U$), but is small
in the inviscid theory (that is $\delta(\partial U/\partial z)_{z=0} \ll U(x)$).

We shall consider a region of the fluid of unit width
and bounded by the surfaces x = X, x = X + ξ, z = Z
and z = z + ζ, where ξ and ζ are small. The
stresses on the surface of this region are as shown in
Figure 6.5. Because the velocity is nearly parallel to
Ox and varies only slowly with x, the only significant
tangential components* are those on the surfaces z = Z

---

* There are, in fact, tangential forces normal to Ox,
  but they have no relevance to the balance of the x-
  component of momentum, and cancel out to the required
  order of small quantities when we consider the balance
  of the z-component of momentum.

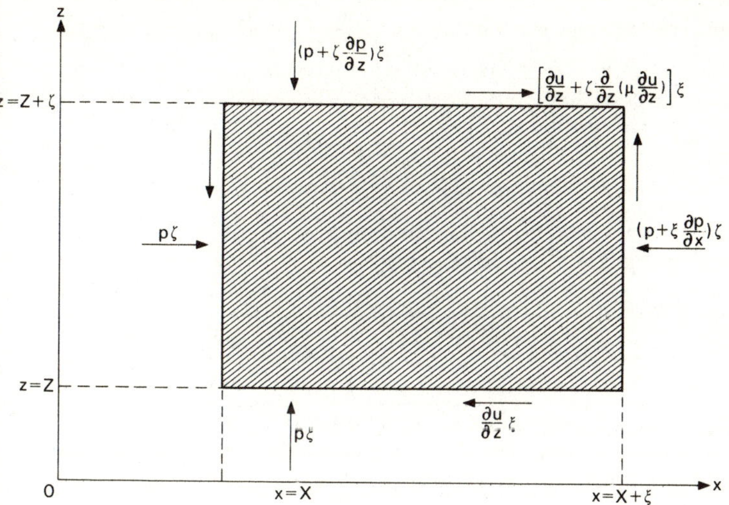

FIGURE 6.5

and $z = Z + \zeta$. Further, the normal components acting on all the surfaces are as shown to the first order in the small quantities ξ, ζ. We consider first the mass flow into the region. Since the flow is steady, the net mass flow must be zero. Hence, using suffices to denote the point where quantities are to be measured, we have

$$\zeta(\rho u)_X + \xi(\rho w)_Z - \zeta(\rho u)_{X+\xi} - \xi(\rho w)_{Z+\zeta} = 0.$$

This gives

$$\xi\zeta\left[\frac{\partial(\rho u)}{\partial x} + \frac{\partial(\rho w)}{\partial z} + \text{terms of the first order in } \xi, \zeta\right] = 0.$$

The quantity in the square brackets is therefore zero, and since we can choose ξ and ζ to be as small as we please, it follows that

$$\frac{\partial(\rho u)}{\partial x} + \frac{\partial(\rho w)}{\partial z} = 0. \tag{103}$$

Next we consider the force components parallel to Oz and restrict ourselves to the case in which there are no external body forces. Then we have that the flux across surfaces of the region of the component in the z-direction of the momentum is $\zeta \frac{\partial p}{\partial z} \xi$, since the motion is steady. Now the flux of this component of momentum is small (since the z-momentum itself is small), and so we have in the boundary layer

$$\frac{\partial p}{\partial z} = 0. \tag{104}$$

Finally we consider the components parallel to Ox. Here the component of force is

$$(p\zeta)_X - (p\zeta)_{X+\xi} - (\mu \frac{\partial u}{\partial z}\xi)_Z + (\mu \frac{\partial u}{\partial z}\xi)_{Z+\zeta} =$$

$$= \xi\zeta \left[- \frac{\partial p}{\partial x} + \frac{\partial}{\partial z}\left(\mu \frac{\partial u}{\partial z}\right) \right] ,$$

and the flux of momentum across the surfaces is

$$(\rho u^2 \zeta)_X - (\rho u^2 \zeta)_{X+\xi} + (\rho u w \xi)_Z - (\rho u w \xi)_{Z+\zeta} =$$

$$= - \xi \frac{\partial}{\partial x}(\rho u^2)\zeta - \zeta \frac{\partial}{\partial z}(\rho u w)\xi$$

$$= - \xi\zeta \left\{ u \left[\frac{\partial(\rho u)}{\partial x} + \frac{\partial(\rho w)}{\partial z} \right] + \rho \left(u \frac{\partial u}{\partial x} + w \frac{\partial u}{\partial z} \right) \right\}$$

$$= - \rho \left(u \frac{\partial u}{\partial x} + w \frac{\partial u}{\partial z} \right) \xi\zeta$$

since the square bracket vanishes, using equation (103). Since the flow is steady this flux of momentum must balance the force on the region, and we have

$$- \xi\zeta \frac{\partial p}{\partial x} + \xi\zeta \frac{\partial}{\partial z}\left(\mu \frac{\partial u}{\partial z}\right) - \xi\zeta\rho \left(u \frac{\partial u}{\partial x} + w \frac{\partial u}{\partial z} \right) = 0$$

to the second order in the small quantities ξ and ζ. This gives

$$u \frac{\partial u}{\partial x} + w \frac{\partial u}{\partial z} = - \frac{1}{\rho} \frac{\partial p}{\partial x} + \frac{1}{\rho} \frac{\partial}{\partial z}\left(\mu \frac{\partial u}{\partial z}\right) . \tag{105}$$

Note that in equation (105), although $\partial u/\partial z \gg \partial u/\partial x$, it is also true that $w \ll u$ and so both terms must be included. The neglect of the corresponding terms in equation (104) really implies that $\partial p/\partial z \ll \partial p/\partial x$ although $\partial u/\partial z \gg \partial u/\partial x$, and this is due to the

assumption that w << u. The set of equations (103),
(104) and (105) is the complete set of equations which
have to be solved. When ρ is constant they are of the
first order in x and of the third* in z: so we need
one boundary condition on x and three on z. We take
these in the form

$$u = U(0) \quad \text{when} \quad x = 0 \quad \text{for all} \quad z > 0, \qquad (106a)$$

$$u = 0 \quad \text{when} \quad z = 0 \quad \text{for all} \quad x > 0, \qquad (106b)$$

$$w = 0 \quad \text{when} \quad z = 0 \quad \text{for all} \quad x > 0, \quad \text{and} \qquad (106c)$$

$$u \to U(x) \quad \text{when} \quad z \to \infty \quad \text{for all} \quad x > 0. \qquad (106d)$$

An alternative form for equation (105) is obtained by
noting that, if viscosity is neglected, the equation
becomes

$$U\frac{\partial U}{\partial x} + W\frac{\partial U}{\partial z} = - \frac{1}{\rho} \frac{\partial P}{\partial x} ,$$

where (U,0,W) is the velocity and P the pressure
corresponding to this fictitious flow. At the surface
the boundary condition on the flow is W - 0, and so
we have

$$U\frac{\partial U}{\partial x} = - \frac{1}{\rho} \frac{\partial P_0}{\partial x} , \qquad (107)$$

where P_0 is the value of the pressure on z = 0. But
in the boundary layer itself $\partial p/\partial z = 0$, using equation
(104). Hence for all z in the boundary layer,
$p = p(x) = P_0(x)$, the value in the inviscid flow for
which equation (107) holds. Hence the pressure term in
equation (105) is

$$- \frac{1}{\rho} \frac{\partial p}{\partial x} = - \frac{1}{\rho} \frac{dP_0}{dx} = U\frac{dU}{dx} .$$

So we may write (105) in the form

$$u\frac{\partial u}{\partial x} + w\frac{\partial u}{\partial z} = U\frac{dU}{dx} + \frac{1}{\rho} \frac{\partial}{\partial z}\left(\mu\frac{\partial u}{\partial z}\right). \qquad (108)$$

We shall, in the rest of this chapter, consider only
incompressible, isothermal flows: that is, we shall treat
μ and ρ as constants. Then equations (103) and (108)

---

* To eliminate w and to obtain an equation for u
  only, we have to differentiate equation (105) with
  respect to z.

become

$$\frac{\partial u}{\partial x} + \frac{\partial w}{\partial z} = 0 \tag{109}$$

and

$$u\frac{\partial u}{\partial x} + w\frac{\partial u}{\partial z} = U\frac{dU}{dx} + \nu\frac{\partial^2 u}{\partial z^2}, \tag{110}$$

where $\nu = \mu/\rho$ is the kinematic viscosity of the fluid. (In many problems concerned with compressible boundary layers, the more general equations can be reduced to these by means of simple transformations; such transformations will not be treated here for lack of space.)

We now have to solve the two equations (109) (110), subject to the boundary conditions (106). It is often convenient to introduce a *stream function* ψ, which is defined by means of its derivatives as follows:

$$u = \frac{\partial \psi}{\partial z}, \qquad w = -\frac{\partial \psi}{\partial x}. \tag{111}$$

Then equation (109) is automatically satisfied. (We shall discuss the physical significance of the stream function in Chapter 8: here we use it purely as a mathematical convenience.) Equation (110) becomes

$$\frac{\partial \psi}{\partial z}\frac{\partial^2 \psi}{\partial x \partial z} - \frac{\partial \psi}{\partial x}\frac{\partial^2 \psi}{\partial z^2} = U\frac{dU}{dx} + \nu\frac{\partial^3 \psi}{\partial z^3}. \tag{112}$$

We have to solve this equation for ψ subject to the boundary conditions

$$\frac{\partial \psi}{\partial z} = U(0) \quad \text{when} \quad x = 0 \quad \text{for all} \quad z > 0, \tag{113a}$$

$$\frac{\partial \psi}{\partial z} = 0 \quad \text{when} \quad z = 0 \quad \text{for all} \quad x > 0, \tag{113b}$$

$$\psi = 0 \quad \text{when} \quad z = 0 \quad \text{for all} \quad x > 0, \quad \text{and} \tag{113c}$$

$$\frac{\partial \psi}{\partial z} \to U(x) \quad \text{when} \quad z \to \infty \quad \text{for all} \quad x > 0. \tag{113d}$$

The third of these conditions comes from the condition (106c) which gives $\partial\psi/\partial x = 0$ when $z = 0$. This clearly means that $\psi = \psi_0$, a constant on $z = 0$. But we can add any constant to ψ without affecting its definition in equation (111); so, adding the constant $-\psi_0$, we obtain the form given in equation (113c).

Before actually finding some solutions of this system, we note that we have not applied a boundary condition at the origin $x = 0, z = 0$. Special care must be taken if the origin is to be included in the flow: we note that it is not usually true that $\partial u/\partial x \ll \partial u/\partial z$ near the origin, so the equations themselves need to be modified. It is fortunate that, in many cases, the conditions near the origin affect the flow only near the origin.

6.3 SIMILARITY SOLUTIONS

There are certain forms of the function $U(x)$ for which the ratio $u(x,y,z)/U(x)$ in the boundary layer is a function of zx^n only, for some value of n. Now we must have

$$\frac{u}{U(x)} = f_1(U,\nu,z,x),$$

where the function f_1 is a dimensionless function of four variables. It follows that U and ν can occur only as a ratio U/ν (otherwise time will appear on the right-hand side). Also we are assuming that z and x occur in the combination zx^n. So we have

$$\frac{u}{U} = f_2\left(\frac{U}{\nu}, zx^n\right),$$

where f_2 is a dimensionless function of two variables. But the right-hand side contains a dimension of length unless it is of the form

$$\frac{u}{U} = f_3\left[zx^n\left(\frac{U}{\nu}\right)^{n+1}\right] = f_3\left[\frac{zU}{\nu}\left(\frac{xU}{\nu}\right)^n\right],$$

where f_3 is a function of a single variable

$$\zeta = \frac{zU}{\nu}\left(\frac{xU}{\nu}\right)^n, \tag{114}$$

and we can write $u = U(x)f_3(\zeta)$. When there is a solution of this form it is called a *similarity solution* because the velocity profiles are similar for all values of x: that is, with suitable scaling (the scaling factor depending only on x), the velocity profiles for all x can all be superimposed.

For simplicity we shall first consider what happens when U is independent of x: this is the problem of a uniform flow past a flat plate along $z = 0$, $x > 0$; this problem was first solved by Blasius in 1908.

Using equation (111) we have

$$\frac{\partial \psi}{\partial z} = U f_3(\zeta),$$

and so, since

$$\frac{\partial}{\partial z} = \frac{U}{\nu}\left(\frac{xU}{\nu}\right)^n \frac{\partial}{\partial \zeta},$$

we have

$$\psi = \nu \left(\frac{\nu}{Ux}\right)^n f(\zeta),$$

where $f_3(\zeta) = f'(\zeta)$ and $f(0) = 0$. Differentiating this, we obtain

$$\frac{\partial \psi}{\partial x} = -\frac{n\nu}{x}\left(\frac{\nu}{Ux}\right)^n f + \frac{\nu n}{x}\left(\frac{zU}{\nu}\right) f',$$

$$\frac{\partial \psi}{\partial z} = U f',$$

$$\frac{\partial^2 \psi}{\partial z^2} = \frac{U^2}{\nu}\left(\frac{xU}{\nu}\right)^n f''$$

$$\frac{\partial^2 \psi}{\partial z \partial x} = nU \frac{zU}{x}\left(\frac{xU}{\nu}\right)^n f'', \quad \text{and}$$

$$\frac{\partial^3 \psi}{\partial z^3} = \frac{U^3}{\nu^2}\left(\frac{xU}{\nu}\right)^{2n} f'''.$$

Substituting these into equation (112) we find that, since $dU/dx = 0$,

$$Uf'nU \frac{zU}{\nu x}\left(\frac{xU}{\nu}\right)^n f'' + \left[\frac{n\nu}{x}\left(\frac{\nu}{Ux}\right)^n f - \frac{nzU}{x} f'\right] \frac{U^2}{\nu}\left(\frac{xU}{\nu}\right)^n f''$$
$$= \frac{U^3}{\nu}\left(\frac{xU}{\nu}\right)^{2n} f'''.$$

This simplifies to

$$\frac{n\nu}{xU} ff'' = \left(\frac{Ux}{\nu}\right)^{2n} f''',$$

or

$$nff'' = \left(\frac{Ux}{\nu}\right)^{2n+1} f'''. \tag{115}$$

Since nff''/f''' is a function of zx^n and $(Ux/\nu)^{2n+1}$ is a function of x only, each must be a constant. This is possible only when $2n + 1 = 0$, that is when $n = -\frac{1}{2}$, and then we have, using equations (114) and (115),

$$u = Uf'(\zeta), \tag{116}$$

where

$$\zeta = z\left(\frac{U}{\nu x}\right)^{\frac{1}{2}} \tag{117}$$

and

$$2f''' + ff'' = 0. \tag{118}$$

Equation (118) is called Blasius's equation, and has to be solved subject to the boundary conditions

$$f(0) = 0, \ f'(0) = 0 \text{ and } f'(\zeta) \to 1 \text{ as } \zeta \to \infty. \tag{119}$$

(Note that the first and last of the conditions (113) are both satisfied when the last of conditions (119) is satisfied, since $U(x) = U(0)$ is a constant in this problem.)

Equation (118) can readily be solved numerically, and the solution is shown in Table 6.1 and Figure 6.6. Note that $u = U$ within 1 per cent when $z(U/\nu x)^{\frac{1}{2}} = 4.8$

TABLE 6.1   The velocity as a function of distance from a flat plate in uniform flow.

$\zeta = z(U/\nu x)^{\frac{1}{2}}$	u/U
0.0	0.0000
0.4	0.1328
0.8	0.2647
1.2	0.3938
1.6	0.5168
2.0	0.6298
2.4	0.7290
2.8	0.8115
3.2	0.8761
3.6	0.9233
4.0	0.9555
4.4	0.9759
4.8	0.9878
5.2	0.9942
5.6	0.9975
6.0	0.9990

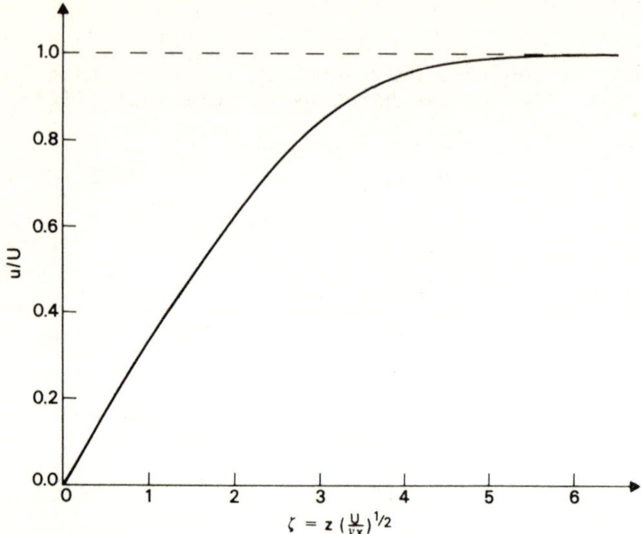

FIGURE 6.6

and within 0.1 per cent when $z(U/\nu x)^{\frac{1}{2}} = 6$. This can be compared with the value of $h(U/\nu x) = 3.5$ given in equation (100) for the simplified model of the boundary layer which we used in Section 6.1.

If we follow the same procedure when U is a function of x, we have

$$\frac{\partial \psi}{\partial x} = -\left(\frac{n\nu}{x} + \frac{n\nu}{U}\frac{dU}{dx}\right)\left(\frac{\nu}{Ux}\right)^n f + \left(\frac{\nu n}{x}\frac{xU}{\nu} + \frac{\nu(n+1)}{U}\frac{zU}{\nu}\frac{dU}{dx}\right)f',$$

and

$$\frac{\partial^2 \psi}{\partial z \partial x} = \frac{dU}{dx}f' + \left[\frac{nU}{x} + (n+1)\frac{dU}{dx}\right]\frac{Uz}{\nu}\left(\frac{xU}{\nu}\right)^n f'';$$

the expressions for $\dfrac{\partial \psi}{\partial z}$, $\dfrac{\partial^2 \psi}{\partial z^2}$ and $\dfrac{\partial^3 \psi}{\partial z^3}$ are unchanged. Substituting into equation (112) we obtain, instead of (115), the relation

$$x\frac{dU}{dx}(f'^2 - 1 + nff'') = U\left[\left(\frac{xU}{\nu}\right)^{2n+1} f''' - nff''\right], \quad (120)$$

where f is a function of ζ as defined in equation (114), and U is a function of x. This can be true only either when $n = -\frac{1}{2}$ and U satisfies the relation

$$\frac{x}{U}\frac{dU}{dx} = m,$$

where m is a constant, or when $U \propto 1/x$ and n is arbitrary. This implies that

$$U \propto x^m, \tag{121}$$

and then equation (120) becomes*

$$2f''' + (m + 1)ff'' + 2m(1 - f'^2) = 0. \tag{122}$$

This is known as the Falkner-Skan equation and it can (like Blasius's equation) be solved numerically subject to the boundary conditions (119) for any particular value of m. For $m > 0$ the solution for which $f' > 0$ for all z (that is, for which there is no backward flow) is unique,† but for $-1 < m < 0$ there is an infinity of solutions, since any solution of equation (122) for such a value of m automatically satisfies the last of the boundary conditions (119). It turns out that only one of these approaches unity exponentially as $\zeta \to \infty$, while all the others differ from unity by an algebraic function of ζ^{-1}. It is usually assumed that the solution which approaches unity exponentially (that is, much faster than the others) is the physically realistic one. Certainly this is true in most cases, both experimentally and when exact solutions are known.

When $m < 0$, of course, the value of u (and of U) when $x = 0$ is undefined. At first sight, therefore, it appears that this is an unrealistic case. However, we can regard the solution as being appropriate only for $x \geq x_0$, where x_0 is some positive finite value of x. The value of $u(x_0,z)$ is not known in this case, but this does not affect the solution of equation (122) subject to the boundary conditions (119). Physically this means that we are ignoring the boundary conditions (113a) which is the one that we noted was included in the last of conditions (119) for the

---

* When $U \propto 1/x$ and n is arbitrary, equation (120) becomes

$$f''' - (\nu/K)^{2n+1}(1 - f'^2) = 0$$

and this reduces to equation (122) with $m = -1$ after a simple change of variable.

† This is also true for $m < -1$, but we shall see later that this is not a physically realistic situation.

Blasius solution. Once this is solved we can, of course, calculate $u(x_0,z)$. The assumption of a similarity

solution therefore implies something about the form of the velocity profile at some starting value of x, and it is only when it is possible to achieve such a profile there that a similarity solution can exist in practice even if the flow $U(x)$ outside the boundary layer is of the right form.

Finally we note the fact that when $m < -0.0904$ the value of $f''(0)$ is negative as shown in Figure 6.7.

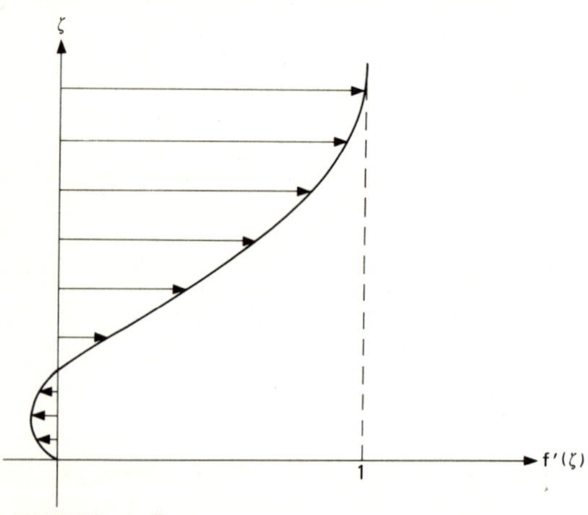

FIGURE 6.7

This implies that near the boundary $\zeta = 0$ the direction of the velocity is opposite to that in the flow outside the boundary layer. The assumptions of boundary-layer theory break down, since the region of backward flow is implicitly excluded from the theory.

6.4 SEPARATION OF THE BOUNDARY LAYER

The phenomenon described at the end of the previous section is associated with the idea of *separation*. It is one of the implicit assumptions of boundary-layer theory that the component of velocity parallel to $x = 0$ is in the direction of increasing x for all z, and this implies that $(\partial u/\partial z)_0 \geq 0$ where $(\partial u/\partial z)_0$ is

the value of $\partial u/\partial z$ when $z = 0$. If there is a similarity solution, as in the previous section, we find that the sign of $(\partial u/\partial z)_0$ is independent of x :

that is, if $m > -0.09$ then $(\partial u/\partial z)_0 > 0$ for all x, and if $m < -0.09$ then $(\partial u/\partial z)_0 < 0$ for all x. In the more general case, however, this is not so and the value of $(\partial u/\partial z)_0$ varies with x in a way depending on the pressure gradient, and may be positive for sufficiently small x but negative for large values of x. If $(\partial u/\partial z)_0$ increases with x, then it cannot become negative (assuming that it is positive for small x) and the pressure gradient is said to be *favourable*. On the other hand, if $(\partial u/\partial z)_0$ decreases as x increases, then it can become negative (though it need not do so) and the pressure gradient is said to be *adverse*. Of course if the pressure gradient is adverse, the boundary-layer solution can apply only as long as $(\partial u/\partial z)_0$ is actually positive. Once separation has occurred the flow far from the boundary may be very significantly altered.

For example, if we consider the flow past a cylinder, we might predict that the inviscid flow would have streamlines of the form shown in Figure 6.8(a). Certainly a solution can be found to the inviscid equations of motion which gives such streamlines. In order to satisfy the boundary conditions at the surface of the cylinder, we have to introduce a boundary layer in the flow as described in this chapter (we take $z = 0$ as the surface of the cylinder and make some allowance for its curvature). At the front of the cylinder the pressure gradient is favourable, but it is adverse behind the cylinder and separation occurs at points represented by S, S in Figure 6.8(a). This has the effect of altering the downstream streamline pattern completely: there is a wake behind the cylinder (shaded, in Figure 6.8(b)) and (except for some mixing near the dividing streamlines) none of the fluid inside this region comes from the region upstream of the cylinder. In practice, the flow is even more complicated than this as it turns out that the wake is not steady, and in all our discussions so far we have assumed that the whole motion is steady.

Another example is the flow along a flat plate up to a step as shown in Figure 6.9. The boundary layer under the inviscid flow suggested in Figure 6.9(a) separates at S, so the streamlines are modified in the way shown in Figure 6.9(b). The fluid in the shaded region does not come from upstream, and moves in small eddy motions.

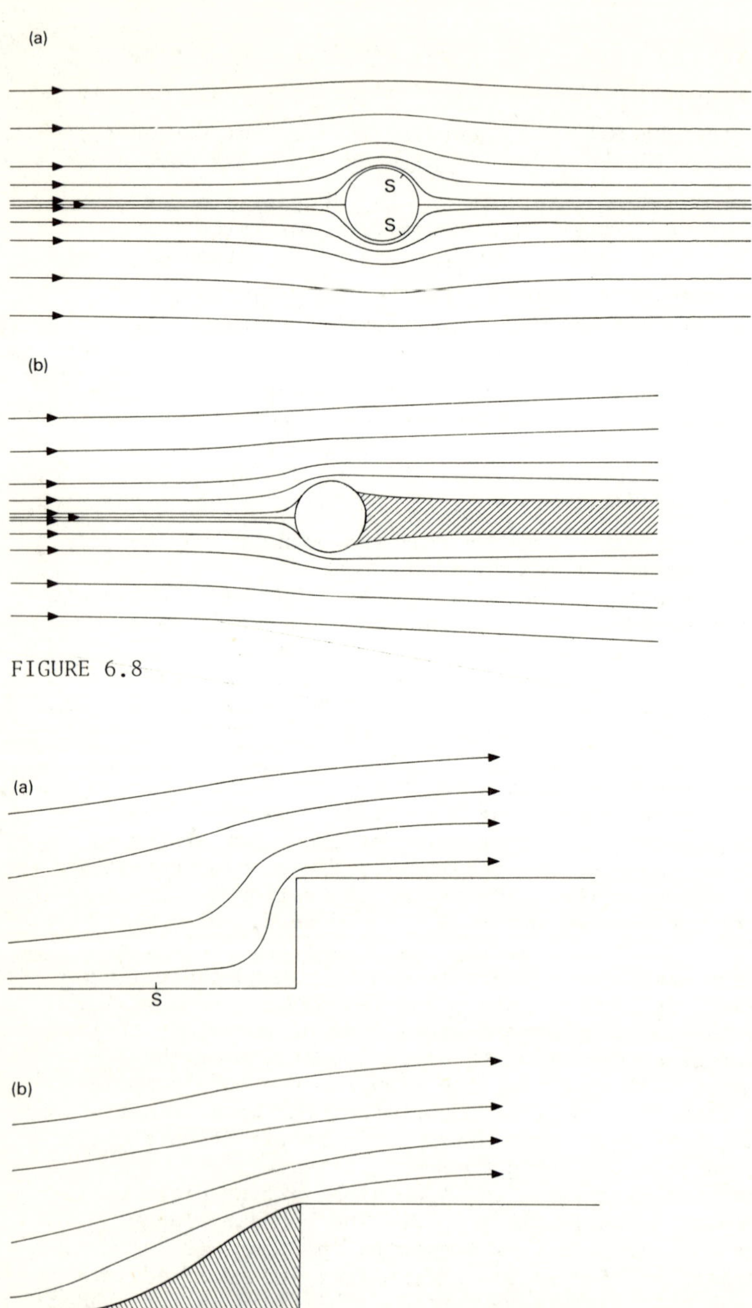

FIGURE 6.8

FIGURE 6.9

Sometimes a boundary layer separates but does not
cause a major change in the outer streamlines and re-
attaches further downstream. This could happen, for
example, in the flow past a corrugated surface as shown
in Figure 6.10, where the shaded area at the rear of

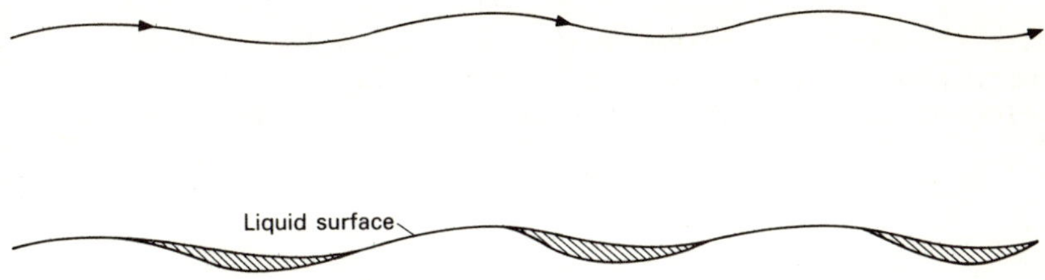

FIGURE 6.10

each corrugation is a region of reverse flow.
     Evidently the flow upstream of the point of
separation will be slightly affected when separation
occurs. We might hope, however, that the point at
which separation occurs (if it does indeed do so) will
be predicted reasonably accurately using boundary-layer
theory. Certainly we would not expect separation to
occur if boundary-layer theory does not predict it.

6.5 TURBULENT BOUNDARY LAYERS

In all our previous discussions we have assumed that
the flow is laminar. Evidently the flow in the boundary
layer can be compared with flow in a pipe (or between
parallel flat plates) and it is well known that for
sufficiently large Reynolds numbers the flow in a pipe
is turbulent. Exactly the same thing can happen in the
boundary layer, and it often happens that the boundary
layer starts by being laminar and then, at some
particular value of x (x_T, say), it suddenly becomes
turbulent. In this case the flow outside the boundary
layer is not affected. The transition to turbulence
occurs at some particular value of R_x - where R_x
is defined in equation (101) - usually about 2×10^6.
But a more appropriate parameter is a Reynolds number
$R_\delta = \rho u \delta / \mu$ based on the 'thickness' of the boundary
layer. What we mean by this we shall discuss more
precisely in the next section.

The theory of the turbulent boundary layer is beyond the scope of this book. We shall merely state here that the thickness of the turbulent boundary layer still increases rather like $(x - x_0)^{\frac{1}{2}}$ on a flat plate, where now x_0 is a virtual origin such that $0 < x_0 < x_T$. This thickness increases rather more rapidly than the corresponding thickness for a laminar layer. The velocity profiles are shown schematically in Figure 6.11.

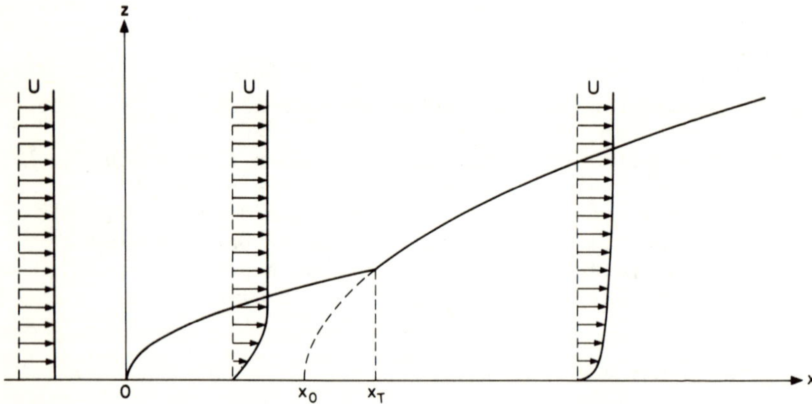

FIGURE 6.11

The value of $(\partial u/\partial z)_0$ for a turbulent layer is, generally speaking, greater than that for a laminar layer. The frictional drag is therefore greater. On the other hand, separation is less likely to occur (for the same reason). Sometimes the flow past a body is such that separation occurs if the boundary layer remains laminar, but not if it becomes turbulent before the laminar separation point. In such a case the total drag on the body is greater in the former case, where the layer remains laminar and separation is allowed to occur, than when the layer is allowed to become turbulent so that there is no separation, even though the frictional drag is greater in the latter case. For this reason turbulence is sometimes artificially introduced when a low drag is required.

It sometimes happens that a laminar layer separates, and then almost immediately becomes turbulent and then re-attaches. Such a case is illustrated in Figure 6.12.

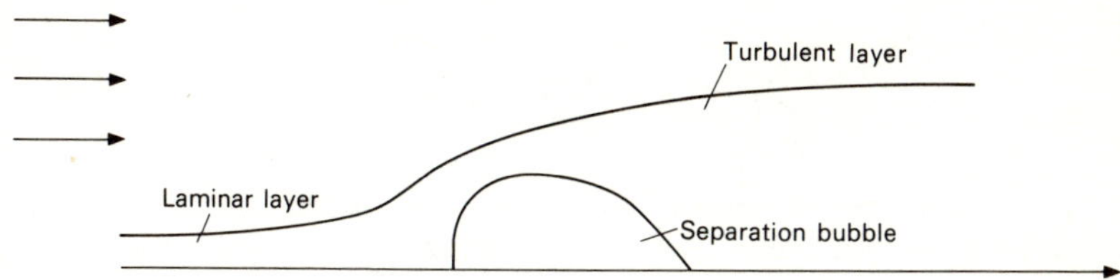

FIGURE 6.12

The small region of reverse flow is called a separation bubble.

6.6 THE DISPLACEMENT THICKNESS AND THE MOMENTUM THICKNESS

In the simplified model of the boundary layer in Section 6.1 we were able to talk without ambiguity about the 'thickness' of the boundary layer as the minimum value of z for which $u = U$. In that model the thickness was, of course, h. In practice, however, the value of u only approaches U as z approaches infinity and there is no clear indication as to what we should regard as the 'edge' of the boundary layer, although we often use the term rather loosely. For example, in Blasius's solution for the flow past a flat plate the thickness of the layer is $\delta = 4.8(\nu x/U)^{\frac{1}{2}}$ if we choose the edge to be where u differs from U by 1 per cent; but, if we choose the edge to be where u differs from U by 0.1 per cent, then the thickness is $6(\nu x/U)^{\frac{1}{2}}$. There is, of course, no problem about this when we are considering orders of magnitude only, but when we are interested in more precise calculations the difficulty arises.

We can, however, define some precise 'thicknesses' which do have a physical significance. Here we shall be concerned only with two of them. The first of these is the *displacement thickness*, which is defined as

$$\delta_1(x) = \int_0^\infty \left(1 - \frac{\rho u}{\rho_0 U}\right) dz , \qquad (123)$$

and the second is the *momentum thickness,* which is
defined as

$$\delta_2(x) = \int_0^\infty \frac{\rho u}{\rho_0 U}\left(1 - \frac{u}{U}\right) dz . \qquad (124)$$

In both these expressions $U(x)$ is the speed in the
inviscid flow just outside the boundary layer and $\rho_0(x)$
is the density in that flow. Hence $u \to U$ and $\rho \to \rho_0$
as $z \to \infty$ for any given x, and the integrals (123)
and (124) are convergent. (This does not follow, of
course, from the fact that the integrands approach zero,
but is, as we shall see, physically necessary.)

The simplest way to think of the displacement thickness is as a measure of the deficit of mass flow due to
the presence of the boundary layer: the loss of mass
occurs through the 'edge' of the layer (that is, $w \neq 0$
there). Thus, if we choose z_∞ to be a fixed value of
z, so large that u is indistinguishable from U for
all values of $x \leq X$, $z \geq z_\infty$, then the rate of mass
flow per unit width across the surface $x = X$ bounded
by $z = 0$ and $z = z_\infty$ is

$$m_1 = \int_0^{z_\infty} \rho u \, dz .$$

If viscous forces could be neglected, the mass flow
would be

$$m_2 = \int_0^{z_\infty} \rho_0 U \, dz .$$

Hence there is a deficit of mass flow whose value is

$$m_2 - m_1 = \int_0^{z_\infty} (\rho_0 U - \rho u) \, dz.$$

In the limit as $z_\infty \to \infty$, this is

$$\int_0^\infty (\rho_0 U - \rho u)\,dz = \rho_0 U \delta_1$$

using the definition (123), since ρ_0 and U are independent of z.

An alternative way of saying the same thing is to note that m_1 is the rate of mass flow if the streamlines in the flow for large values of z were displaced through a distance $\delta_1(x)$. If the streamline which passes through the point $x = X$, $z = z_\infty + \Delta_1(X)$ also passes through the point $x = 0$, $z = z_\infty$, then the flow rates across the surfaces $x = 0$ and $x = X$ are the same. Hence

$$\int_0^{z_\infty} \rho_0 U\,dz = \int_0^{z_\infty + \Delta_1(X)} \rho u\,dz = \int_0^{z_\infty} \rho u\,dz + \rho_0 U \Delta_1$$

since, for $z > z_\infty$, we have $\rho = \rho_0$ and $u = U$ to all intents and purposes. It follows that

$$\Delta_1 = \int_0^{z_\infty} \left(1 - \frac{\rho u}{\rho_0 U}\right)dz \to \delta_1 \quad \text{as} \quad z_\infty \to \infty.$$

The momentum thickness, δ_2, has a similar connection with the momentum deficit due to the presence of the bounday layer. There is a deficit of momentum involved with the mass deficit, and δ_2 is related directly to the excess deficit. The rate of flow of momentum across the surface $x = X$ between $z = 0$ and $z = z_\infty$ is

$$M_1 = \int_0^{z_\infty} \rho u^2\,dz,$$

that across $x = 0$ is

$$M_2 = \int_0^{z_\infty} \rho_0 U^2\,dz,$$

and the rate of loss of momentum due to the mass deficit is

$$M_3 = U(m_2 - m_1) = \int_0^{z_\infty} (\rho_0 U^2 - \rho u U)\,dz.$$

Hence the excess rate of loss of momentum is

$$M_2 - M_1 - M_3 = \int_0^{z_\infty} (\rho u U - \rho u^2)\,dz \rightarrow \rho_0 U^2 \delta_2$$

as $z_\infty \rightarrow \infty$, using the definition (124).

These two quantities have many uses in boundary-layer theory. Here we shall be mostly concerned with the displacement thickness δ_1. When drag forces have to be calculated, however, the momentum thickness δ_2 is more useful.

Either thickness can be used when we wish to discuss transition to turbulence. We define

$$R_{\delta_1} = \frac{\rho_0 U \delta_1}{\mu_0}, \quad R_{\delta_2} = \frac{\rho_0 U \delta_2}{\mu_0}, \tag{125}$$

where μ_0 is the viscosity of the fluid at the edge of the boundary layer. Transition to turbulence occurs when R_{δ_1} is approximately 3×10^3 or when R_{δ_2} is approximately 6×10^2, and these parameters are more accurate than the parameter R_x defined in the previous paragraph.

In the Blasius solution, all the quantities defined in this section can be calculated, and it is found that

$$\delta_1 = 1.72 \left(\frac{\mu_0 x}{\rho_0 U} \right)^{\frac{1}{2}}, \quad R_{\delta_1} = 1.72\, R_x^{\frac{1}{2}},$$

$$\delta_2 = 0.66 \left(\frac{\mu_0 x}{\rho_0 U} \right)^{\frac{1}{2}}, \quad R_{\delta_2} = 0.66\, R_x^{\frac{1}{2}}.$$

6.7 THE PROCEDURE FOR CALCULATING THE FLOW AT HIGH REYNOLDS NUMBERS

We do this by an iterative process as follows:

STEP 1 Find a solution to the inviscid equations of motion past the given body, subject to the boundary condition that the normal component of relative velocity is zero at the rigid boundaries. Chapters 7, 8 and 9 will be concerned with some methods of doing this.

STEP 2 (a) Find a laminar boundary-layer solution as described in this chapter.

(b) Check that turbulence does not occur: if it does, find the turbulent boundary layer for $x > x_T$ where transition to turbulence occurs at $x = x_T$.

(c) Check that separation does not occur: if it does, find a new inviscid solution to allow for the reverse flow which occurs in the fluid and return to Step 2(a).

(d) Calculate the displacement thickness, δ_1.

STEP 3   Find a solution of the inviscid equations of motion subject to the boundary condition that the normal component of relative velocity is zero, not on the surface $(z = 0)$ of the body itself, but on the surface $z = \delta_1$.

The effect of this (to the first order) is to displace the streamlines through a distance δ_1, which we have already seen to be the effect of the boundary layer.

STEP 4   Repeat Step 2, using the inviscid solution of Step 3 instead of that of Step 1.

Steps 3 and 4 can be repeated indefinitely in principle. In practice it is rarely necessary to go further than Step 3, and we often stop at Step 2. This degree of approximation is found, in many cases, to agree with experiment taking account of experimental accuracy of measurement.

EXERCISES

1   An incompressible fluid whose kinematic viscosity is ν occupies the region $z > 0$ and is bounded below by the rigid, horizontal flat plate $z = 0$ which moves in its own plane with velocity $U \cos \omega t$ parallel to the x-axis. The fluid is at rest far from the plate and it may be assumed that its velocity is everywhere parallel to the x-axis. By looking for a solution to the equations of question 4 in Chapter 5 of the form $\mathcal{R}\{U\,e^{i\omega t}\,f(z)\}$, find the velocity distribution in the fluid.

If the amplitude of the displacement of the plate is 50 cm and its maximum speed is 1.5 m/s, find the value of z at which the amplitude of the fluid velocity is $U/100$, given that $\nu = 1.5 \times 10^{-6} \mathrm{m}^2/\mathrm{s}$.

2 (Rayleigh's problem) An incompressible fluid whose
kinematic viscosity is ν occupies the region $z > 0$
and is bounded by the horizontal, rigid flat plate
$z = 0$. Initially everything is at rest, but at time
$t = 0$ the plate begins to move with velocity U
parallel to the x-axis, and this motion is maintained
for all $t > 0$. It may be assumed that the velocity of
the fluid is everywhere parallel to the x-axis. Show
that at time t the speed of the fluid at distance z
from the plate is $U\{1 - \text{erf}[z/2\sqrt{(\nu t)}]\}$, where

$$\text{erf } y = \frac{2}{\sqrt{\pi}} \int_0^y \exp(-\eta^2)\,d\eta.$$

(Hint: solve the equations of question 4 in Chapter 5
using Laplace transforms; it may be assumed that the
Laplace transform of $\{1 - \text{erf}(\alpha/\sqrt{t})\}$ is $s^{-1}\exp(2\alpha\sqrt{s})$.)

3 An incompressible fluid whose kinematic viscosity
is ν occupies the region $z > 0$ and is bounded by a
rigid horizontal plate in the plane $z = 0$. The plate
is rotating about the z-axis with angular velocity ω
and, far from the plate, the velocity of the fluid is
the superposition of a solid-body rotation like that of
the plate and, relative to Cartesian axes Oxyz
rotating with the plate, a uniform velocity U parallel
to the x-axis. The velocity of the fluid may be assumed
to be everywhere parallel to the plate. The rotation
of the fluid is so great that it can be assumed that
there is a balance between the Coriolis forces and the
viscous forces: thus, if the fluid velocity is $(u,v,0)$
referred to the rotating axes Oxyz,

$$- 2\omega v = \nu\frac{\partial^2 u}{\partial z^2}, \quad 2\omega u = 2\omega U + \nu\frac{\partial^2 v}{\partial z^2}.$$

Find the velocity of the fluid as a function of z and
show that the end-points of the velocity vector at
different values of z for given x,y lie on a spiral
(the Ekman spiral).

4 A rigid flat plate $z = 0, x > 0$ disturbs the flow
of a stream of incompressible fluid whose kinematic
viscosity is ν and which, far from the plate, has
velocity $(U,0,0)$ referred to fixed Cartesian axes
Oxyz. The velocity of the fluid is $(u,0,w)$ and a
possible approximation to the value of u is

$$u = \begin{cases} Uz(2\delta - z)/\delta^2, & 0 \leq z \leq \delta(x) \\ U, & z \geq \delta(x). \end{cases}$$

By considering the mass and momentum balances, show that

$$\delta(x) = (30\nu x/U)^{\frac{1}{2}}.$$

Find also the displacement thickness and the momentum thickness for this distribution of u.

5 Alternative approximations to the value of u in the previous question are

(a) $u = \begin{cases} U \sin(\pi z/2\delta), & 0 \leq z \leq \delta(x) \\ U, & z \geq \delta(x); \end{cases}$

(b) $u = U \tanh(z/\delta)$, $z \geq 0$.

In each case find $\delta(x)$ as a function of x and also find the displacement and momentum thicknesses.

6 Air at 20°C and atmospheric pressure flows with speed 5 m/s between two parallel flat plates whose distance apart is 1 cm. How far from the entrance will the boundary layers meet, if this is supposed to happen when the speed of the fluid midway between the plates is reduced by (a) 1 per cent, (b) 10 per cent?

7 An incompressible fluid, whose density is ρ and whose viscosity is μ, occupies the region $z > 0$; the fluid velocity is $(u,0,w)$ referred to fixed Cartesian axes. At the boundary $z = 0$, this velocity is zero and $u \rightarrow U(x)$ when $z \rightarrow \infty$. By integrating the equation of continuity with respect to z, show that

$$\lim_{z \to \infty} \left(w + z\frac{dU}{dx} \right) = U\frac{d\delta_1}{dx} + \delta_1 \frac{dU}{dx}$$

where $\delta_1 = \displaystyle\int_0^{\infty} (1 - u/U)dz$ is the displacement thickness.

Assuming that the Reynolds number of the flow is high enough for the boundary-layer approximation to hold, integrate the momentum equation with respect to z to show that

$$\frac{\tau_0}{\rho} = U^2\frac{d\delta_2}{dx} + (\delta_1 + 2\delta_2) U\frac{dU}{dx},$$

where $\tau_0 = -\mu(\partial u/\partial z)_{z=0}$ is the tangential stress at the wall and $\delta_2 = \int_0^\infty (u/U - u^2/U^2)dz$ is the momentum thickness.

8 Derive the expression given in the previous question for the skin friction τ_0 from first principles (that is, without using either the equation of continuity or the momentum equation).

9 For a turbulent boundary layer under a uniform flow, a common approximation for the mean velocity profile is

$$u = \begin{cases} U(z/\delta)^{1/7}, & 0 \le z \le \delta(x) \\ U, & z \ge \delta(x). \end{cases}$$

Show that the momentum thickness is $\delta_2 = 7\delta/72$.

Experiments by Blasius for flow in smooth pipes at moderate Reynolds numbers indicated that the skin friction is given $0.023\rho U^2 (U\delta/\nu)^{-1/4}$, where ν is the viscosity of the fluid. Show that the approximation assumed above gives

$$\delta = 0.38(\nu x^4/U)^{1/5},$$

where x is measured from a suitable origin.

10 A perfect gas whose viscosity μ is directly proportional to the absolute temperature has velocity $(u,0,w)$, density ρ and pressure p. These quantities satisfy the boundary-layer equations

$$\frac{\partial(\rho u)}{\partial x} + \frac{\partial(\rho w)}{\partial z} = 0,$$

$$u\frac{\partial u}{\partial x} + w\frac{\partial u}{\partial z} = \frac{1}{\rho}\frac{\partial}{\partial z}\left(\mu\frac{\partial u}{\partial z}\right),$$

$$\frac{\partial p}{\partial z} = 0.$$

The boundary conditions are (a) $u = 0$, $v = 0$ when $z = 0$, and (b) $\rho \to \rho_1$, $u \to u_1$ as $z \to \infty$, where ρ_1, u_1 are given constants. By defining a stream function ψ and a new variable

$$\zeta = \int_0^z \frac{\rho}{\rho_1}\, dz \, ,$$

show that the equations and boundary conditions reduce
to those for an incompressible fluid.

Chapter 7

EULER'S EQUATIONS OF MOTION: VORTICITY

7.1 THE EQUATION OF CONTINUITY

In this chapter we find differential equations which hold everywhere in the fluid and deduce some results from them without trying to solve them completely. We take within the fluid an arbitrary fixed domain V (see Figure 7.1) whose boundary is the surface ∂V. We consider the rate of change of mass, momentum and energy of the fluid in this domain.

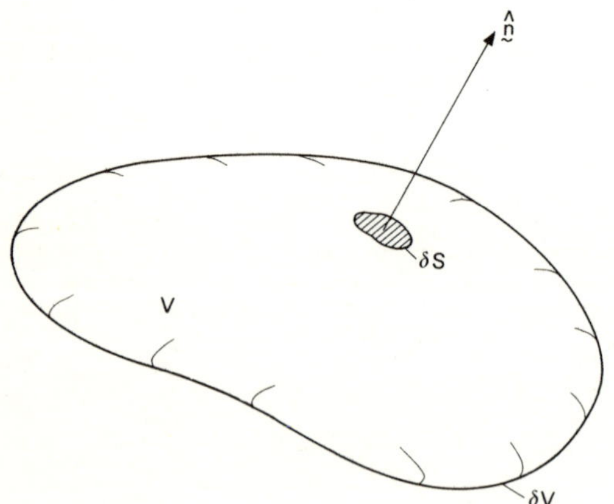

FIGURE 7.1

The rate of increase of mass is

$$\frac{d}{dt} \iiint_V \rho \, dV = \iiint_V \frac{\partial \rho}{\partial t} \, dV,$$

since V is a fixed domain, independent of time.

When there are mass sources in the region, which
contribute a mass m per unit volume per unit time
(usually m = 0), then the increase of mass is due
partly to these sources and partly to the inflow of
fluid across the surface ∂V. The rate of mass increase
due to the sources is $\iiint_V m\, dV$, and that due to the
inflow across the surface ∂V (which is assumed to be
piecewise smooth) is

$$- \iint_{\partial V} \rho \underset{\sim}{u} . \hat{\underset{\sim}{n}}\; dS \;=\; - \iiint_V \operatorname{div}(\rho \underset{\sim}{u})\; dV,$$

where $\hat{\underset{\sim}{n}}$ is the unit vector in the direction of the
outward normal of the surface ∂V, and we have used
the divergence theorem.* We therefore have that

$$\iiint_V \frac{\partial \rho}{\partial t}\, dV \;=\; \iiint_V m\, dV \;-\; \iiint_V \operatorname{div}(\rho \underset{\sim}{u})\, dV,$$

which may be rearranged in the form

$$\iiint_V \left[\frac{\partial \rho}{\partial t} + \operatorname{div}(\rho \underset{\sim}{u}) - m \right] dV = 0.$$

Now the domain V is arbitrary, and this integral can
be zero for all domains V only if the integrand (which
is assumed to be continuous) is zero everywhere. Hence

$$\frac{\partial \rho}{\partial t} + \operatorname{div}(\rho \underset{\sim}{u}) = m \qquad\qquad (126)$$

at every point in the fluid. Usually m = 0, and we
shall assume that this is true in future unless otherwise stated. So we have

$$\frac{\partial \rho}{\partial t} + \operatorname{div}(\rho \underset{\sim}{u}) = 0. \qquad\qquad (127)$$

---

* This is the theorem that, for a continuously
  differentiable vector field $\underset{\sim}{R}$, we have

$$\iiint_V \operatorname{div} \underset{\sim}{R}\, dV \;=\; \iint_{\partial V} \underset{\sim}{R} . \hat{\underset{\sim}{n}}\; dS .$$

  It is sometimes known as Gauss's theorem or as
  Green's lemma.

Using the definition of $\dfrac{D\rho}{Dt}$ given in equation (9), we may write this equation in the alternative form

$$\frac{D\rho}{Dt} + \rho \ \text{div} \ \underset{\sim}{u} = 0. \tag{128}$$

For an incompressible fluid we can (by definition) neglect the fractional change of density of a given element of fluid as it moves about: that is, $\dfrac{1}{\rho}\dfrac{D\rho}{Dt}$ is negligibly small. Hence in this case equation (128) reduces to

$$\text{div} \ \underset{\sim}{u} = 0. \tag{129}$$

In general, the quantity $\dfrac{1}{\rho}\dfrac{D\rho}{Dt}$ is equal to $-\dfrac{1}{v}\dfrac{Dv}{Dt}$ where $v = \dfrac{1}{\rho}$ is the specific volume of the fluid. Hence

$$\Delta = -\frac{1}{\rho}\frac{D\rho}{Dt} \tag{130}$$

is the rate of increase of volume per unit volume of a given element of fluid (whose mass, of course, remains constant) and we see from equation (128) that

$$\Delta = \text{div} \ \underset{\sim}{u}, \tag{131}$$

the divergence of the velocity. The scalar quantity Δ is often simply called the *divergence* of the flow; it is exactly analogous to the dilatation in the linear theory of elasticity. It is evident from equation (129) that, for incompressible flow, the divergence is negligible.

7.2 THE MOMENTUM EQUATIONS

We shall in this section assume that the internal forces in the fluid are normal to the surface on which they act (that is, we neglect viscosity). We consider the momentum balance of the fluid in the domain V of the previous section. (When viscous forces are also included, the resulting equations are called the Navier-Stokes equations; these are derived in Appendix 2.)
The rate of increase of momentum of the fluid in the domain is

$$\frac{d}{dt} \iiint_V \rho \underset{\sim}{u} \ dV = \iiint_V \frac{\partial}{\partial t}(\rho \underset{\sim}{u}) dV,$$

since, as before, the domain V is independent of time. We shall assume that any mass sources in the fluid contribute no momentum to the fluid (that is, that they are symmetrical); an extra term can easily be included in the equations if this is not the case. Any source of momentum in the fluid which contributes no mass is indistinguishable from a body force, and so need not be included specifically as a source of momentum. Thus there are three mechanisms for increasing the momentum of the fluid in the domain V : the first is any body force (for example, that due to gravity) which may be acting on the fluid, and this has value $\iiint_V \rho \underset{\sim}{F}\, dV$,

where $\underset{\sim}{F}$ is the force per unit mass on the fluid. The second mechanism is inflow of fluid across ∂V; this fluid carries momentum with it, and the resulting rate of momentum inflow is

$$- \iint_{\partial V} \rho \underset{\sim}{u}(\underset{\sim}{u}.\hat{\underset{\sim}{n}})\, dS = - \iiint_V \left[(\underset{\sim}{u}.\nabla)(\rho\underset{\sim}{u}) + \rho\underset{\sim}{u}\,\mathrm{div}\,\underset{\sim}{u}\right] dV, \quad (132)$$

where we have used the divergence theorem for each component of the left-hand side of equation (132), and the identity

$$\mathrm{div}(\rho u_i \underset{\sim}{u}) = \underset{\sim}{u}.\nabla(\rho u_i) + \rho u_i\,\mathrm{div}\,\underset{\sim}{u}, \quad i = 1,2,3.$$

The negative sign in equation (132) occurs because we want an expression for the *inflow* of momentum, and $\hat{\underset{\sim}{n}}$ is directly *outward* from V.

Finally, there are the internal forces on the fluid; since we are here assuming them to be normal, their resultant is the force due to the pressure on V, and this has magnitude

$$- \iint_{\partial V} p\,\hat{\underset{\sim}{n}}\, dS = - \iiint_V \mathrm{grad}\,p\, dV, \quad (133)$$

where we have used an alternative form of the divergence theorem. The negative sign occurs here because the force on the fluid within V due to pressure from the fluid outside acts *inwards*, whereas $\hat{\underset{\sim}{n}}$ is in the opposite direction.

Putting all these together, we have

$$\iiint_V \frac{\partial}{\partial t}(\rho u)\, dV = I_1 + I_2 + I_3$$

where

$$I_1 = \iiint_V \rho \underset{\sim}{F} \, dV,$$

$$I_2 = - \iiint_V \left[(\underset{\sim}{u}.\nabla)(\rho\underset{\sim}{u}) + \rho\underset{\sim}{u} \, \mathrm{div} \, \underset{\sim}{u} \right] dV, \text{ and}$$

$$I_3 = - \iiint_V \mathrm{grad} \, p \, dV.$$

This may be rearranged in the form

$$\iiint_V \left[\frac{\partial}{\partial t}(\rho\underset{\sim}{u}) + (\underset{\sim}{u}.\nabla)(\rho\underset{\sim}{u}) + \rho\underset{\sim}{u} \, \mathrm{div}\,\underset{\sim}{u} + \mathrm{grad} \, p - \rho\underset{\sim}{F} \right] dV = 0.$$

As in the previous section, the domain V is arbitrary and we assume that the integrand is continuous; it follows that the integrand is zero everywhere in the fluid. It follows that

$$\frac{\partial}{\partial t}(\rho\underset{\sim}{u}) + (\underset{\sim}{u}.\nabla)(\rho\underset{\sim}{u}) + \rho\underset{\sim}{u} \, \mathrm{div}\,\underset{\sim}{u} = \rho\underset{\sim}{F} - \mathrm{grad} \, p.$$

This may be rearranged in the form

$$\underset{\sim}{u}\left[\frac{\partial\rho}{\partial t} + (\underset{\sim}{u}.\nabla)\rho + \rho\,\mathrm{div}\,\underset{\sim}{u} \right] + \rho\left[\frac{\partial\underset{\sim}{u}}{\partial t} + (\underset{\sim}{u}.\nabla)\underset{\sim}{u} \right] = \rho\underset{\sim}{F} - \mathrm{grad} \, p,$$

and the first square bracket is zero, using the equation of continuity (128). So, using the expression for $\frac{D\underset{\sim}{u}}{Dt}$ (given in equation (9)), we have

$$\frac{D\underset{\sim}{u}}{Dt} = \underset{\sim}{F} - \frac{1}{\rho} \, \mathrm{grad} \, p. \tag{134}$$

An alternative way of expressing this is

$$\frac{\partial\underset{\sim}{u}}{\partial t} + \tfrac{1}{2}\,\mathrm{grad}\,(\underset{\sim}{u}.\underset{\sim}{u}) - \underset{\sim}{u}\wedge\mathrm{curl}\,\underset{\sim}{u} = \underset{\sim}{F} - \frac{1}{\rho}\,\mathrm{grad}\,p,$$

since $\mathrm{grad}(\underset{\sim}{u}.\underset{\sim}{u}) = 2(\underset{\sim}{u}.\nabla)\underset{\sim}{u} + 2\underset{\sim}{u}\wedge\mathrm{curl}\,\underset{\sim}{u}$. When $\underset{\sim}{F}$ is a conservative field of force with potential Ω, so that $\underset{\sim}{F} = -\,\mathrm{grad}\,\Omega$, this may be written

$$\frac{\partial\underset{\sim}{u}}{\partial t} + \mathrm{grad}(\tfrac{1}{2}\underset{\sim}{u}.\underset{\sim}{u} + \Omega) + \frac{1}{\rho}\,\mathrm{grad}\,p = \underset{\sim}{u}\wedge\mathrm{curl}\,\underset{\sim}{u}. \tag{135}$$

We shall use this form for manipulation later in this chapter.

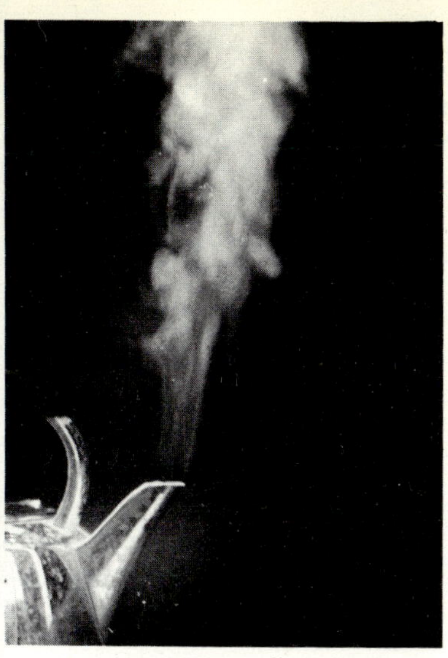

1 Laminar and turbulent
flow in steam emerging
from a kettle

2 The Severn bore

3a   Bow waves behind swimming ducks

3b   Bow waves behind minesweepers

7.3 THE ENERGY EQUATION

We shall consider what happens when the only forms of energy are potential, kinetic and internal energy. There are, of course, other possible sources of energy which may occur, the most common being the introduction of heat directly into the fluid and, when the fluid is an electrical conductor and magnetic fields are present, electromagnetic energy. The presence of heat sources can readily be accounted for by including an extra term analogous to that which we introduced for the mass source when we derived the equation of continuity. The only case we shall consider in which electromagnetic energy is important is in Chapter 11, question 8. In a real fluid, there will be dissipative forces present, but we are neglecting these in the present discussion (their effect is to transform kinetic energy into heat energy) and it follows that $\underset{\sim}{F}$ is a conservative field of force with a potential energy per unit mass Ω, as given at the end of the previous section.

The kinetic energy per unit mass is $\frac{1}{2}\underset{\sim}{u}.\underset{\sim}{u}$ and the internal energy per unit mass will be written as E ; for an ideal gas, $E = c_v T$ where c_v is the specific heat at constant volume (which is constant over a large range of temperatures) and T is the absolute temperature. For a liquid, $E = cT$ where c is the specific heat (and this, too, is constant over a large range of temperatures for most liquids) and T is measured from some suitable origin. So the rate of increase of the energy contained in the domain V is

$$\frac{d}{dt}\iiint_V \rho\,(\Omega + \tfrac{1}{2}\underset{\sim}{u}.\underset{\sim}{u} + E)\,dV = \iiint_V \frac{\partial}{\partial t}\left[\rho\,(\Omega + \tfrac{1}{2}\underset{\sim}{u}.\underset{\sim}{u} + E)\right]dV,$$

since the domain V is independent of time. This is due partly to inflow of energy with the fluid as it crosses the surface ∂V, partly to the work done by the internal forces in the fluid (which we are still assuming to be normal ones) and partly to heat conduction across ∂V. The rate of inflow of energy which is carried by the fluid as it crosses V is given by

$$-\iint_{\partial V} \rho\,(\Omega + \tfrac{1}{2}\underset{\sim}{u}.\underset{\sim}{u} + E)\underset{\sim}{u}.\hat{\underset{\sim}{n}}\,dS =$$

$$= -\iiint_V \text{div}\left[\rho\,(\Omega + \tfrac{1}{2}\underset{\sim}{u}.\underset{\sim}{u} + E)\underset{\sim}{u}\right]dV,$$

where we have used the divergence theorem in its usual

form. The rate of working of the internal forces is

$$-\iint_{\partial V} p\underset{\sim}{u}.\hat{\underset{\sim}{n}}\,dS = -\iiint_V \operatorname{div}(p\underset{\sim}{u})\,dV.$$

In both these expressions the minus sign is due to the fact that we are concerned with *inflow*, whereas $\hat{\underset{\sim}{n}}$ is directed *outwards* from V. The rate of increase of heat due to conduction of heat is

$$\iint_{\partial V} k\,\hat{\underset{\sim}{n}}.\operatorname{grad} T\,dS = \iiint_V \operatorname{div}(k\operatorname{grad} T)\,dV,$$

where k is the thermal conductivity of the fluid. It follows that

$$\iiint_V \frac{\partial}{\partial t}\left[\rho(\Omega + \tfrac{1}{2}u.u + E)\right]dV = I_1 + I_2 + I_3$$

where

$$I_1 = -\iiint_V \operatorname{div}\left[\rho(\Omega + \tfrac{1}{2}u.u + E)u\right]dV,$$

$$I_2 = -\iiint_V \operatorname{div}(pu)\,dV, \quad \text{and}$$

$$I_3 = \iiint_V \operatorname{div}(k\operatorname{grad} T)\,dV.$$

This may be re-arranged to give

$$\iiint_V \left\{\frac{\partial}{\partial t}\left[\rho(\Omega + \tfrac{1}{2}\underset{\sim}{u}.\underset{\sim}{u} + E)\right] + \operatorname{div}\left[\rho(\Omega + \tfrac{1}{2}\underset{\sim}{u}.\underset{\sim}{u} + E)\underset{\sim}{u}\right]\right.$$
$$\left. + \operatorname{div}(p\underset{\sim}{u}) - \operatorname{div}(k\operatorname{grad} T)\right\} dV = 0.$$

As before, the integrand must be zero everywhere since it is continuous and the domain V is arbitrary.

Assuming that $\dfrac{\partial \Omega}{\partial t} = 0$, which is usually the case, this may be rearranged in the form

$$(\Omega + \tfrac{1}{2}\underset{\sim}{u}.\underset{\sim}{u} + E)\left[\frac{\partial \rho}{\partial t} + \operatorname{div}(\rho\underset{\sim}{u})\right] + \rho\left[\tfrac{1}{2}\frac{\partial}{\partial t}(\underset{\sim}{u}.\underset{\sim}{u}) + \frac{\partial E}{\partial t}\right]$$

$$+ \rho(\underset{\sim}{u}.\nabla)(\Omega + \tfrac{1}{2}\underset{\sim}{u}.\underset{\sim}{u} + E) + \underset{\sim}{u}.\operatorname{grad} p + p\operatorname{div}\underset{\sim}{u} =$$

$$= \operatorname{div}(k\operatorname{grad} T).$$

The first term vanishes, using the equation of continuity (128), and since $\underset{\sim}{F} = -\,\text{grad}\,\Omega$ we can write the rest as

$$\rho\left(\frac{\partial E}{\partial t} + \underset{\sim}{u}.\nabla E\right) + p\,\text{div}\,\underset{\sim}{u} +$$

$$+ \rho\underset{\sim}{u}.\left[\frac{\partial \underset{\sim}{u}}{\partial t} - \underset{\sim}{F} + (\underset{\sim}{u}.\nabla)\underset{\sim}{u} - \frac{1}{\rho}\,\text{grad}\,p\right] = \text{div}(k\,\text{grad}\,T).$$

The square bracket of the left-hand side is zero, using equation (134), and so

$$\rho\frac{DE}{Dt} + p\,\text{div}\,\underset{\sim}{u} = \text{div}(k\,\text{grad}\,T), \qquad (136)$$

where, as usual, the operator $\dfrac{D}{Dt} = \dfrac{\partial}{\partial t} + \underset{\sim}{u}.\nabla$. Usually the thermal conductivity k is uniform throughout the fluid, so that the right-hand side may be written as $k\,\nabla^2 T$.

For an incompressible fluid $\text{div}\,\underset{\sim}{u} = 0$, and we have

$$\frac{DT}{Dt} = \kappa\,\nabla^2 T, \qquad (137)$$

where $\quad \kappa = \begin{cases} k/\rho c & \text{for a liquid} \\ k/\rho c_v & \text{for an incompressible ideal gas.} \end{cases}$

The constant κ is called the thermometric conductivity.

For a compressible ideal gas, the left-hand side of equation (136) may be written, using successively equations (1), (128), (5) and (1) again,

$$\rho\left(c_v\frac{DT}{Dt} + RT\,\text{div}\,\underset{\sim}{u}\right) = \rho\left(c_v\frac{DT}{Dt} - \frac{RT}{\rho}\frac{D\rho}{DT}\right)$$

$$= \rho\,c_p\frac{DT}{Dt} - \rho R\frac{DT}{Dt} - RT\frac{D\rho}{Dt}$$

$$= \rho\,c_p\frac{DT}{Dt} - \frac{Dp}{Dt}.$$

Hence we can write equation (136) in the form

$$\rho\,c_p\frac{DT}{Dt} - \frac{Dp}{Dt} = k\nabla^2 T. \qquad (138)$$

When $\dfrac{\partial p}{\partial t} = 0$ and the external forces are conservative with potential Ω such that $\dfrac{\partial\Omega}{\partial t} = 0$, we have

$$\frac{Dp}{Dt} = \underset{\sim}{u}.\nabla p = -\rho\underset{\sim}{u}.\frac{D\underset{\sim}{u}}{Dt} - \rho\underset{\sim}{u}.\text{grad}\,\Omega = -\rho\frac{D}{Dt}\left(\tfrac{1}{2}\underset{\sim}{u}.\underset{\sim}{u} + \Omega\right).$$

Hence for steady motion with conservative external forces, we have

$$\rho \frac{D}{Dt} \left(c_p T + \tfrac{1}{2} \underset{\sim}{u} \cdot \underset{\sim}{u} + \Omega \right) = k \nabla^2 T.$$

When, in addition, heat conduction is negligible, this may be written

$$\frac{D}{Dt} \left(c_p T + \tfrac{1}{2} \underset{\sim}{u} \cdot \underset{\sim}{u} + \Omega \right) = 0. \tag{139}$$

Hence the quantity $c_p T + \tfrac{1}{2} \underset{\sim}{u} \cdot \underset{\sim}{u} + \Omega$ is constant for a given element of fluid - that is (since the motion is steady) along a streamline, as we found in this particular case in equation (19a) of Chapter 2.

7.4 BERNOULLI'S EQUATIONS FOR INCOMPRESSIBLE FLOW

Although Bernoulli's equation in the form (19a) can, as we have just seen, be derived from the energy equation (136) for a compressible gas, the form (17) for incompressible flow cannot be obtained from the same equation. This is due to the fact that, when the flow is incompressible, the energy equation reduces to the form (137) and this is entirely a heat balance and is only implicitly affected (through terms in $\frac{DT}{Dt}$) by the velocity and pressure fields of the flow. Further the equations (128) and (134) can, at least in principle, be solved for p and $\underset{\sim}{u}$ without using the energy equation, since ρ is constant, and then equation (137) is used to find T. (For compressible flow, we cannot dissociate the energy equation from the mass and momentum equations, and have to solve the equations (128), (134), (136) and (1) for the variables p, ρ, T and $\underset{\sim}{u}$.) This means that the kinetic and potential energy balance is independent of the internal energy, and can be obtained directly from equation (134).

As in the previous section, we consider only the case in which the external forces are conservative and so we can write $\underset{\sim}{F} = -$ grad Ω. We consider two different cases: (a) steady flow and (b) irrotational flow (as defined below).

CASE (a) We write $\overset{\wedge}{\underset{\sim}{t}}$ as the unit vector in the direction of the tangent to the streamline through a point, and multiply equation (135) scalarly by $\overset{\wedge}{\underset{\sim}{t}}$ (that is, we consider only its component along the streamline). Then, since $\frac{\partial \underline{u}}{\partial t} = \underline{0}$ and $\operatorname{grad} \rho = \underline{0}$, we have

$$\hat{\underline{t}}.\operatorname{grad}\left(\tfrac{1}{2}\underline{u}.\underline{u} + \Omega + \frac{p}{\rho}\right) = \hat{\underline{t}}.\underline{u} \wedge \operatorname{curl} \underline{u}$$

and this is identically zero since $\underline{u}$ and $\hat{\underline{t}}$ are in the same direction. But $\hat{\underline{t}}.\operatorname{grad}(\ ) = \frac{\partial}{\partial s}(\ )$, where $\frac{\partial}{\partial s}$ denotes differentiation along the stream-line. Hence

$$\frac{\partial}{\partial s}\left(\tfrac{1}{2}\underline{u}.\underline{u} + \Omega + \frac{p}{\rho}\right) = 0$$

and

$$\tfrac{1}{2}\underline{u}.\underline{u} + \Omega + \frac{p}{\rho} = \text{constant along a streamline.} \quad (140)$$

This is the form of Bernoulli's equation which we derived earlier in equation (17) of Chapter 2.
CASE (b) It often happens that there exists a function ψ such that the velocity field $\underline{u}$ is given by $\underline{u} = \operatorname{grad} \phi$. In this case the flow is said to be *irrotational*. The function ϕ is not defined uniquely, but only within an additive function of time, t. It is immediately clear that, when the flow is irrotational, we have $\operatorname{curl} \underline{u} = \operatorname{curl} \operatorname{grad} \phi = \underline{0}$. Since, in addition, we have $\rho = \text{constant}$, equation (135) becomes

$$\operatorname{grad}\left(\frac{\partial \phi}{\partial t} + \tfrac{1}{2}\underline{u}.\underline{u} + \Omega + \frac{p}{\rho}\right) = \underline{0}.$$

Hence in this case, the quantity $\frac{\partial \phi}{\partial t} + \tfrac{1}{2}\underline{u}.\underline{u} + \Omega + \frac{p}{\rho}$ is a function of t only. By a suitable addition to the function ϕ, we can arrange that this function of time is in fact a constant.* Hence we have

$$\frac{\partial \phi}{\partial t} + \tfrac{1}{2}\underline{u}.\underline{u} + \Omega + \frac{p}{\rho} = \text{constant everywhere.} \quad (141)$$

This form of Bernoulli's equation is valid for irrotational flow, whether it is steady or not.

---

* In fact we could arrange for it to be zero, but
  we don't often bother to do so.

It is worth noting particularly that the form (140) holds only for steady flow, that the 'constant' on the right-hand side may vary from streamline to streamline, and that the flow may be, but need not be, irrotational. On the other hand, the form (141) holds only for irrotational flow, the constant on the right-hand side is constant throughout the fluid and the flow may be, but need not be, steady. When the flow is both irrotational and steady, the two forms are clearly identical, and the 'constant' on the right-hand side of (140) really is a constant throughout the fluid.

7.5 THE VORTICITY EQUATION

We define the *vorticity* of the flow at a given point in the fluid as

$$\underset{\sim}{\omega} = \text{curl } \underset{\sim}{u} . \tag{142}$$

This quantity is very closely related to angular velocity, as we can see by considering a rigid body rotating about a fixed point with angular velocity $\underset{\sim}{\omega}_0$. Then, if

$\underset{\sim}{u}(\underset{\sim}{r})$ is the velocity of an element of the rigid body, we have (by the definition of angular velocity) $\underset{\sim}{u} = \underset{\sim}{\omega}_0 \wedge \underset{\sim}{r}$. Hence, for the rigid body (since $\underset{\sim}{\omega}_0$ is constant),

$$
\begin{aligned}
\text{curl } \underset{\sim}{u} &= \text{curl}(\underset{\sim}{\omega}_0 \wedge \underset{\sim}{r}) \\
&= \underset{\sim}{\omega} \text{ div } \underset{\sim}{r} - (\underset{\sim}{\omega}_0 . \nabla)\underset{\sim}{r} \\
&= 3\underset{\sim}{\omega}_0 - \underset{\sim}{\omega}_0 \\
&= 2\underset{\sim}{\omega}_0 ,
\end{aligned}
$$

and $\frac{1}{2}\text{curl } \underset{\sim}{u}$ is the angular velocity of the body. A mass of fluid does not normally move like a rigid body in the sense that it does not have an angular velocity $\underset{\sim}{\omega}_0$ whose value is independent of the point in the fluid. However the quantity $\frac{1}{2}$ curl $\underset{\sim}{v}$ (where $\underset{\sim}{v}$ is the local velocity of the fluid) exists at every point and is, like the angular velocity for a solid, a measure of the rotation of the fluid. (The factor $\frac{1}{2}$ is omitted since its only purpose is to show the analogy with angular velocity.)

If we take the curl of equation (135), we find that, when the external force field is conservative

$$\frac{\partial \underset{\sim}{\omega}}{\partial t} - \text{curl}(\underset{\sim}{u} \wedge \underset{\sim}{\omega}) = - \text{curl}\left(\frac{1}{\rho}\text{grad}\,p\right) = - \text{grad}\left(\frac{1}{\rho}\right) \wedge \text{grad } p .$$

If we assume in addition that *either* ρ = constant (so
that the flow is incompressible), *or* $p = p(\rho)$ (in
which case the flow is said to be *barotropic**), the
right-hand side of this expression is easily seen to be
zero; often (but not always) one or other of these
assumptions is a good one. Now we have the standard
vector identities

$$\text{curl}(\underset{\sim}{u}\wedge\underset{\sim}{\omega}) = \underset{\sim}{u}\,\text{div}\,\underset{\sim}{\omega} - \underset{\sim}{\omega}\,\text{div}\,\underset{\sim}{u} + (\underset{\sim}{\omega}.\nabla)\underset{\sim}{u} - (\underset{\sim}{u}.\nabla)\underset{\sim}{\omega},$$

$$\text{div}\,\underset{\sim}{\omega} = \text{div}(\text{curl}\,\underset{\sim}{u}) = 0.$$

So, using equation (128), we have

$$\text{curl}(\underset{\sim}{u}\wedge\underset{\sim}{\omega}) = \underset{\sim}{\omega}\,\frac{1}{\rho}\,\frac{D\rho}{Dt} + (\underset{\sim}{\omega}.\nabla)\underset{\sim}{u} - (\underset{\sim}{u}.\nabla)\underset{\sim}{\omega}. \qquad (143)$$

Hence, for incompressible and for barotropic motions,

$$\frac{\partial\underset{\sim}{\omega}}{\partial t} - \frac{\underset{\sim}{\omega}}{\rho}\,\frac{D\rho}{Dt} - (\underset{\sim}{\omega}.\nabla)\underset{\sim}{u} + (\underset{\sim}{u}.\nabla)\underset{\sim}{\omega} = 0,$$

and this may be written in the form

$$\frac{1}{\rho}\,\frac{D\underset{\sim}{\omega}}{Dt} - \frac{\underset{\sim}{\omega}}{\rho^2}\,\frac{D\rho}{Dt} = \frac{1}{\rho}\,(\underset{\sim}{\omega}.\nabla)\underset{\sim}{u}\;;$$

after a trivial rearrangement, we obtain

$$\frac{D}{Dt}\left(\frac{\underset{\sim}{\omega}}{\rho}\right) = \left(\frac{\underset{\sim}{\omega}}{\rho}\,.\,\nabla\right)\underset{\sim}{u}\;. \qquad (144)$$

It follows from this equation, as we shall see below,
that if an element of fluid has zero vorticity at some
instant, then it can never acquire any vorticity unless
it passes through a region of infinite velocity gradient.
(This result was originally due to Helmholtz.) In a
real fluid, of course, infinite velocity gradients do
occur: what happens, as we have seen in Chapter 6, is
that viscous forces become important and then equation
(144) no longer holds. A formal proof of the result,
which can be obtained directly from equation (144) for
an inviscid fluid, is given in Appendix 4, but here we
merely point out its plausibility. For if at some
instant, $\underset{\sim}{\omega}$ is zero, then so is $\frac{D}{Dt}\left(\frac{\underset{\sim}{\omega}}{\rho}\right)$; further,
differentiating equation (144) following the motion,
we obtain for the ith component

---

* A compressible flow which is not barotropic is said
  to be *baroclinic*.

152 Chapter 7

$$\frac{D^2}{Dt^2}\left(\frac{\omega_i}{\rho}\right) = \left[\frac{D}{Dt}\left(\frac{\underline{\omega}}{\rho}\right) \cdot \nabla\right]u_i + \frac{\underline{\omega}}{\rho} \cdot \frac{D}{Dt}\left(\nabla u_i\right)$$

$$= \left[\left(\frac{\underline{\omega}}{\rho} \cdot \nabla\right)\underline{u}\right] \cdot \nabla u_i + \frac{\underline{\omega}}{\rho} \cdot \frac{D}{Dt}\left(\nabla u_i\right),$$

and this, too, is zero. Similarly for higher derivatives.
Hence, as long as the vorticity of the element can be
expressed as a convergent Taylor series with respect to
time, its value remains zero always.
There are, of course, many flows in which all the
fluid comes from a region where the velocity is uniform -
and in such a region the vorticity is certainly zero.
Hence, in such a' flow, the vorticity is zero everywhere
except, perhaps, in the neighbourhood of any obstacles
in the flow. But this means that curl $\underline{u}$ = $\underline{0}$, and so
there exists a function ϕ such that $\underline{u}$ = grad ϕ.
We note, for future reference, that there is no need
for ϕ to be a single-valued function unless the region
of zero vorticity is simply connected. This is the
definition of *irrotational* flow, as given in the previous
section, and we now see that it can often occur. The
scalar function ϕ is called the *velocity potential*
(presumably because it is a function from which the
velocity can be derived).
We should emphasize that this result holds only when
viscous forces are neglected. This is not very
surprising since it is not implausible that 'angular
velocity' can be produced or destroyed only by
tangential forces, and in the present discussion we
are ignoring these.
When we consider the flow of a fluid past a model in
a wind tunnel, it is evident that all the fluid comes
from a region of zero vorticity. So it is only fluid
which has passed somewhere near the model (or, of
course, near the wall of the tunnel) which can possess
vorticity. This fluid is either in the boundary layer
or in the wake behind the body. To all intents and
purposes, the vorticity in the flow can be taken to be
zero, everywhere except in the boundary layers and in
the wake. In fact, a very useful way of thinking of
the boundary layer is as a layer of vorticity rather than
as a layer of sharp velocity gradient (the two things
are, of course, equivalent). Analogous with this
result is another that if, at some instant, the
vorticity of an element of fluid is non-zero, then it is
always non-zero - but when it is non-zero we cannot say,
in general, that it is constant.

However, in the special case of two-dimensional flow (that is, when $\underset{\sim}{u} = \underset{\sim}{i}\,u_1(x,y) + \underset{\sim}{j}\,u_2(x,y)$), the right-hand side of equation (144) is zero, for

$$\underset{\sim}{\omega} = \underset{\sim}{k}\left(\frac{\partial u_2}{\partial x} - \frac{\partial u_1}{\partial y}\right)$$

and so

$$(\underset{\sim}{\omega}.\nabla)\underset{\sim}{u} = \left(\frac{\partial u_2}{\partial x} - \frac{\partial u_1}{\partial y}\right)\frac{\partial u_3}{\partial z}\underset{\sim}{k} = \underset{\sim}{0}$$

since $\underset{\sim}{u}$ is independent of z. It follows that in the case of two-dimensional flow $\frac{D}{Dt}\left(\frac{\omega}{\rho}\right) = 0$, and so the quantity $\frac{\omega}{\rho}$ remains constant for a given element of fluid.

It is clear from equation (129) that the velocity potential ϕ for the irrotational flow of an incompressible fluid satisfies Laplace's equation $\nabla^2\phi = 0$. It is a well-known result of vector calculus that a function ϕ which

   (i) satisfies Laplace's equation in a simply-connected region,

  (ii) has continuous first derivatives throughout the region and its boundaries, and

(iii) has a given normal derivative on all the boundaries

is unique within an additive arbitrary constant. Now the normal derivative of the velocity potential at a surface is exactly equal to the normal component of the fluid velocity there and, as we saw in Section 1.6, this is specified for the flow of an inviscid fluid whose boundaries are known. It follows that, if we can find a solution ϕ to Laplace's equation, and its derivatives satisfy the boundary conditions of a particular problem, then this is the velocity potential for the unique irrotational flow which exists for this problem. We note that the method used to find an appropriate solution to Laplace's equation is unimportant: we may use a systematic technique, we may have a flash of inspiration, or we may (and this is a very common method) have sufficient experience to build up a suitable velocity potential from a number of other problems which partially correspond to the given one.

7.6 VORTEX LINES AND TUBES

A *vortex line* is a curve which is at every point on it tangential to the direction of the vorticity at that

point. There is an obvious analogy with streamlines,
but it is important to remember that a vortex line does
not give the direction of flow, but rather a direction
of swirl. For example, if a cup of tea is stirred with
a vertical spoon moving horizontally, the streamlines
are horizontal and the vortex lines are vertical.
Another example, which we shall discuss in more detail
later on, is a smoke ring: here, in the motion relative

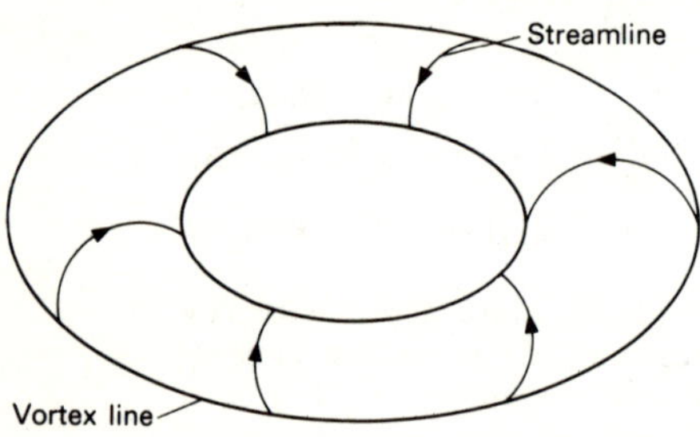

FIGURE 7.2

to the ring, the streamlines are circles passing through
the ring, as shown in Figure 7.2, and the vortex lines
are circles going round the ring.
    A *vortex tube* is the bundle of vortex lines which go
through a closed curve C (see Figure 7.3). Suppose

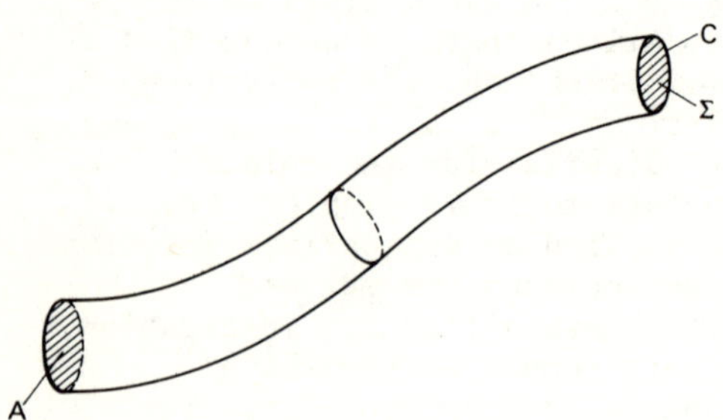

FIGURE 7.3

that A is the cross-sectional area of a vortex tube
at an arbitrary point on its length, and that Σ is
some surface entirely within the tube bounded by C.
Then we consider the domain V bounded by the surface
∂V consisting of A, Σ and the part of the curved

surface of the tube between A and Σ. Then, using the divergence theorem, we obtain,

$$\iiint_V \operatorname{div} \underset{\sim}{\omega}\, dV = \iint_{\partial V} \underset{\sim}{\omega}.\hat{\underset{\sim}{n}}\, dS = \iint_A \underset{\sim}{\omega}.\hat{\underset{\sim}{n}}\, dS + \iint_\Sigma \underset{\sim}{\omega}.\hat{\underset{\sim}{n}}\, dS$$

where $\hat{\underset{\sim}{n}}$ is, as usual, the outward-pointing normal to the surface ∂V, since $\underset{\sim}{\omega}.\hat{\underset{\sim}{n}} = 0$ on the curved part of the surface V. Now $\operatorname{div} \underset{\sim}{\omega} = \operatorname{div}(\operatorname{curl} \underset{\sim}{u}) = 0$, and if we suppose that $\underset{\sim}{\omega}$ is directed from Σ towards A, we have $\iint_A \underset{\sim}{\omega}.\hat{\underset{\sim}{n}}\, dS = \iint_A \omega\, dS$. It follows that

$$\iint_A \omega\, dS = - \iint_\Sigma \underset{\sim}{\omega}.\hat{\underset{\sim}{n}}\, dS ,$$

whatever the position of A. (If A is the other side of Σ, or if it intersects it, we obtain the same result by taking an alternative surface Σ' further along the tube, so that $\underset{\sim}{\omega}$ is in the direction from Σ' towards A, and repeating the argument.) Another way of saying this is that the quantity

$$\kappa = \iint \omega\, dS , \tag{145}$$

where the integration is over an area of cross-section of the tube, is constant along the tube. The quantity κ is called the *strength* of the vortex tube.

It follows at once that a vortex tube cannot end within the fluid, for, suppose that in Figure 7.4 it ended at Q. We continue the tube a little further to

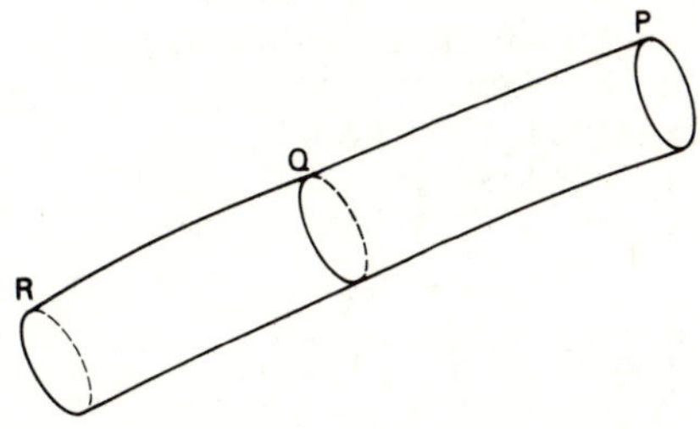

FIGURE 7.4

R, noting that there is no vorticity between Q and R (or else there would still be a vortex tube there), and apply the same argument. We find, as before, that

$$\iint_P \omega \, dS = \iint_R \omega \, dS$$

and the right-hand side is zero because $\omega = 0$ on R. Hence $\kappa = 0$, where κ is the strength of the tube defined by (145), and this is nonsense in general. By shrinking the cross-sectional area at P down to a point, the same result holds for vortex lines.

There are, therefore, two possibilities: either the vortex lines begin and end on the boundaries of the fluid (in which case the extension QR to the tube in Figure 7.4 does not exist within the fluid), or they form closed curves like those in the smoke ring.

Another important result for an inviscid fluid is that vortex lines move with the fluid if this is inviscid. For a fluid which has all its vorticity concentrated in a number of discrete vortex tubes, this is a direct result of the fact that the fluid which is not in a vortex tube at some instant can never be in a vortex tube (since it can never possess vorticity), and the fluid which is in a vortex tube must always be in one (since it must always possess vorticity). The result (originally due to Kelvin) can, however, be proved in general, and we shall proceed to do this.

First we introduce the concept of *circulation*. If C is a simple closed curve in the fluid, then we define the circulation round C as

$$\kappa = \int_C \underset{\sim}{u} . \hat{\underset{\sim}{t}} \, dc , \tag{146}$$

where $\hat{\underset{\sim}{t}}$ is the unit vector tangential to C at a point on it, and dc is the length of an element of C at this point. We have used the same symbol for the strength of the vortex tube (145) through C and the circulation round C (146), because if S is any surface within the fluid bounded by C we have from Stokes's theorem

$$\int_C \underset{\sim}{u} . \hat{\underset{\sim}{t}} \, dc = \iint_S (\text{curl } \underset{\sim}{u}) . \hat{\underset{\sim}{n}} \, dS = \iint_S \underset{\sim}{\omega} . \hat{\underset{\sim}{n}} \, dS .$$

Hence the circulation round C is equal to the strength of the vortex tube passing through C. In particular, of course, if there is no vortex tube passing through C then there is no circulation round C.

If the flow happens to be irrotational everywhere in the neighbourhood of C, so that we can write $\underset{\sim}{u} = \text{grad } \phi$, we have

$$\kappa = \int_C \text{grad } \phi . \hat{\underset{\sim}{t}} \, dc = \int_C \frac{\partial \phi}{\partial c} \, dc = \Delta\phi \, ,$$

the change in ϕ in going once round the curve C. It follows that, if there is vorticity anywhere in the flow (so that κ is not identically zero everywhere), the velocity potential ϕ is not single-valued. Conversely, if ϕ is not single-valued, there must be vorticity somewhere in the flow, though it may be confined to an extremely small region.

For the irrotational flow of an incompressible fluid in a multiply-connected region, the conditions discussed at the end of Section 7.5 are insufficient to determine the flow uniquely. In addition, we need to know the circulation round every irreducible curve in the region; it can be shown that an irrotational flow satisfying both the original boundary conditions and the conditions on the circulation is unique.

Suppose, then, that the fluid which at time t lies on the curve C has moved at time $t + \tau$ (where τ is small) to the neighbouring curve C'. The circulation round C at time t is κ, and that round C' at

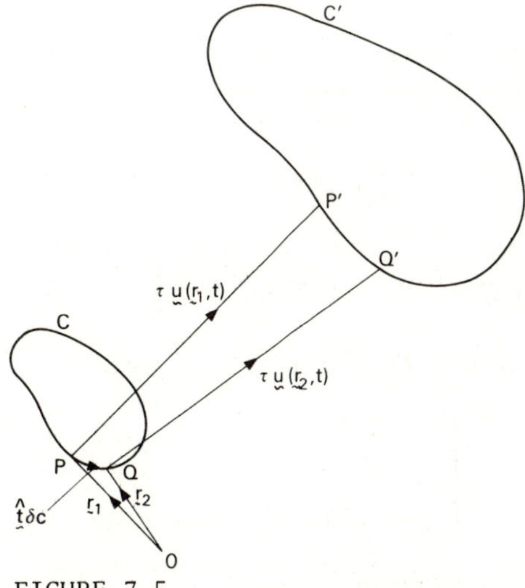

FIGURE 7.5

time $t + \tau$ is κ'. We shall therefore write

$$\frac{D\kappa}{Dt} = \lim_{\tau \to 0} \left(\frac{\kappa' - \kappa}{\tau} \right)$$

for the rate of change of circulation round C moving with the fluid. Suppose that PQ is an element of length δc on C, and that the fluid at P moves to P' on C', and that at Q to Q' on C', as shown in Figure 7.5. We write $\hat{\underset{\sim}{t}}\,\delta c = \overrightarrow{PQ}$, $\hat{\underset{\sim}{t}}'\delta c' = \overrightarrow{P'Q'}$, and we let P, Q have vector coordinates $\underset{\sim}{r}_1$, $\underset{\sim}{r}_2$ respectively referred to some fixed origin O, and we write $\underset{\sim}{u}$ for the velocity of the fluid at P at time t and $\underset{\sim}{u}'$ for that at P' at time $t + \tau$. Then, by definition,

$$\kappa = \int_C \underset{\sim}{u}.\hat{\underset{\sim}{t}}\,dc \quad \text{and} \quad \kappa' = \int_{C'}\underset{\sim}{u}'.\hat{\underset{\sim}{t}}'dc' \,. \quad \text{Now} \quad \overrightarrow{PP'} = \underset{\sim}{u}\tau$$

to the first order in τ, and so, to this order,

$$\underset{\sim}{u}' = \underset{\sim}{u}\left[\underset{\sim}{r}_1 + \tau\underset{\sim}{u}(\underset{\sim}{r}_1,t), \ t + \tau\right] = \underset{\sim}{u}(\underset{\sim}{r}_1,t) + \tau\frac{D\underset{\sim}{u}}{Dt} \,,$$

where $D\underset{\sim}{u}/Dt$ is evaluated at P. Also, to the first order in δc and τ, we have

$$\overrightarrow{QQ'} = \tau\underset{\sim}{u}(\underset{\sim}{r}_2,t) = \tau\underset{\sim}{u}(\underset{\sim}{r}_1,t) + \tau(\delta c\,\hat{\underset{\sim}{t}}.\nabla)\underset{\sim}{u}$$

$$= \overrightarrow{PP'} + \tau(\delta c\,\hat{\underset{\sim}{t}}.\nabla)\underset{\sim}{u} \,.$$

It follows that, to the same order,

$$\hat{\underset{\sim}{t}}'\,\delta c' = \overrightarrow{P'Q} = \overrightarrow{PQ} + \overrightarrow{QQ'} - \overrightarrow{PP'}$$

$$= \hat{\underset{\sim}{t}}\,\delta c + \tau(\delta c\,\hat{\underset{\sim}{t}}.\nabla)\underset{\sim}{u} \,,$$

and so

$$\underset{\sim}{u}'.\hat{\underset{\sim}{t}}'\delta c' - \underset{\sim}{u}.\hat{\underset{\sim}{t}}\,\delta c = \tau\left[\frac{D\underset{\sim}{u}}{Dt}.\hat{\underset{\sim}{t}}\,\delta c + \underset{\sim}{u}.(\delta c\,\hat{\underset{\sim}{t}}.\nabla)\underset{\sim}{u}\right] \,.$$

Now, using equation (134), the right-hand side of this expression is

$$\left[(\underset{\sim}{F} - \frac{1}{\rho}\,\text{grad}\,p).\hat{\underset{\sim}{t}} + \tfrac{1}{2}(\hat{\underset{\sim}{t}}.\nabla)u^2\right]\tau\,\delta c =$$

$$= \hat{\underset{\sim}{t}}.\text{grad}(-\Omega - \frac{p}{\rho} + \tfrac{1}{2}u^2)\tau\,\delta c$$

when $\underset{\sim}{F} = -\text{grad}\,\Omega$ and ρ is constant. Hence

$$\kappa' - \kappa = \int_{C'} \underset{\sim}{u'} \cdot \underset{\sim}{\hat{t}'} dc' - \int_{C} \underset{\sim}{u} \cdot \underset{\sim}{\hat{t}} \, dc$$

$$= \tau \int_{C} \underset{\sim}{\hat{t}} \cdot \operatorname{grad}(-\Omega - \frac{p}{\rho} + \tfrac{1}{2}u^2) dc$$

to the first order in τ, and this integral is
identically zero. It follows that

$$\frac{D\kappa}{Dt} = \lim_{\tau \to 0} \left(\frac{\kappa' - \kappa}{\tau} \right) = 0 \; .$$

This result is also true for compressible barotropic
motion, since in this case also $\dfrac{1}{\rho}$ grad p can be
expressed as the gradient of a scalar quantity.
     Hence the strength of the vortex tube of the fluid
which at time t passes through C does not vary with
time. This result is true for any vortex tube, and in
particular for one with very small cross-section. Hence
it is true for vortex lines as well. That is, we have
shown that for an incompressible inviscid fluid (and for
a compressible barotropic, inviscid fluid) the vortex
lines move with the fluid. A corollary of this result
is that due to Helmholtz, which we have already discussed
in Section 7.5.

7.7 LINE VORTICES

It sometimes happens that all the vorticity is confined
to a number of discrete tubes, the diameters of which
are much smaller than the distances between them. Under
these circumstances, it is a good approximation to
regard the tubes as having infinitesimally small cross-
sectional areas, but finite strength - this implies, of
course, that the vorticity itself must be infinitely
large in the tubes, but although this is physically
unrealistic it is an acceptable mathematical model.
These model vortex tubes are called *line vortices*. We
shall be concerned with situations in which the vorticity
is zero (and the flow irrotational) everywhere except on
the line vortices. When we are looking for solutions of
the inviscid equations of motion we must expect to find
singularities along the line vortices and we have to
apply the condition that the circulation round them has
the correct value.

We shall discuss here two types of line vortex - the rectilinear line vortex and the circular ring vortex. The first lies, as its name implies, along a straight line, and we shall restrict our discussion to cases in which all of them are parallel to the z-axis. We shall consider first a fluid whose motion is caused completely by a single line vortex of strength κ which, at time $t = 0$, lies along the z-axis. From symmetry, the velocity of the fluid is a transverse velocity and

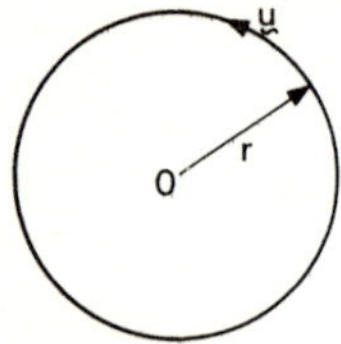

FIGURE 7.6

depends only on r, where (r,θ,z) are cylindrical polar coordinates (see Figure 7.6). It is shown in Appendix 5 that, for $r > a$, this flow is identical with that in a viscous fluid caused by the rotation at constant angular velocity about its axis of a right circular cylinder of radius a, which is within the fluid. In both the idealized and the real flow, the fluid moves in circles round the z-axis and the circulation, round a circle of radius r whose centre is on the z-axis, is

$$\kappa = \int_0^{2\pi} ur \, d\theta = 2\pi ur,$$

and this is constant. Hence the speed of the fluid at distance r from the vortex is

$$u = \frac{\kappa}{2\pi r}, \tag{147}$$

and this is the flow described in equation (11) of Chapter 2. The corresponding velocity potential is

$$\phi = \frac{\kappa}{2\pi} \log r. \tag{148}$$

In the idealized flow, the fluid on the z-axis remains there (although it is spinning very fast). But the vortex lines move with the fluid and so the line vortex too remains along the z-axis for all time; we have therefore established the principle that a rectilinear

line vortex cannot cause its own translational motion.
(This is to be expected, too, for the rotating cylinder
in a real fluid of which the line vortex is a model:
when all the motion is due to the rotation of the
cylinder, there is no preferred direction, and so there
can be no sideways force on the cylinder, which will
therefore remain in position unless there is some other
force acting on it.)

If, however, we have n rectilinear line vortices
parallel to the z-axis of strengths $\kappa_1, \kappa_2, \ldots, \kappa_n$ which
at time t are respectively through the points
$A_1, A_2, \ldots, A_n$ in the x-y plane, then it is easily
verified that at a point P which is different from all
the points A_i, the velocity potential is

$$\phi = \sum_{s=1}^{n} \phi_s = \sum_{s=1}^{n} \frac{\kappa_s}{2\pi} \log r_s , \qquad (149)$$

where r_s is the distance of P from the line vortex
through A_s. To verify this we note that the circulation
round any single vortex has the right value, and also
that the velocity approaches zero as $r \to \infty$: hence all
the boundary conditions are satisfied and the expression
(149) does give the flow due to the n vortices alone.
It follows that at any point the velocity of the fluid
at any point in the neighbourhood of one of the vortices
(that at A_1, say) is not necessarily zero: it is, in
fact, the vector sum of the separate velocities due to
each of the other vortices. Instead of trying to express
this in a precise mathematical formulation, we shall
consider a number of special cases.

EXAMPLE 7.7.1 Two vortices with equal strengths in
opposite senses.
Suppose we have line vortices parallel to the z-axis
which, at time $t = 0$, pass through the points
A(0,a,0), B(0,-a,0) and that their strengths are
respectively κ and $-\kappa$. This is shown in Figure
7.7. At the point P the velocity is the vector
sum of $\kappa/2\pi AP$ perpendicular to AP in the sense
shown and $\kappa/2\pi BP$ perpendicular to BP in the
sense shown. At A, however, the translational
velocity of the fluid due to the vortex at A is
zero, but there will be a velocity because of the
motion due to the vortex at B - and this has
magnitude $\kappa/2\pi AB = \kappa/4\pi a$ and is in the direction
parallel to Ox. Similarly, the velocity at B is

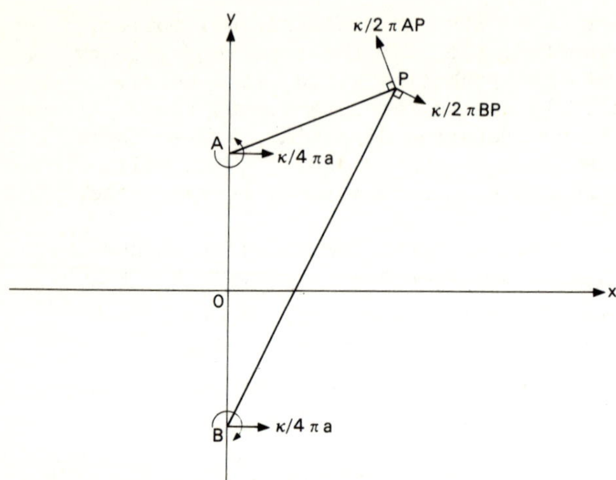

FIGURE 7.7

due entirely to the vortex at A, and this is also
$\kappa/4\pi a$ parallel to Ox. So the two vortices move,
at time t = 0, parallel to each other with speed
$\kappa/4\pi a$. They therefore remain the same distance
apart and go on moving parallel to each other.

COROLLARY: From symmetry, we see that there is no flow
across the plane y = 0. Hence the flow in the region
y > 0 due to the two vortices is the same as that due
to the single vortex at A in the presence of a rigid
wall in the surface y = 0. The vortex at B is said
to be the *image* of that at A in the wall.

EXAMPLE 7.7.2 Two vortices with equal strengths in
the same sense.
Here we shall consider two vortices which, at time
t = 0, pass through the same points A and B as
before, but this time each has strength κ. The
velocity at P is as shown in Figure 7.8. That at
A is $\kappa/4\pi a$ in the negative direction of the
x-axis, and that at B is $\kappa/4\pi a$ in the positive
direction of the x-axis. Hence the vortices both
move in directions perpendicular to the line
joining the points AB with the same speed, but
in the opposite sense. The effect is therefore
that of rotation about the z-axis, and in fact the
vortices remain generators of the cylinder
$x^2 + y^2 = a^2$.

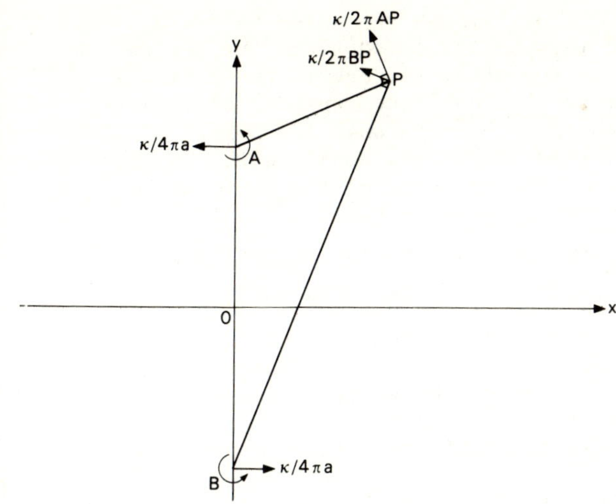

FIGURE 7.8

EXAMPLE 7.7.3 An infinite row of vortices of equal
strength in the same sense.
Here we consider an infinite set of vortices (as
shown in Figure 7.9) all parallel to the z-axis,

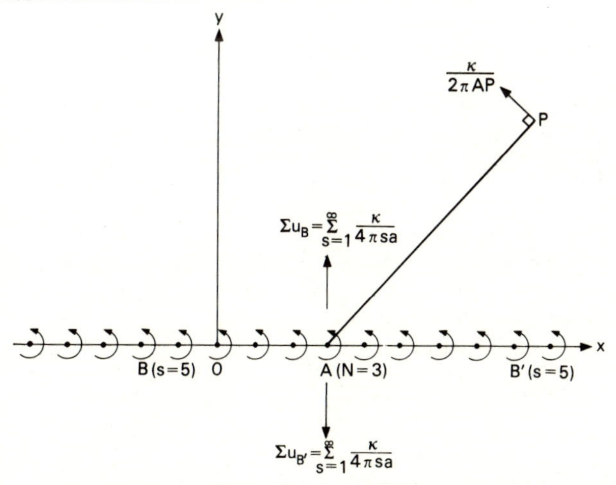

FIGURE 7.9

all of the same strength κ (in the same sense)
which, at time t = 0, pass through the points
(na,0,0) where n is a positive or a negative
integer or zero. We concentrate for the moment on

a particular vortex through the point A $(Na,0,0)$. The motion of this vortex is equal to the motion of the fluid at the point A due to the presence of all the other vortices. We consider the other vortices in pairs - those through the point $B[(N-s)a,0,0]$ and $B'[(N+s)a,0,0)]$ with $s = 1,2,3,\ldots$. The velocity of the fluid at A due to the vortex through B is $\left(0, \dfrac{\kappa}{2\pi sa}, 0\right)$ and that due to the vortex through B' is $\left(0, -\dfrac{\kappa}{2\pi sa}, 0\right)$. Hence the velocity at A due to the vortices through B and B' together is zero - and this is true for each value of s. It follows that the vortex through A does not move - and this is true for each of the vortices. So we see that the pattern of vortices remains unchanged.

To find the motion of the fluid due to this row of vortices, we note that the velocity at the point $P(x,y,z)$ due to the vortex through the point A is $(u_N,v_N,0)$, where

$$u_N = -\frac{\kappa}{2\pi AP}\sin P\hat{A}x = -\frac{\kappa y}{2\pi[(x-Na)^2 + y^2]}$$

and

$$v_N = \frac{\kappa}{2\pi AP}\cos P\hat{A}x = \frac{\kappa(x-Na)}{2\pi[(x-Na)^2 + y^2]} .$$

The velocity at P due to all the vortices is therefore $(u,v,0)$ where

$$u = \sum_{N=-\infty}^{N=\infty} u_N, \quad v = \sum_{N=-\infty}^{N=\infty} v_N .$$

Using the well-known identities

$$\sum_{N=-\infty}^{N=\infty} \frac{\beta}{(\alpha - N)^2 + \beta^2} = \frac{\pi\sinh\pi\beta\cosh\pi\beta}{\sinh^2\pi\beta + \sin^2\pi\alpha} ,$$

$$\sum_{N=-\infty}^{N=\infty} \frac{\alpha - N}{(\alpha - N)^2 + \beta^2} = \frac{\pi\sin\pi\alpha\cos\pi\alpha}{\sinh^2\pi\beta + \sin^2\pi\alpha} ,$$

it is easy to show that

$$u = -\frac{\kappa}{2a}\frac{\sinh(\pi y/a)\,\cosh(\pi y/a)}{\sinh^2(\pi y/a) + \sin^2(\pi x/a)},$$

$$v = \frac{\kappa}{2a}\frac{\sin(\pi x/a)\,\cos(\pi x/a)}{\sinh^2(\pi y/a) + \sin^2(\pi x/a)}.$$

The streamlines are given by the differential equation

$$\frac{dy}{dx} = \frac{v}{u} = \frac{\sin(2\pi x/a)}{\sinh(2\pi y/a)},$$

and this integrates to give

$$\cosh(2\pi y/a) - \cos(2\pi x/a) = C, \text{ a constant.}$$

We see immediately that

$$-1 + c \le \cosh(2\pi x/a) \le 1 + c, \ \cos(2\pi x/a) \le 1 + C.$$

Hence, for a streamline on which $C > 2$, there are no real values of x for which $y = 0$, and such a streamline is a wavy line roughly parallel to the x-axis. Also, for a streamline on which $C < 2$, there are some values of x for which no real value of y exists, and such a streamline is closed and encircles one of the vortices. The full pattern is shown in Figure 7.10.

FIGURE 7.10

Note in particular that, for large enough values of y, the streamlines are almost straight lines parallel to the x-axis. In fact, it can easily be seen from the general expressions for u and v that

$$u \to -\kappa/2a, \quad v \to 0 \quad \text{as } y \to \infty$$

and

$$u \to \kappa/2a, \quad v \to 0 \quad \text{as } y \to -\infty ;$$

so we see that, far enough away from the vortices, the flow for $y > 0$ becomes indistinguishable from a uniform flow with velocity $\kappa/2a$ in the negative direction of the x-axis, while for $y < 0$ it becomes indistinguishable from that with a velocity of the same magnitude but in the opposite direction. The limiting case in which $\kappa \to 0$, $a \to 0$, but where the ratio κ/a remains finite, is discussed in Section 7.9 and question 13.

EXAMPLE 7.7.4 Two infinite rows of vortices: the Kármán vortex street.
Here we consider two parallel rows of vortices like those described in Example 7.7.3 which lie in the planes $y = \pm b$. Those in the plane $y = b$ are all of strength κ (all in the same sense) and at time $t = 0$ pass through the points $(na,b,0)$, where n is a positive or negative integer or zero; those in the plane $y = -b$ are all of strength κ in the opposite sense and, at time $t = 0$ pass through the points $[(n+\tfrac{1}{2})a,-b,0]$, where n is again a positive or negative integer or zero. This system is shown in Figure 7.11. We concentrate for the moment on a particular vortex through A $(Na,b,0)$, and consider the effect on it of the other vortices. We have already seen that those in the same plane have no effect on it. We now consider the effect on A of the pair of vortices through $C[(N-s-\tfrac{1}{2})a,-b,0]$ and $C'[(N+s+\tfrac{1}{2})a,-b,0]$ with $s = 0,1,2,\ldots$. It is readily seen that the velocity at A due to these vortices together is $(u_s,0,0)$ where

$$u_s = \frac{\kappa}{\pi AC} \cdot \frac{2b}{AC}$$

$$= \frac{2\kappa b}{\pi[4b^2 + (s+\tfrac{1}{2})^2 a^2]}$$

$$= \frac{8\kappa b}{\pi[16b^2 + (2s+1)^2 a^2]}.$$

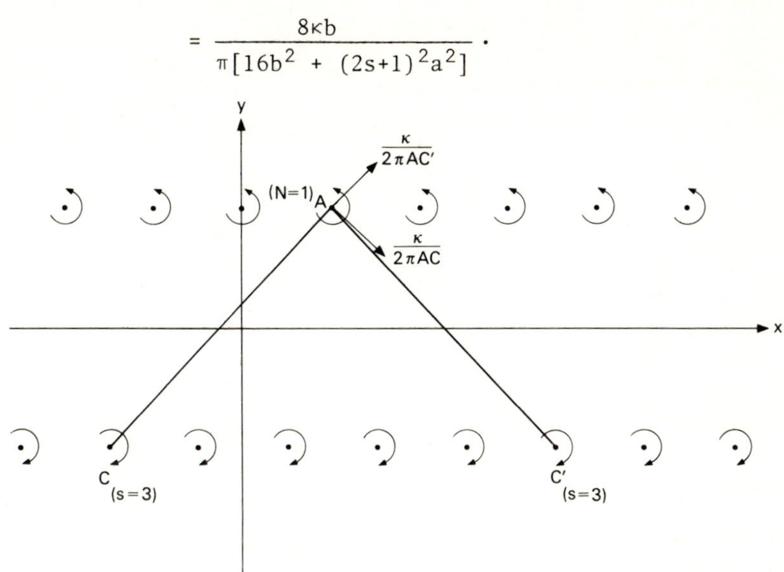

FIGURE 7.11

Hence the velocity at A due to all the other vortices is   (U,0,0)   where

$$U = \frac{2\kappa}{\pi a} \sum_{s=0}^{\infty} \frac{4(b/a)}{16(b/a)^2 + (2s+1)^2} = \frac{\kappa}{2a} \tanh(2\pi b/a),$$

using the well-known result that

$$\sum_{s=0}^{\infty} \frac{c}{c^2 + (2s+1)^2} = \frac{\pi}{4} \tanh(\tfrac{1}{2}\pi c).$$

It follows that each of the vortices in the plane
y = b moves with velocity U = (κ/2a) tanh(2πb/a)
parallel to the x-axis, and it is easy to verify
that the same is true for each vortex in the plane
y = -b. Hence the pattern of the vortices moves as
a whole with this velocity.

Using the result obtained in the last example for the
separate rows of vortices, it is easy to see that the
velocity of the fluid for large values of y (either
positive or negative) is negligible, since each row
produces equal and opposite velocities there.

The system is called a Kármán vortex street and occurs in practice behind a bluff body such as a circular cylinder placed across a uniform stream. This is shown schematically in Figure 7.12. Vortices

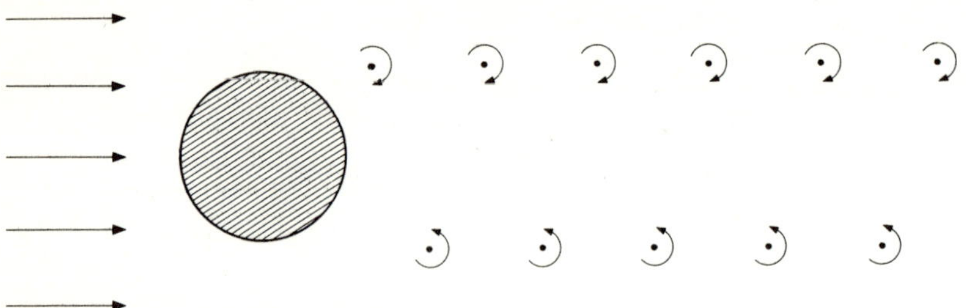

FIGURE 7.12

are cast off alternately on opposite sides, presumably (in a real fluid) due to boundary layer separation as discussed in the previous chapter. Several diameters downstream of the cylinder, the system is indistinguishable from the infinite street discussed above.

7.8 RING VORTICES

The second type of line vortex we shall consider is the circular ring vortex: this is a particular case of an idealized closed vortex tube. We suppose that, at time $t = 0$, the vorticity is all concentrated in the circle $x^2 + y^2 = a^2$, $z = 0$ as in Figure 7.13.

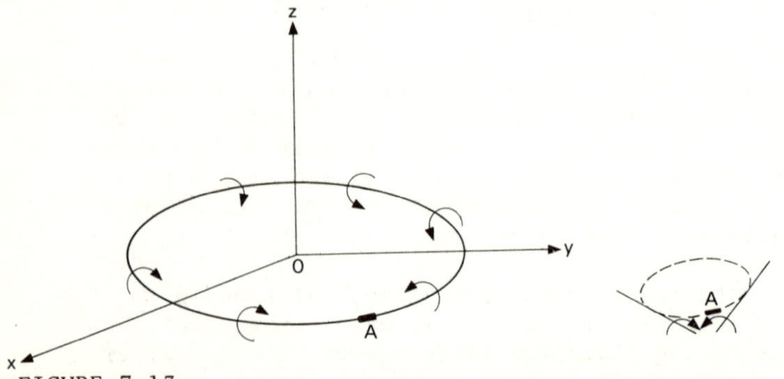

FIGURE 7.13

We shall not develop the theory of the subsequent
motion in detail, but merely give a qualitative account
of what happens. Suppose we consider an element A of
the vortex ring; its motion will be the same as that
due to the rest of the ring of the fluid there. From
symmetry it is easy to see that this must be in the
plane containing Oz and A. Inspection of the inset
diagram in Figure 7.13 suggests that the motion is in
fact parallel to the z-axis and, for a ring whose
vorticity is in the sense shown, in the negative
direction of that axis. The whole effect, therefore,
might be expected to be that the ring does not change
shape or size, but travels parallel to its axis of
symmetry at constant speed - and this is in fact what
is observed to happen* (although, in a real fluid, of
course, the vorticity in the ring is affected by the
viscosity of the fluid, and the ring gradually
dissipates - and so slows down.)
    It can be shown that the motion of the fluid due to
the presence of the vortex ring is identical with that
due to a distribution of doublets over the circle
$x^2 + y^2 < a^2$, $z = 0$. The formal expression for the
velocity field involves elliptic integrals and will not
be discussed here.

7.9 VORTEX SHEETS

In Example 7.7.3 we discussed the motion due to a row
of line vortices and found that, far enough away from
the row the velocity is nearly uniform. In fact, if
the distance between the vortices in the row is a,
then the velocity of the fluid differs by less than
0.05 per cent from a uniform stream at a distance 1.5a
from the row itself. The situation can be idealized
still further by letting the distance a between the
vortices approach zero and, at the same time, letting
the strength κ of each vortex approach zero in such a
way that the quantity $K = \kappa/a$ remains constant. We
call this new idealization a *vortex sheet* of strength
K: it is evident from the results of the earlier
example (or from first principles - see question 13)
that it is a surface of discontinuity separating two
uniform flows.

---

* A smoke ring is an example of such a vortex, made
  visible by the presence of the smoke.

By superposing a uniform velocity equal and opposite to one of them on the whole flow, we obtain a uniform flow of fluid separated from a fluid at rest by the vortex sheet - or a uniform flow of fluid over a rigid surface in the plane of the sheet. But this is precisely the situation in an inviscid fluid when we neglect the boundary layer. In fact, what we are doing when we consider the inviscid flow past a rigid body is to idealize the boundary layer as a vortex sheet. This explains why we can so often take the tangential velocity of the fluid at the surface to be non-zero, the apparent infinite velocity gradient which results being absorbed into the vortex sheet which can be treated within the framework of the theory of inviscid fluids. Of course, the uniform vortex sheet represents a boundary layer for a uniform flow: but it is easy to modify the theory (and take $K = K(x)$) to allow for a flow which is not uniform.

As a second approximation (Step 3 in Section 6.7) we replace the boundary layer by a vortex sheet along the surface $z = \delta_1(x)$, where $\delta_1(x)$ is the displacement thickness, the fluid in the region $0 \le z < \delta_1(x)$ having zero velocity. This is, at least from a mathematical point of view, a rather more satisfactory way of looking at the problem than the approach we used earlier.

Physically, too, it is useful to consider the boundary layer as a region of vorticity rather than of velocity gradient. We can think of the vorticity as being produced by friction at the surface; this vorticity diffuses out through the fluid because of the viscous forces and is also swept along by the oncoming stream. Provided that the speed at which it is swept along by the oncoming stream is large compared with the speed with which it is diffused outwards by the viscous forces (that is, as long as the Reynolds number is large) we would expect the layer containing the vorticity to be comparatively thin, as we found was indeed the case in Chapter 6, so that boundary-layer theory can be applied.

EXERCISES

1 A fluid has density ρ and pressure p, both of which depend on time t and on position $\underset{\sim}{r}$. When $p = p(\rho)$ show that

$$\mathrm{grad} \int_{p_0}^{p} \frac{1}{\rho} \, dp = \frac{1}{\rho} \, \mathrm{grad} \, p \, ,$$

where p_0 is a reference pressure.

By integrating Euler's equations of motion, show that, if $\underset{\sim}{u}$ is the velocity at the point $\underset{\sim}{r}$ at time t, and if there is a conservative field of force whose potential is Ω, then

(a) for steady flow, $\frac{1}{2}\underset{\sim}{u}.\underset{\sim}{u} + \Omega + \int \frac{dp}{\rho} =$ constant along a streamline, and

(b) for irrotational flow, $\frac{\partial \phi}{\partial t} + \frac{1}{2}\underset{\sim}{u}.\underset{\sim}{u} + \Omega + \int \frac{dp}{\rho}$ is

constant everywhere, with $u = \mathrm{grad} \, \phi$.

In case (a) show that the equation reduces to that in Section 2.5 when the fluid is an ideal gas and the flow is adiabatic.

2 The motion of an incompressible fluid is axially symmetric about the z-axis. The radial component of velocity is u and the axial component is w. Show that, if r is distance measured from the z-axis, there exists a function Ψ such that

$$u = \frac{1}{r} \frac{\partial \Psi}{\partial z} \, , \quad w = - \frac{1}{r} \frac{\partial \Psi}{\partial r} \, ,$$

and that surfaces on which Ψ is constant are stream surfaces. (The function Ψ is Stokes's stream function.)

Show that, for a uniform stream whose velocity is U parallel to the z-axis, $\Psi = - \frac{1}{2}Ur^2$ and for a point source of fluid of strength m (that is, it emits a volume 4πm per unit time) at the origin,

$$\Psi = mz/(r^2 + z^2)^{\frac{1}{2}}.$$

3 A sink of strength m is at the point z = a and a source of strength m is at z = -a in a uniform stream U parallel to the z-axis of an incompressible fluid. An element of fluid starting in the uniform stream at a distance b from the z-axis crosses the plane z = 0 at a distance c from the z-axis. Show that

$$b^2 = c^2 - \frac{4ma}{U(c^2 + a^2)^{\frac{1}{2}}} \, ,$$

and sketch the streamlines of the flow.

4 If $\underset{\sim}{v}$ is a given vector function of position in a simply connected domain D and it is known that curl $\underset{\sim}{v}$ = 0 everywhere in D , show that there exists a scalar function ϕ such that $\underset{\sim}{v}$ = grad ϕ everywhere in D.

5 A single-valued function ϕ of position is known to be a solution of Laplace's equation everywhere in a simply-connected volume V, both ϕ and its first derivatives being continuous throughout V and its bounding surface   V.   Further, the normal derivative of ϕ is specified everywhere on ∂V. Show that, if a second function ϕ' satisfies the same conditions, then $\phi - \phi'$ is constant everywhere in V.
  By introducing a suitable dividing surface into a doubly-connected volume V, prove a corresponding result for a multi-valued function ϕ which satisfies the same conditions together with the additional condition that the change in ϕ in one circuit of each irreducible curve in V has the same specified value.

6 A circular cylinder, whose radius is a, moves with uniform velocity U in a direction perpendicular to its generators in an incompressible inviscid fluid. The resulting motion of the fluid is two-dimensional and irrotational, there are no singularities in the flow and there is a circulation κ round the cylinder. Transform the problem to one of steady flow and show that the velocity distribution of Example 2.2.3 is appropriate to the transformed problem. Find the velocity potential of this steady motion.
  For the original motion, show that
  (i) when $|\kappa|$ < $4\pi aU$, there are two points in each
      plane perpendicular to the axis of the cylinder
      at which the fluid has the same velocity as the
      cylinder and that both these points are on the
      surface of the cylinder; and
  (ii) when $|\kappa|$ > $4\pi aU$, there is only one such point
      and that it is not on the cylinder. Deduce that,
      in this case, the fluid in the neighbourhood of
      the cylinder is carried along with it.

7 For small reversible adiabatic disturbances of a compressible fluid at rest, show that the density perturbation ρ' satisfies the equation

$$\frac{\partial^2 \rho'}{\partial t^2} = a_0^2 \nabla^2 \rho',$$

where a_0 is the speed of sound in the fluid at rest.
When the motion is known to have spherical symmetry,
show that

$$\frac{\partial^2 (r\rho')}{\partial t^2} = a_0^2 \frac{\partial^2 (r\rho')}{\partial r^2}$$

Find the equation determining the frequencies of the
normal modes of vibration inside a rigid spherical shell
of radius b and show that, for high frequencies, the
shell behaves like a pipe of length b which is open at
one end and closed at the other.

8   A sphere is immersed in an otherwise unbounded,
compressible inviscid fluid; its centre is fixed and its
radius is $R(1 + \varepsilon \cos \omega t)$, where R, ε, ω are constants
such that $\varepsilon \ll 1$ and $\omega R \ll a_0$, the speed of sound in
the fluid. Assuming that all the disturbances are small,
reversible and adiabatic, find the density distribution
in the fluid.

9   The flow of an unbounded incompressible fluid is
caused by two vortices, each parallel to the z-axis, of
equal strength and, at time t = 0, passing through the
points (a,0,0), (-a,0,0). Show that the elements of
fluid which, at time t = 0, are at the points (b,0,0),
(-b,0,0) remain in the plane containing the vortices as
along as $b^2 = 5a^2$.

10 An unbounded incompressible fluid has, at some
instant, vortices parallel to the z-axis which are of
strength κ and pass through each of the points (a,0,0),
(-a,0,0) and a third of strength $- \kappa/2$ through the
origin. Show that the vortices are stationary and find
the stagnation points of the flow.

11 An incompressible inviscid fluid is bounded by the
planes x = 0, y = 0 and, at some instant, a line
vortex parallel to the z-axis is of strength κ and
passes through the point (a,b,0) in the fluid. Find
a suitable image system for the motion and calculate the
subsequent path of the vortex.

12 Line vortices parallel to the z-axis are of equal
strength κ and pass through the points $(\pm na,b,0)$
and similar vortices of strength $- \kappa$ pass through the
points $(\pm na,0,0), n = 0,1,2,\ldots$. Show that each
vortex moves along its row with speed $\dfrac{\kappa}{2a} \coth\left(\dfrac{\pi b}{a}\right)$.

13 A uniform vortex sheet consists of a distribution in the plane $y = 0$ of line vortices, all parallel to the z-axis, with strength κ per unit length of the x-axis. Show that the velocity of the fluid at the point (x,y,z) is $(u,v,0)$, where

$$u = -\frac{\kappa}{2\pi} \int_{-\infty}^{\infty} \frac{y \, d\xi}{y^2 + (x - \xi)^2},$$

$$v = \frac{\kappa}{2\pi} \int_{-\infty}^{\infty} \frac{(x - \xi)d\xi}{y^2 + (x - \xi)^2}.$$

Evaluate these integrals and describe the flow.

Chapter 8

TWO-DIMENSIONAL FLOW

8.1 THE STREAM FUNCTION

We shall consider in this chapter motion which is independent of the z-coordinate and perpendicular to the z-axis. Hence the streamline pattern (and everything else) is the same in every plane parallel to the x-y plane. It is convenient to refer to the motion as 'in the plane z = 0' even when we mean 'parallel to the plane z - 0', and sometimes the equation f(x,y) = 0 will be taken as representing a curve in this plane and sometimes it will be taken as representing a cylindrical surface whose generators are parallel to the z-axis. The context will indicate which of the two meanings is appropriate. The reason for this is that, later in this chapter and in the next, we wish to use the symbol z for a different purpose. The velocity of the fluid is (u,v,0), but we usually write this as (u,v) for short, since the third component is always zero.

   For flows of this kind, equation (127) may be written in the form

$$\frac{\partial \rho}{\partial t} + \frac{\partial (\rho u)}{\partial x} + \frac{\partial (\rho v)}{\partial y} = 0 \ . \tag{150}$$

There are two special cases of particular interest. First we consider the case of steady compressible flow, so that equation (150) becomes

$$\frac{\partial (\rho u)}{\partial x} + \frac{\partial (\rho v)}{\partial y} = 0 \ ,$$

and it is clear that a function ψ' exists such that

$$\rho u = \rho_0 \frac{\partial \psi'}{\partial y} \ , \quad \rho v = - \rho_0 \frac{\partial \psi'}{\partial x} \ , \tag{151}$$

where ρ_0 is some (constant) reference density. Such a
function is called the *stream function,* and is determined
only within an additive constant. To discover the
physical meaning of the function ψ', we consider the
flow across the fixed curve AB in the x-y plane as
shown in Figure 8.1. The rate of mass flow across AB *

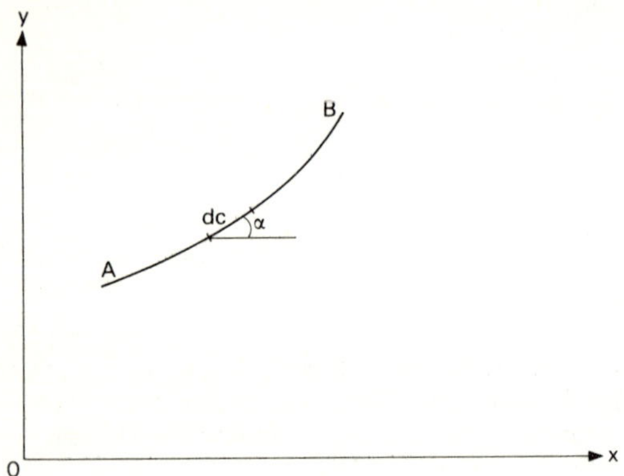

FIGURE 8.1

from left to right (looking from A towards B) is
$\int_{AB} \rho (u \sin \alpha - v \cos \alpha)\, dc$ where a typical element of
AB has length dc and makes an angle α with the
x-axis. Now dc has components dx, dy parallel to
the axes so that $dx = dc \cos \alpha$ and $dy = dc \sin \alpha$.
Hence the rate of mass flow, using equation (151) is

$$\int_{AB} \rho (u\, dy - v\, dx) = \rho_0 \int_{AB} \left(\frac{\partial \psi'}{\partial y}\, dy + \frac{\partial \psi'}{\partial x}\, dx \right)$$

$$= \rho_0 \int_{AB} d\psi'$$

$$= \rho_0 (\psi'_B - \psi'_A) ,$$

So ψ' is a measure of the rate of mass flow of the
fluid.

   The second case, which we shall consider in more
detail, is that of incompressible flow (which may, or

---

* We really mean the rate of mass flow across unit
  length of the cylinder whose cross-section is AB,
  of course.

may not, be steady). Then equation (150) becomes

$$\frac{\partial u}{\partial x} + \frac{\partial v}{\partial y} = 0 , \qquad (152)$$

and there exists a function ψ such that

$$u = \frac{\partial \psi}{\partial y} , \quad v = - \frac{\partial \psi}{\partial x} . \qquad (153)$$

It is determined within an additive arbitrary function of time t by this definition. Once again it is called the stream function, and the analogy with the stream function of equation (151) is obvious; in fact, when the flow is steady and incompressible, the two functions are identical. This time ψ is a measure of the rate of volume flow (which is, of course, directly proportional to the rate of mass flow for an incompressible fluid) across the curve AB in Figure 8.1, for this is

$$\int_{AB} (u \sin\alpha - v \cos\alpha)\,dc = \int_{AB} \left(\frac{\partial \psi}{\partial y}\,dy + \frac{\partial \psi}{\partial x}\,dx \right)$$

$$= \int_{AB} d\psi$$

$$= \psi_B - \psi_A .$$

In particular, if AB is part of a streamline, there is no flow across it and so $\psi_B = \psi_A$. Conversely if $\psi_B = \psi_A$ for all points A, B on a curve, then there is no flow across it and the curve is a streamline. We find, therefore, that the streamlines are given by the curves $\psi = $ constant: the use of this result is often a better way of finding the streamlines when the flow is two-dimensional than the methods discussed in Chapter 2, but it cannot be applied to three-dimensional flow, of course.

As examples we shall discuss the flows described in Chapter 2 by equations (10), (11) and (14).

EXAMPLE 8.1.1

$$u = U\left[1 - \frac{a^2(x^2 - y^2)}{(x^2 + y^2)^2} \right], \quad v = - \frac{2Ua^2xy}{(x^2 + y^2)^2} .$$

We have in this case

$$\frac{\partial \psi}{\partial x} = - v = \frac{2Ua^2 xy}{(x^2 + y^2)^2} ,$$

and so

$$\psi = - \frac{Ua^2 y}{x^2 + y^2} + f(y) ;$$

this may be differentiated to give

$$\frac{Ua^2 (y^2 - x^2)}{(x^2 + y^2)^2} + f'(y) = \frac{\partial \psi}{\partial y}$$

$$= u$$

$$= U\left[1 - \frac{a^2 (x^2 - y^2)}{(x^2 + y^2)^2} \right] .$$

Hence $f'(y) = U$ and $f(y) = Uy + a$ constant, which
we can ignore. It follows that

$$\psi = Uy - \frac{Ua^2 y}{x^2 + y^2} = \frac{Uy[(x^2 + y^2) - a^2]}{x^2 + y^2} , \qquad (154)$$

and the streamlines are given by

$$Uy[(x^2 + y^2) - a^2] = k(x^2 + y^2) \qquad (155)$$

where k is a constant. In particular, on the
streamline $k = 0$ we have either $y = 0$ or
$x^2 + y^2 = a^2$. That is, the y-plane and the surface
of the cylinder $x^2 + y^2 = a^2$ form part of the same
stream surface. Equation (155) gives the equation
for the streamlines in the x-y plane in Cartesian
coordinates: in polar coordinates (r,θ) the
equation becomes

$$U \sin \theta = \frac{kr}{r^2 - a^2} .$$

It is easily verified that, for $r > a$, this gives
a streamline pattern of the form shown in Figure 2.1,
the upper half corresponding to positive values of
k and the lower to negative values.

EXAMPLE 8.1.2

$$u = - \frac{\kappa y}{2\pi(x^2 + y^2)}, \quad v = \frac{\kappa x}{2\pi(x^2 + y^2)}.$$

We have

$$\frac{\partial \psi}{\partial x} = - v = - \frac{\kappa x}{2\pi(x^2 + y^2)}, \quad \frac{\partial \psi}{\partial y} = u = - \frac{\kappa y}{2\pi(x^2 + y^2)}.$$

By inspection we see immediately that these integrate to give

$$\psi = - \frac{\kappa}{4\pi} \log(x^2 + y^2), \tag{156}$$

and the streamlines are given by $x^2 + y^2 =$ constant. That is, they are circles whose centres are all on the z-axis. In fact, equation (156) describes the motion due to the line vortex in Example 7.7.1.

EXAMPLE 8.1.3

$$u = U e^{ky} \sin[k(x - ct)], \quad v = - U e^{ky} \cos[k(x - ct)].$$

We have

$$\frac{\partial \psi}{\partial x} = - v = U e^{ky} \cos[k(x - ct)],$$

$$\frac{\partial \psi}{\partial y} = u = U e^{ky} \sin[k(x - ct)]$$

and, again by inspection, we see that

$$\psi = \frac{U}{k} e^{ky} \sin[k(x - ct)] + f(t), \tag{157}$$

where $f(t)$ is an arbitrary function of t which can usually be ignored. The streamlines are therefore given by

$$e^{ky} \sin[k(x - ct)] = \text{constant}$$

and this agrees, as it should, with the result we obtained in Example 2.2.5.

8.2 TWO-DIMENSIONAL IRROTATIONAL FLOW: THE COMPLEX POTENTIAL

In the rest of this chapter we shall consider what happens when the flow is irrotational as well as two-dimensional and incompressible. The stream function ψ of equation (153) exists for two-dimensional flow whether it is irrotational or not; when it is irrotational, however, we can define (as in Chapter 7) a velocity potential ϕ by the relation

$$u = \frac{\partial \phi}{\partial x}, \quad v = \frac{\partial \phi}{\partial y}, \tag{158}$$

and ϕ is determined, like ψ, only within an arbitrary function of t. By comparing equations (153) and (158) we see that

$$\frac{\partial \phi}{\partial x} = \frac{\partial \psi}{\partial y}, \quad \frac{\partial \phi}{\partial y} = - \frac{\partial \psi}{\partial x},$$

and these are the well-known Cauchy-Riemann relations of complex variable theory. Since they hold it follows that $\phi + i\psi$ is an analytic function of the complex variable $z = x + iy$. (The reader is reminded that here z is a complex number and has no connection whatsoever with a third physical coordinate in a Cartesian system of axes.) So we can define the *complex potential*

$$w(z) = \phi(x,y) + i\,\psi(x,y), \quad z = x + iy, \tag{159}$$

for a two-dimensional, irrotational incompressible flow. We note that

$$\frac{dw}{dz} = \frac{\partial \phi}{\partial x} + i\,\frac{\partial \psi}{\partial x} = u - iv, \tag{160}$$

and so, once $w(z)$ is known, both components of velocity are immediately derivable in a particularly simple way. An alternative form of equation (160) is obtained if we write $u = q \cos\theta$, $v = q \sin\theta$: it then becomes

$$\frac{dw}{dz} = q\, e^{-i\theta}. \tag{161}$$

Here, of course, q is the magnitude of the velocity and θ is the angle it makes with the x-axis. The quantity $\frac{dw}{dz}$ is sometimes referred to as the *complex velocity*.

8.3 SOME SIMPLE EXAMPLES OF THE COMPLEX POTENTIAL

If we are given sufficient boundary conditions to
determine a two-dimensional irrotational motion in a
given region, the general problem is to find the
appropriate complex potential. It is often easier to
do this if we already have available some standard
results. These are in the first instance most easily
obtained by proceeding in the reverse order; that is,
by considering particular forms for the complex
potential and finding out what flows they represent.
In this section we consider seven such cases.

EXAMPLE 8.3.1

$$w = Uz \, , \quad u \text{ real.} \tag{162}$$

Then

$$\frac{dw}{dz} = U = u - iv \, .$$

Hence equation (162) is the complex potential of a
uniform flow whose velocity is U parallel to the
x-axis.

EXAMPLE 8.3.2

$$w = Uz \, e^{-i\beta}, \quad U, \beta \text{ real.} \tag{163}$$

Here

$$\frac{dw}{dz} = U \, e^{-i\beta} = U \cos \beta - iU \sin \beta \, .$$

So equation (163) represents a uniform flow whose
velocity is U in a direction which makes an angle
β with Ox (see Figure 8.2).

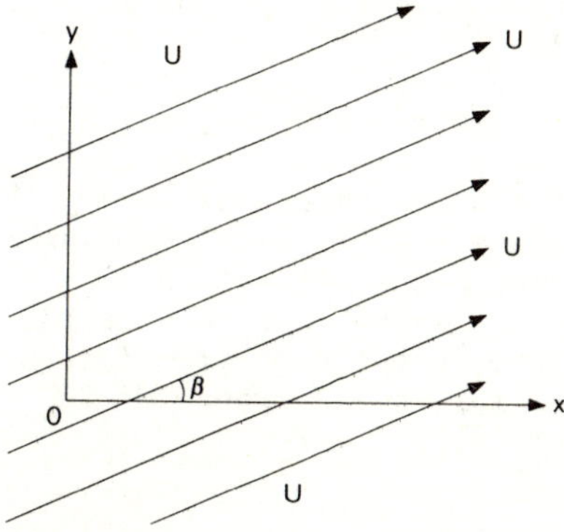

FIGURE 8.2

EXAMPLE 8.3.3
$$w = C z^n, \quad n \text{ a positive integer}, \quad C \text{ real}. \qquad (164)$$

Writing $z = r e^{i\theta'}$, we have
$$w = C r^n e^{in\theta'} = C r^n (\cos n\theta' + i \sin n\theta').$$

Here $\psi = C r^n \sin n\theta'$ and the streamlines are given by
$$r^n \sin n\theta' = \text{constant}.$$

This means that the flow is separated into independent sectors by the planes $\theta' = \dfrac{k\pi}{n}$ $(k = 0,1,2,\ldots,n-1)$,

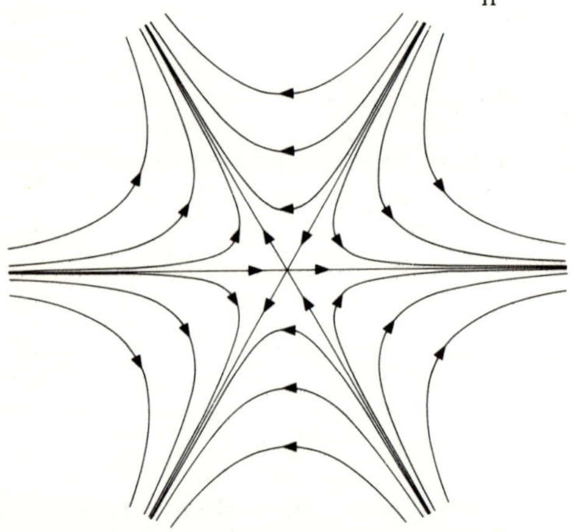

FIGURE 8.3

there being no flow across these planes. To find the direction of the flow along a given streamline we note that, using equation (161),
$$q e^{-i\theta} = \frac{dw}{dz} = nC z^{n-1} = nCr^{n-1} e^{i(n-1)\theta'}.$$

Hence on the planes $\theta' = k\pi/n$ we have, for C positive, $\theta = - (n - 1)k\pi/n = - k\pi + \theta'$. It follows that the flow in the dividing plane $\theta' = k\pi/n$ is away from the origin when k is even and towards the origin when k is odd. (The opposite is true for C negative.) The streamlines for $n = 3$ when C is positive are shown in Figure 8.3.

EXAMPLE 8.3.4

$$w = C z^{-n}, \quad n \text{ a positive integer, } C \text{ real.} \qquad (165)$$

Writing $z = r e^{i\theta'}$, we have

$$w = C r^{-n} e^{-in\theta'} = C r^{-n}(\cos n\theta' - i \sin n\theta').$$

Hence $\psi = - C r^{-n} \sin n\theta'$, and the streamlines are given by

$$r^n = k \sin n\theta',$$

where k is a constant. As in the previous example, the flow is divided into sectors. The streamline

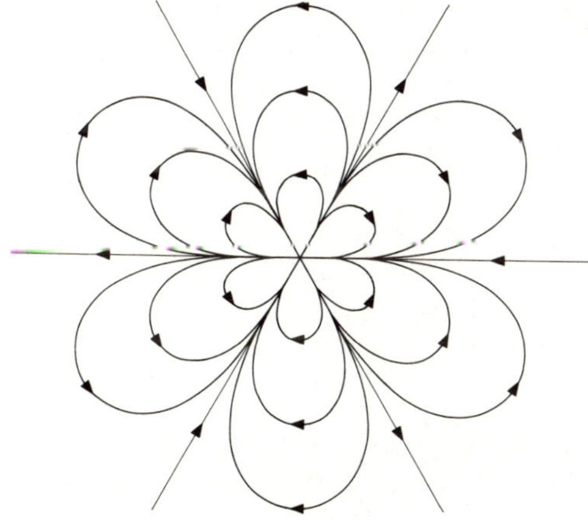

FIGURE 8.4

pattern for $n = 3$ when C is positive is shown in Figure 8.4.

EXAMPLE 8.3.5

$$w = m \log z, \quad m \text{ real.} \qquad (166)$$

Writing $z = r e^{i\theta'}$, we have

$$w = m(\log r + i\theta').$$

Hence $\psi = m\theta'$, and the streamlines are given by $\theta' = $ constant; that is they are straight lines through the origin in the x-y plane.

Further, if we go once round any curve which encloses the origin, the change in ψ is $2\pi m$.

Hence there is a volume flow of fluid across any cylinder (not necessarily circular) whose axes are parallel to the line $x = 0, y = 0$ and which contains the origin; the rate of flow is $2\pi m$ per unit length of the cylinder and this is independent of the particular cylinder chosen. Clearly, then, we have a source of fluid along the line $x = 0$, $y = 0$ which emits a volume $2\pi m$ of fluid per unit length per unit time. This is called a *line source* of strength m. (If $m < 0$, we have a *line sink* of strength $|m|$.)

EXAMPLE 8.3.6

$$w = -\frac{i\kappa}{2\pi} \log z, \quad \kappa \text{ real}. \tag{167}$$

Writing $z = r\,e^{i\theta'}$, we have

$$w = -\frac{i\kappa}{2\pi} (\log r + i\theta'),$$

and so

$$\psi = -\frac{\kappa}{2\pi} \log r, \quad \phi = \frac{\kappa\theta'}{2\pi}.$$

Hence the streamlines are the circles $r = $ constant. Further the circulation round any curve C which encircles the line $x = 0, y = 0$ is

$$\int_C \mathbf{u}.\hat{\mathbf{t}}\, dc = \int_C \mathrm{grad}\, \phi.\hat{\mathbf{t}}\, dc = \int_C \frac{\partial\phi}{\partial c}\, dc$$

the change in ϕ as we go once round C, which is κ, and this is independent of the particular curve chosen. Hence the flow represented by equation (167) is that due to a line vortex of the kind described in Section 7.7.

EXAMPLE 8.3.7

$$w = C\,e^{kz}, \quad C, k \text{ real}.$$

We have $\dfrac{dw}{dz} = Ck\,e^{kz}$, and so

$$u = Cke^{kx} \cos ky, \quad v = -Cke^{kx} \sin ky,$$

and this is identical with the flow in Example 8.1.3, except that x and y are interchanged and the dependence on t is absent. The streamline pattern is therefore similar to that shown in Figure 2.6, except that the whole system must be turned through a right angle.

Of these seven complex potentials, all are useful in finding potentials for more complicated flow systems, although some of them do not give realistic flow in their own right.

8.4 FURTHER EXAMPLES OF THE COMPLEX POTENTIAL

Here we superimpose some of the flows of the previous section and find the corresponding complex potential by addition.

EXAMPLE 8.4.1
A line source of strength m through $x = a$, $y = 0$ together with a line sink of strength m through $x = -a$, $y = 0$.

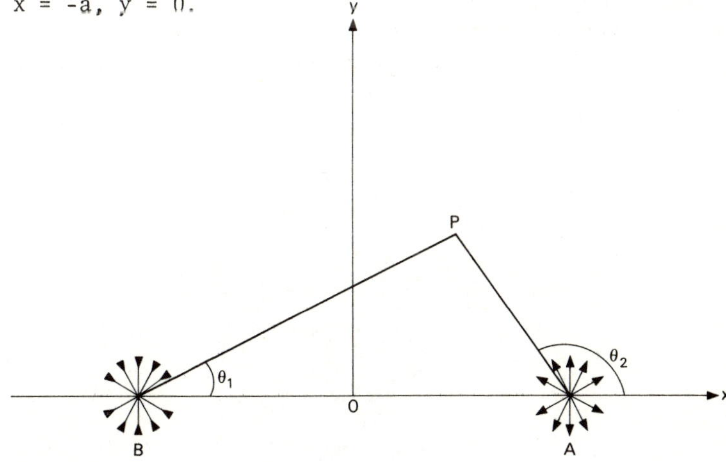

FIGURE 8.5

The complex potential due to the source is

$$w_1(z) = m \log(z - a),$$

and that due to the sink is

$$w_2(z) = -m \log(z + a).$$

Now consider the complex potential $w = w_1 + w_2$, so that

$$w = m \log \frac{z - a}{z + a} . \qquad (168)$$

Near the source, $|w_2| \ll |w_1|$ and so $w \sim w_1$; similarly near the sink $|w_1| \ll |w_2|$ and so there $w \sim w_2$. Further, when z is large, the velocity is negligibly small, since

$$\frac{dw}{dz} = \frac{2ma}{z^2 - a^2} .$$

Hence the complex potential of equation (168) represents the required flow, since it gives the right flow in the neighbourhood of both singularities and at infinity, and since it gives no new singularities.

EXAMPLE 8.4.2
A *line doublet* of strength μ through $x = 0$, $y = 0$ and directed in the positive direction of the x-axis. This is, by definition, the flow of the previous example when $m = \mu/2a$ and $a \to 0$.
    In this case we have

$$w(z) = \lim_{a \to 0} \left[\frac{\mu}{2a} \left(\log \frac{z - a}{z + a} \right) \right]$$

$$= \lim_{a \to 0} \left[- \frac{\mu}{a} \left(\frac{a}{z} + \frac{a^3}{z^3} + \cdots \right) \right]$$

$$= - \frac{\mu}{z} \lim_{a \to 0} \left(1 + \frac{a^2}{z^2} + \cdots \right) .$$

That is

$$w(z) = - \frac{\mu}{z} . \qquad (169)$$

This is therefore a special case of equation (165) of the previous section (with $n = 1$). The streamlines are given by

$$r = k \sin \theta' ,$$

and these are circles (as shown in Figure 8.6) touching the plane $y = 0$ where $x = 0$.

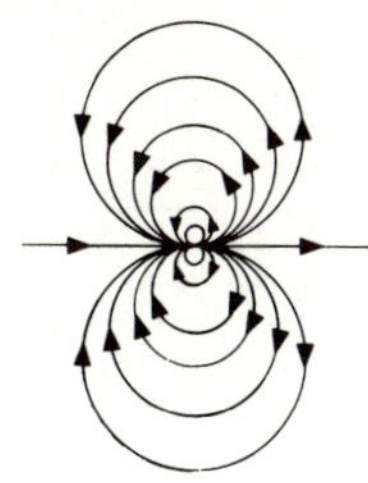

FIGURE 8.6

EXAMPLE 8.4.3
A line source of strength m through $x = 0$, $y = 0$
together with a stream which, far from the source,
is uniform with velocity U parallel to the x-axis.
     The complex potential due to the source is
$w_1 = m \log z$, and that due to the uniform stream is

$w_2 = Uz$. Consider the complex potential $w = w_1 + w_2$
so that

$$w = m \log z + Uz, \quad m, U \text{ real.} \tag{170}$$

It is easy to verify that near the source $w \sim w_1$

and that, far from the source, $w \sim w_2$; further

there are no singularities (other than the given
source) in the flow. Hence the complex potential of
equation (170) represents the required flow.
     The stream function is $\psi = m\theta' + U r \sin \theta'$
where $z = re^{i\theta'}$, and so the streamlines are given
by $m\theta' + Ur \sin \theta' = $ constant, which may be written

$$\theta' + \frac{U}{m} y = k ,$$

where k is a constant. The curve given by $k = \pi$
is the negative part of the x-axis together with the
curve ABC in Figure 8.7. Note that the curve ABC
is asymptotic to the planes $y = \pm m\pi/U$. The cylinder
through this curve divides the flow into two regions -
that to the left contains fluid originating from the
uniform stream and that to the right contains fluid
originating from the source. Note that, since

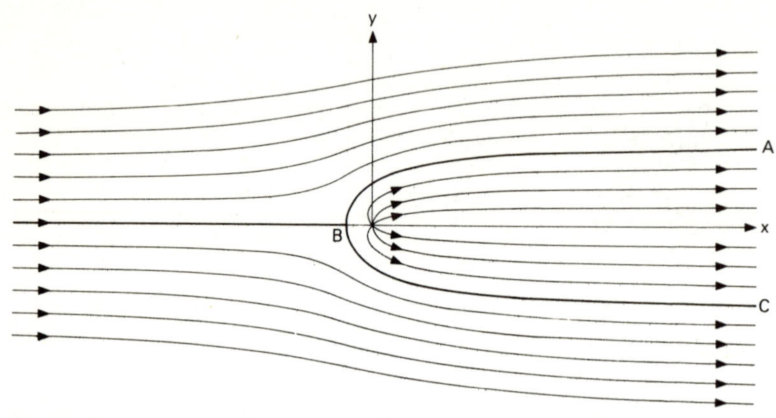

FIGURE 8.7

$$u = U + \frac{m}{r} \cos \theta', \quad v = \frac{m}{r} \sin \theta',$$

the flow at large distances from the origin is indistinguishable from a uniform stream of velocity U parallel to Ox, except that the fluid in the domain $x > 0$, $|y| < \frac{m\pi}{U}$ has all originated from the source and the rest has originated from far upstream. The point B is, of course, a stagnation point: it has coordinates $\left(-\frac{m}{U}, \, 0\right)$.

Streamlines having $k < \pi$ originate from the source, and those with $k > \pi$ originate from the uniform stream.

EXAMPLE 8.4.4
A line doublet of strength μ through $x = 0$, $y = 0$ and directed along the negative direction of the x-axis, together with a stream which, far from the doublet, is uniform with velocity U parallel to the positive direction of the x-axis.
Here the complex potentials due to the doublet and the uniform stream are respectively $w_1 = \frac{\mu}{z}$ and $w_2 = Uz$. We consider the complex potential

$w = w_1 + w_2,$ so that

$$w = Uz + \frac{\mu}{z}, \quad U, \mu \text{ real.} \tag{171}$$

It is easy to verify that this satisfies all the conditions of the problem (and this is left as an exercise for the reader). The stream function is

$\psi = \left(Ur - \frac{\mu}{r}\right) \sin \theta',$ and so the streamlines are given by

$$\left(Ur - \frac{\mu}{r}\right) \sin \theta' = k ,$$

where k is a constant. In particular, when $k = 0$ the streamline is the x-axis ($\theta' = 0$ and $\theta' = \pi$) together with the circle $r^2 = \frac{\mu}{U}$. So the circular cylinder $r = \left(\frac{\mu}{U}\right)^{\frac{1}{2}}$ divides the region into two independent parts, as shown in Figure 8.8. The part

FIGURE 8.8

outside the cylinder contains only fluid from the uniform stream, and that inside only fluid from the doublet.

EXAMPLE 8.4.5
A line doublet of strength μ through $x = 0, y = 0$ directed along the negative direction of the x-axis inside the cylinder $x^2 + y^2 = a^2.$

The complex potential given by equation (171) satisfies all the boundary conditions (since we are ignoring viscosity) if $\left(\frac{\mu}{U}\right)^{\frac{1}{2}} = a$. Hence the required solution is

$$w(z) = \mu\left(\frac{1}{z} + \frac{z}{a^2}\right), \qquad \mu, \ a \ \text{real}. \tag{172}$$

The streamline pattern is identical with that inside the circle in Figure 8.8.

EXAMPLE 8.4.6
The flow past the cylinder $x^2 + y^2 = a^2$ which, far from the cylinder, is a uniform stream whose velocity is U parallel to the positive direction of the x-axis.

Once again the complex potential given by (171) satisfies all the boundary conditions if $\left(\frac{\mu}{U}\right)^{\frac{1}{2}} = a$. Hence the required solution is

$$w(z) = U\left(z + \frac{a^2}{z}\right), \qquad U, \ a \ \text{real}. \tag{173}$$

The streamline pattern is identical with that outside the circle in Figure 8.8.

This solution is, of course, a formal solution to the problem which satisfies all the boundary conditions for inviscid flow. We have seen in Chapter 6, however, that for large Reynolds numbers (for small Reynolds numbers the inviscid flow is not even a good approximation anywhere) there will be a boundary layer, and that this will separate some-where behind the cylinder. The flow defined by equation (173) for $|z| > a$ is therefore not physically realistic everywhere, although for negative values of x it is quite a good approx-imation to the flow. However, we shall find this model useful as a stepping stone to the flow past an aerofoil, and we shall discuss this in the next chapter.

EXAMPLE 8.4.7
The flow of the previous example together with a circulation κ about the cylinder. (This circulation could physically be caused *either* by spinning the cylinder and taking account of the viscous forces near the cylinder *or* by the casting off of a line vortex of strength $-\kappa$ behind the cylinder.

We shall use it chiefly, however, as a stepping stone
in aerofoil theory.)
   It is easy to verify that the complex potential

$$w(z) = U\left(z + \frac{a^2}{z}\right) - \frac{i\kappa}{2\pi} \log z \qquad (174)$$

satisfies the boundary conditions since, if
$z = re^{i\theta'}$, then

(i) $\psi = U\left(r - \frac{a^2}{r}\right)\sin \theta' - \frac{\kappa}{2\pi} \log r$, and so $r = a$

   is the streamline $\psi = -\frac{\kappa}{2\pi} \log a$;

(ii) $\phi = U\left(r + \frac{a^2}{r}\right)\cos \theta' + \frac{\kappa}{2\pi} \theta'$, and so the

   circulation round any curve enclosing the
   cylinder is κ;

(iii) $\frac{dw}{dz} \to U$ as $z \to \infty$, and so the flow far from

   the cylinder is a uniform stream with velocity
   U parallel to the x-axis.

   The flow defined by equation (174) is the same as
that discussed in the Example 2.2.3, as can be
verified by differentiating equation (174).

8.5 THE FORCE ON A CYLINDER IN STEADY FLOW

If a cylinder is moving through a fluid at rest with
constant velocity, there will be forces on the cylinder
due to the presence of the fluid. The component normal
to the direction of the motion of the cylinder is called
the *lift* and that in the direction opposite to the motion
of the cylinder is the *drag*. As usual, we superpose a
velocity on the system in such a way that the cylinder
is at rest and then the resulting flow is steady.
   We suppose that the force per unit length necessary
to keep the cylinder stationary has components (X,Y)
parallel to the axes Ox, Oy. We assume that the
cylinder is a stream surface for the inviscid flow past
it (this will be approximately true for the flow of a
real fluid past a sufficiently smooth slender body) and
that the complex potential of the flow is $w(z)$. We
consider an element δS of the cylinder which is of
unit length perpendicular to the x-y plane and whose
outward-pointing normal has direction cosines (ℓ,m);
then the pressure force on the cylinder has components

$-\iint_S p\ell \, dS$, $-\iint_S pm \, dS$, as can be seen from Figure 8.9.

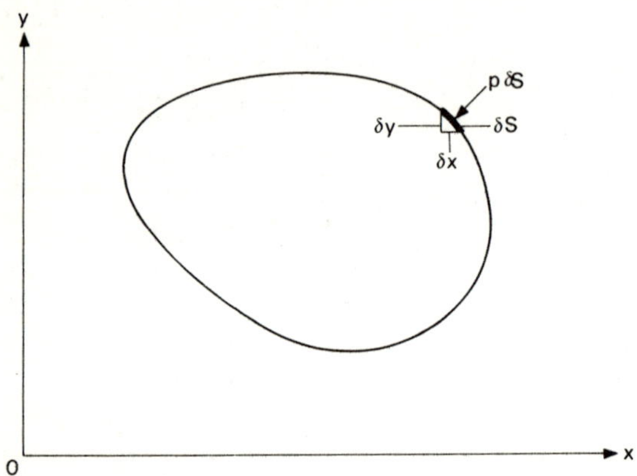

FIGURE 8.9

It follows that the force per unit length on the body
due to the pressure of the surrounding fluid is given by

$$X = - \iint_S p\ell \, dS, \quad Y = - \iint_S pm \, dS,$$

where all body forces on the cylinder are neglected.
If the element δS has directed components δx, δy
parallel to Ox, Oy respectively, then it is clear from
Figure 8.9 that $\delta y = \ell \, \delta S$ and $\delta x = - m \, \delta S$. Hence we
can write

$$X - iY = - \int_B p \, dy - i \int_B p \, dx$$

$$= - i \int_B p(dx - i \, dy),$$

where B is the intersection of the cylindrical surface
with the x-y plane. Now, still neglecting body forces,
we have from Bernoulli's equation (17) that on the
surface of the cylinder (which is a stream surface)
$p = - \frac{1}{2}\rho q^2 + $ constant, where q is the magnitude of
the velocity. Now the integral of any constant round
the closed curve B is zero; hence

$$X - iY = \tfrac{1}{2} i\rho \int_B q^2 (dx - i\,dy)$$

$$= \tfrac{1}{2} i\rho \int_B \frac{dw}{dz} \overline{\left(\frac{dw}{dz}\right)} \overline{dz} \; .$$

Now $\overline{\left(\dfrac{dw}{dz}\right)} = u + iv$ represents a vector in the same direction as dz on the streamlines (by definition of a streamline); it follows that $\overline{\left(\dfrac{dw}{dz}\right)}\,\overline{dz}$ is real and equal to its complex conjugate, which is $\dfrac{dw}{dz}\,dz$. It follows that we can write

$$X - iY = \tfrac{1}{2} i\rho \int_B \left(\frac{dw}{dz}\right)^2 dz \; .$$

Suppose now that C is any closed curve around the cylinder. Then, since the flow is irrotational every-where between B and C, we have

$$\int_B \left(\frac{dw}{dz}\right)^2 dz = \int_C \left(\frac{dw}{dz}\right)^2 dz \; ,$$

using Cauchy's first theorem in complex variable theory. Hence

$$X - iY = \tfrac{1}{2} i\rho \int_C \left(\frac{dw}{dz}\right)^2 dz \; . \tag{175}$$

Also, if the moment about the line $x = 0$, $y = 0$ of this force is G, then

$$G = \iint_S (p\ell y - pmx)\,dS = \int_B p(y\,dy + x\,dx)$$

$$= \mathcal{R} \int_B p\; z \; \overline{dz}$$

$$= \mathcal{R} \int_B (\text{constant} - \tfrac{1}{2}\rho q^2) z\; \overline{dz},$$

using Bernoulli's theorem, as before. (The symbol $\mathcal{R}$ signifies the real part of the following complex number.)

The integral of the first term is zero, and so

$$G = - \tfrac{1}{2}\rho \times \mathcal{R} \int_B z \, \frac{dw}{dz} \overline{\left(\frac{dw}{dz}\right)} \, \overline{dz}$$

$$= - \tfrac{1}{2}\rho \times \mathcal{R} \int_B z \left(\frac{dw}{dz}\right)^2 dz \ ,$$

using again the result that, on B, $\overline{\left(\dfrac{dw}{dz}\right)}\,\overline{dz} = \dfrac{dw}{dz}\,dz$.
Since the flow is irrotational between the curves B and C, we finally have

$$G = \mathcal{R}\left[- \tfrac{1}{2}\rho \int_C \left(\frac{dw}{dz}\right)^2 z \, dz \right] \ , \tag{176}$$

where C is any closed curve round the cylinder.

The expressions (175) and (176) are due to Blasius, and the results of this section are often known as *Blasius's theorem*.

As an example we consider the flow described by equation (174). Here we have

$$\frac{dw}{dz}^2 = \left[U\left(1 - \frac{a^2}{z^2}\right) - \frac{i\kappa}{2\pi z} \right]^2$$

$$= U^2 - \frac{i\kappa U}{z} - \frac{\kappa^2 + 8\pi^2 U^2 a^2}{4\pi^2 z^2} + \frac{i\kappa U a^2}{\pi z^3} + \frac{U^2 a^4}{z^4}$$

and the residue of this at the origin (its only pole) is $-\dfrac{i\kappa U}{\pi}$. Hence

$$X - iY = \tfrac{1}{2}i\rho \int_C \left(\frac{dw}{dz}\right)^2 dz$$

$$= - \pi\rho \times \text{residue of} \ \left(\frac{dw}{dz}\right)^2 \ \text{at the origin}$$

$$= i\rho\kappa U \ .$$

Hence $X = 0$ and $Y = -\rho\kappa U$, and the force per unit length on the cylinder is $\rho\kappa U$, in the negative direction of the y-axis if κ and U are both positive. For this problem, therefore, the drag is zero and the lift has magnitude $\rho\kappa U$.

We note that, if there is no circulation, there is no force on the cylinder (this is an example of d'Alembert's paradox). When κ and U are both positive (see Figure 8.10), the speed on the upper half $(y > 0)$ of

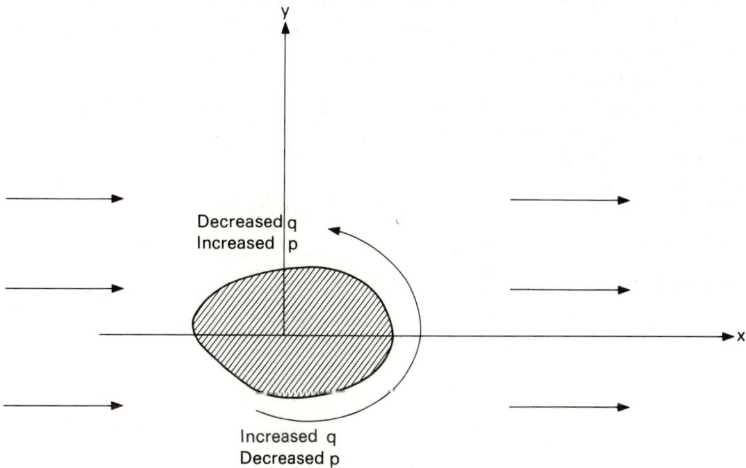

FIGURE 8.10

the cylinder is reduced due to the circulation, and that on the lower half is increased; using Bernoulli's equation, we see that the pressure on the upper half of the cylinder is increased when circulation is present and that on the lower half is decreased. Hence we would expect there to be a resultant 'downward' force on the cylinder. So the result is predictable qualitatively, and can be extended to a more generally shaped cylinder. The more detailed discussion of d'Alembert's paradox is postponed to Section 8.7.

For the flow defined by equation (174) the residue

of $\left(\dfrac{dw}{dz}\right)^2 z$ is $-\left(\dfrac{\kappa^2 + 8\pi^2 U^2 a^2}{4\pi^2}\right)$,   and so

$$G = \mathcal{R}\left[\pi i \rho\left(\frac{\kappa^2 + 8\pi^2 U^2 a^2}{4\pi^2}\right)\right],$$

which is zero, as we would expect from symmetry.

8.6 THE KUTTA-JOUKOWSKI THEOREM

We now prove the same result for the force on an arbitrary right cylinder whose cross-section in the x-y plane is the curve B, when there is a flow past the cylinder which, far from it, has velocity U parallel to the positive direction of the x-axis. (There is no loss of generality in choosing this direction for the velocity, since we can always rotate the axes so that it is true.) We suppose further that there is a circulation κ round the cylinder and no flow across the surface of the cylinder (that is, B is a streamline).

We choose a closed curve C in the x-y plane which contains B completely and assume that the flow is irrotational between B and C. When $|z|$ is large, we have that $\dfrac{dw}{dz} \sim U$. There are no singularities in the flow outside the cylinder, so we can certainly write the analytic function $\dfrac{dw}{dz}$ in the form

$$\frac{dw}{dz} = U + \sum_{n=1}^{\infty} \frac{a_n}{z^n},$$

and so

$$\int_C \frac{dw}{dz}\, dz = 2\pi i a_1.$$

But

$$\int_C \frac{dw}{dz}\, dz = \int_B \frac{dw}{dz}\, dz = \int_B dw = \int_B d\phi + i\int_B d\psi = \kappa,$$

since $\int_B d\phi = \kappa$, the circulation round B, and ψ is a constant on the streamline B. Hence $a_1 = -\dfrac{i\kappa}{2\pi}$.

Further

$$\left(\frac{dw}{dz}\right)^2 = U^2 + \sum_{n=1}^{\infty} \frac{b_n}{z^n},$$

where $b_1 = 2Ua_1 = -i\kappa U/\pi$. Hence

$$X - iY = -\pi\rho \times \text{residue of } \left(\frac{dw}{dz}\right)^2 = i\rho\kappa U.$$

It follows that $X = 0$, $Y = -\rho\kappa U$ for *any cylinder* whose generators are perpendicular to the x-y plane in a flow whose velocity at infinity is U parallel to the positive direction of the x-axis, and whose circulation round the cylinder is κ. This is the Kutta-Joukowski theorem.

It is interesting to note (i) that there is no force at all if there is no circulation and (ii) that when there is a force it acts in a direction perpendicular to the stream. The first of these points is a statement of d'Alembert's paradox for the uniform flow past a cylinder: this is discussed in more detail in the next section. The second point is true only as long as a_1 is purely

imaginary - that is, as long as there is no resultant source of fluid inside the surface B. Of course, if there were such a resultant source, B could not be a streamline. There may be any number of sources as long as they are balanced by an equivalent number of sinks, so that the resultant source strength is zero: in particular, there could be doublets inside the boundary B, and we saw in the special case discussed in Example 8.4.4 that a single doublet in a uniform stream can represent the flow of a uniform stream past a circular cylinder. Referring to Example 8.4.3, we can see that a source (and, by implication, a distribution of sources) in a uniform stream represents the flow of a uniform stream past an infinite bluff body. We now see that a distribution of doublets in a uniform stream can represent (at least in an idealized flow) the flow of a uniform stream past a finite bluff body. In a real fluid the resulting flow breaks down behind the cylinder because of the effect of viscous forces which cause a wake to be formed.
In the idealized flow, we can also calculate the moment of the forces due to the fluid on the cylinder, using equation (176). It is easy to show that the moment is

$$G = - 2\pi\rho U \times \mathfrak{J} a_2,$$

where the symbol $\mathfrak{J}$ denotes the imaginary part of the following complex number; we need more information about the flow to determine this, and we cannot do so without knowing something about the shape of the cylinder.

8.7 D'ALEMBERT'S PARADOX

It is not difficult nowadays to see the fallacy in d'Alembert's paradox, but it is possible to understand the problem when we remember that the idea of a boundary layer was not developed until after the turn of the century. Before that it seemed to be true that inviscid theory gave good estimates of the velocities and pressures in a steady uniform flow (with no circulation)

past an obstacle (and we know this to be true for high
Reynolds number flows except in the boundary layer and
in the wake), but it did not give a good estimate of the
forces on the body.

That this must be so as far as drag is concerned is
clear if we take axes moving with the steady uniform
flow. Evidently the kinetic energy of the motion is
the same for all time, since there is nothing to
distinguish one instant from another. Now if there were
a resultant force on the body which had a component in
the direction of the uniform flow, this would do work on
the moving body and so increase the kinetic energy. As
this does not happen, no force with such a component can
exist.

There are three main reasons why this argument breaks
down - all connected with the boundary layer.* In the
first place, the presence of the boundary layer
introduces a tangential drag on the body in its own
right, as we saw in Chapter 6: this is called the
friction drag, and is evidently proportional to the
viscosity. Second, the boundary layer often separates
and a wake is produced behind the body. The effect of
this is effectively to change the body from one of finite
size to one which extends to infinity downstream (the
external flow sees the body and wake together as a
single obstacle). A special case is discussed briefly
in Section 9.7 for a two-dimensional flow, and we shall
see then that a finite drag force is predicted: this
is called the *form drag.* Finally, the flow downstream
of the obstacle is often unsteady, and this in turn
results in a further drag on the body: because many
unsteady flows can be represented as wave motions,
this is called *wave drag,* and some examples of this will
be discussed in Chapter 10. In each of these types of
drag a mechanism exists for the dissipation of the
energy produced by the force acting on the body, so that
this can now do work.

EXERCISES

1 An incompressible inviscid fluid is in motion due to
the presence of a cylinder, whose generators are perpendicular to the x-y plane, whose axis moves with velocity
(U,V,0) and which has angular velocity Ω about the axis.

---

* For flows with a low Reynolds number, where boundary-
  layer theory is inappropriate, there is no question of
  inviscid theory being valid, of course.

Show that a stream function ψ exists.

At the instant when the axis of the cylinder passes through the point (x_0, y_0) show that, on the surface of the cylinder, the quantity

$$\psi + Vx - Uy + \tfrac{1}{2}\Omega\left[(x - x_0)^2 + (y - y_0)^2\right]$$

is constant.

2 An incompressible inviscid fluid is in steady, two-dimensional motion in the x-y plane and the vorticity has magnitude ω which is a function of position. Show that a stream function ψ exists and that

$$\frac{\partial^2\psi}{\partial x^2} + \frac{\partial^2\psi}{\partial y^2} = -\omega.$$

If the fluid is contained within the elliptic cylinder $\frac{x^2}{a^2} + \frac{y^2}{b^2} = 1$, and the vorticity is uniform everywhere, show that the streamlines are similar ellipses and that each element of fluid makes one circuit in time $2\pi(a^2 + b^2)/\omega ab$. Find also the position where the fluid pressure is a minimum.

3 A line source of strength m passes through the point $(a,0)$ and a line sink of equal strength through $(-a,0)$. Find the stream function for the motion and sketch the streamlines.

4 Discuss the nature of the two-dimensional, irrotational flow of an incompressible, inviscid fluid in the neighbourhood of the origin when the complex potential is given by

(a) $\dfrac{dw}{dz} = \dfrac{A}{z} + f(z)$,

(b) $\dfrac{dw}{dz} = \dfrac{iA}{z} + f(z)$,

(c) $\dfrac{dw}{dz} = \dfrac{A}{z^2} + f(z)$,

where, in each case, A is real and $f(z)$ is a function of z which is analytic at $z = 0$.

5 A cylindrical log of diameter $2a$ lies on the horizontal bed of a deep ocean. Assuming that there is no separation behind the log, vertify that the complex potential $w(z) = a\pi V \coth(a\pi/z)$ represents a uniform flow past the log.

6   An inviscid fluid is in two-dimensional, irrotational motion in the space between two circular cylinders whose cross-sections satisfy the equations

$$(x - a_1)^2 + y^2 = b_1^2, \quad (x - a_2)^2 + y^2 = b_2^2, \quad \text{where}$$

$a_1^2 - b_1^2 = a_2^2 - b_2^2 = c^2 > 0.$ Verify that the complex potential of the flow is

$$w = \frac{i\kappa}{2\pi} \log\left(\frac{z - c}{z + c}\right),$$

where κ is the circulation round the innder cylinder.
   Show also that the force on the inner cylinder has

magnitude $\rho\kappa^2/4\pi c,$ where ρ is the density of the fluid.

7   Show that the complex potential

$$w(z) = -\frac{i\Omega(a + b)}{4(a - b)}\left[z - \sqrt{(z^2 - a^2 + b^2)}\right]^2$$

represents the flow of an incompressible fluid which has been set in motion by an elliptic cylinder whose axis is fixed, which rotates about the axis with angular velocity ω, and whose cross-section is an ellipse whose axes have lengths $2a, 2b$ and are instantaneously coincident with the coordinate axes.

8   An incompressible, inviscid fluid, whose density is ρ, is in two-dimensional, irrotational motion inside a cylindrical vessel, the velocity potential of the flow being ϕ. The unit vector $\underset{\sim}{k}$ is parallel to the generators, $\underset{\sim}{H}$ is the angular momentum of the fluid (per unit length of the cylinder) about a point 0, the curve Γ encloses the cross-section through 0 of the cylinder, and r is distance measured from the line through 0 parallel to $\underset{\sim}{k}$. Show that

$$\underset{\sim}{H} = -\rho\underset{\sim}{k}\int_\Gamma \phi r \, dr.$$

9   An incompressible inviscid fluid is in irrotational motion in simply connected space bounded by the surface S_n, each of which is bounded. If ϕ is the velocity potential of the flow, show that its kinetic energy is

$$\tfrac{1}{2}\rho \sum_n \iint_{S_n} \phi \operatorname{grad} \phi \cdot \hat{\underset{\sim}{n}} \, dS,$$

where $\hat{\underset{\sim}{n}}$ is the unit normal to the surface S_n directed out of the fluid.

When the fluid is contained between two fixed cylinders, whose generators are parallel, one being inside the other, show that the kinetic energy of the fluid per unit length of the cylinders is $\frac{1}{2}\rho\kappa(\psi_1 - \psi_2)$, where κ is the circulation of the flow round the inner cylinder, ψ_2 is the value of the stream function on the outer cylinder and ψ_1 is that on the inner.

Chapter 9

THE USE OF CONFORMAL MAPPINGS

9.1 MORE GENERAL TWO-DIMENSIONAL FLOWS

In the previous chapter we found the complex potential for a number of two-dimensional flows, but the methods we used were rather hit-or-miss ones. If we want to find the flow past a body of given shape, or to find the free streamlines in a separated flow, we need a more systematic theory.

The determination of the flow past a cylindrical body of given shape (which is not usually circular) is of particular use in aerofoil theory. In practice a subsonic aircraft consists of wings of finite length together with a fuselage, tail and other appendages. It turns out that (as long as the wings are of uniform cross-section and not swept back*) a good approximation to the real flow past such a body is to think of the aircraft as being primarily a cylinder of infinite length whose cross-section is the same as that of the wings. (It is then necessary to make corrections to the resultant flow for the finite length of the wings and for the presence of the appendages like the fuselage and the tail.) Such an infinite cylinder is called an *aerofoil:* its cross-section is usually fairly slender and tapers towards the trailing edge. The correction for finite wings can be made most easily by introducing trailing vortices at the ends of the wings as shown in Figure 9.1(b). There will, of course, be circulation round the wings when the aircraft is flying (it is this which causes the necessary lift), so that the wing itself can be idealized as a line vortex parallel to its generators; for an infinite wing, this vortex would, of course, be of infinite length.

---

* The theory can, however, be modified to deal with swept-back wings.

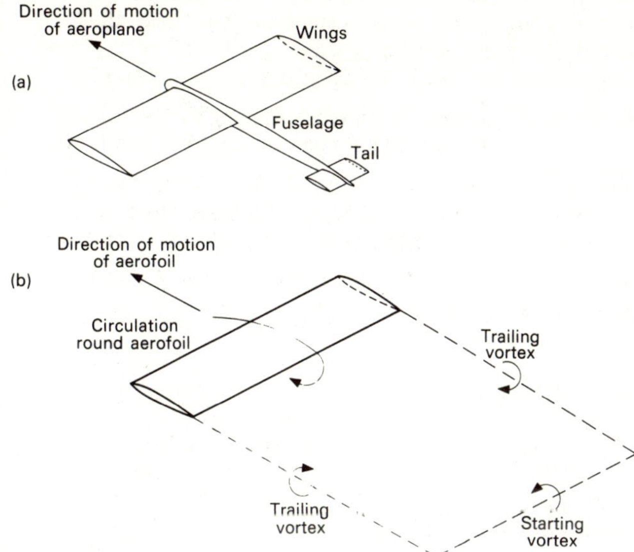

FIGURE 9.1

For a finite wing, it cannot continue much beyond the end
of the wing, nor can it end in the fluid, so it continues
as a pair of trailing vortices as shown in Figure 9.1(b).
The trailing vortices would, in an inviscid fluid,
continue right back to the ground and join up with the
starting vortex to complete a ring. In a real fluid, the
vortices are a good approximation to the flow near the
aerofoil itself, but viscous forces tend to dissipate
them farther back and so they die out. The effect of the
fuselage (and the other appendages) can usually be
included by introducing a few more (smaller) vortices
into the flow. Usually, however, we require to know the
basic flow past the infinite aerofoil in more detail than
this brief account gives, and the theory of the present
chapter will help to determine this for a given case.

The necessity for discussing free streamline theory
has been indicated in the previous chapter and will be
discussed further in Section 9.7.

We shall suppose that we wish to find the complex
potential $w(z)$ of a steady, two-dimensional, inviscid,
incompressible flow having certain cylindrical surfaces
(whose generators are perpendicular to the x-y plane)
as stream surfaces. We suppose that a typical surface
intersects the x-y plane in the curve C, which may be
either a simple closed curve or may be an infinite curve

(it cannot end in the fluid). Then C will be a streamline of the flow. We shall suppose further that there is a number of line sources in the fluid, a typical one being of strength m at A $(z = z_0)$, a number of line vortices, a typical one being of strength κ at $z = z_1$, and a number of doublets (which may be regarded as combinations of sources and sinks). Finally, if the fluid is infinite in extent, we shall suppose that as $|z| \to \infty$, $w(z) \sim w_\infty(z)$, a known function of z: in many cases this asymptotic flow will be uniform and parallel to the x-axis, and in this case $w_\infty(z) = Uz$.

9.2 CONFORMAL MAPPINGS

Suppose now that $\zeta(z)$ is a function of z, analytic for all values of $z = x + iy$ (except possibly at a number of discrete values* of z) within the fluid in the z-plane, and we shall write $\zeta = \xi + i\eta$. Then the function $\zeta(z)$ provides a mapping from each point P of the z-plane to a corresponding point P' of the ζ-plane. Thus if P is the point z in the z-plane, then P' is the point $\zeta(z)$ in the ζ-plane, and this is unique since the function $\zeta(z)$ is analytic. (Note that there may be more than one point P which correspond to a given point P' in the ζ-plane, but this does not concern us for the moment.) Further, neighbouring points in the z-plane become neighbouring points in the ζ-plane, so the closed curve C in the z-plane becomes a closed curve C' in the ζ-plane.

Suppose that P and Q are two neighbouring points in the z-plane such that $PQ = r_1 e^{i\theta_1}$, and that the corresponding points in the ζ-plane are P' and Q' respectively, as shown in Figure 9.2. If we write $P'Q' = \rho_1 e^{i\phi_1}$, we have

$$\rho_1 e^{i\phi_1} = \zeta(z + r_1 e^{i\theta_1}) - \zeta(z) = \frac{d\zeta}{dz} r_1 e^{i\theta_1} \qquad (177)$$

to the first order in r_1. Hence $\rho_1 = \left|\frac{d\zeta}{dz}\right| r$, which

---

* Mathematically this can be expressed by saying that $\zeta(z)$ is a function of z which is analytic almost everywhere.

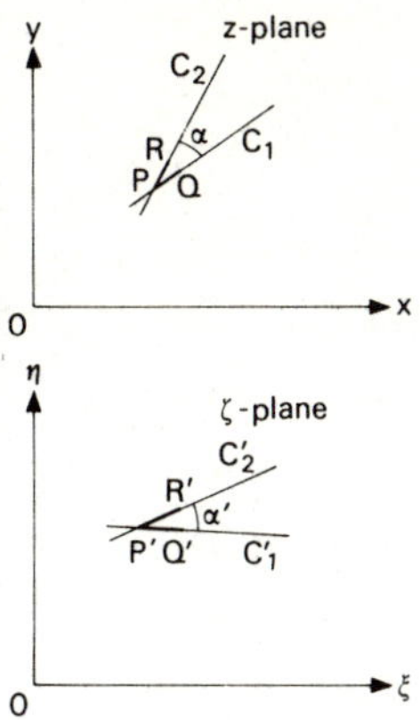

FIGURE 9.2

may be written

$$|P'Q'| = |PQ|\left|\frac{d\zeta}{dz}\right|. \tag{178}$$

Suppose that C_1 is a curve through P and Q, and that C_2 is another curve through P which intersects C_1 at an angle α. We suppose that the corresponding curves C_1', C_2' intersect in the ζ-plane at an angle α', as shown in Figure 9.2. If R is a neighbouring point to P on C_2 and R' is the corresponding point on C_2' (so that P' and R' are neighbouring points on C_2') then the angle QPR is α and the angle $Q'P'R'$ is α'. Writing $PR = r_2 e^{i\theta_2}$ and $P'R' = \rho_2 e^{i\phi_2}$, we have $\alpha = \theta_2 - \theta_1$ and $\alpha' = \phi_2 - \phi_1$. Using equation (177) for the points P, R, P', R', instead of P, Q, P', Q', we have

$$\rho_2 e^{i\phi_2} = \frac{d\zeta}{dz} r_2 e^{i\theta_2}$$

and so, unless $\frac{d\zeta}{dz} = 0$ (in which case $\rho_1 = \rho_2 = 0$), we have

$$\frac{\rho_2}{\rho_1} e^{i(\phi_2 - \phi_1)} = \frac{r_2}{r_1} e^{i(\theta_2 - \theta_1)}$$

or

$$\frac{\rho_2}{\rho_1} \, e^{i\alpha'} = \frac{r_2}{r_1} \, e^{i\alpha}. \tag{179}$$

It follows that, when $\dfrac{d\zeta}{dz} \neq 0$,

$$\alpha' = \alpha \tag{180}$$

and

$$\frac{\rho_2}{\rho_1} = \frac{r_2}{r_1} \, .$$

The fact that the angles are preserved at nearly every point in the mapping is the reason for calling it conformal.

We shall next consider a simple example of conformal mappings. We shall in fact take $\zeta = z^2$, so that $\dfrac{d\zeta}{dz} = 2z$. In this case, we have

$$\xi + i\eta = x^2 - y^2 + 2ixy,$$

$$\xi = x^2 - y^2, \quad \eta = 2xy.$$

Hence

$$\xi = x^2 - \frac{\eta^2}{4x^2} = \frac{\eta^2}{4y^2} - y^2.$$

So the lines $x = $ constant in the z-plane become the parabolas $\xi = x^2 - \dfrac{\eta^2}{4x^2}$ in the ζ-plane and the lines $y = $ constant in the z-plane become the parabolas $\xi = \dfrac{\eta^2}{4y^2} - y^2$ in the ζ-plane. The result of equation (180) shows that the two families of parabolas in the ζ-plane, as sketched in Figure 9.3, intersect at right angles (except at the origin, where $\dfrac{d\zeta}{dz} = 0$). This result may, of course, be verified directly but it is unnecessary to do so in view of the general theory given above.

This example illustrates what happens when more than one point of the z-plane map into a single point of the ζ-plane. For if the point P_1 (a,b) maps into the point P' in the ζ-plane, then so does the point P_2 $(-a,-b)$. (The point P' is, of course,

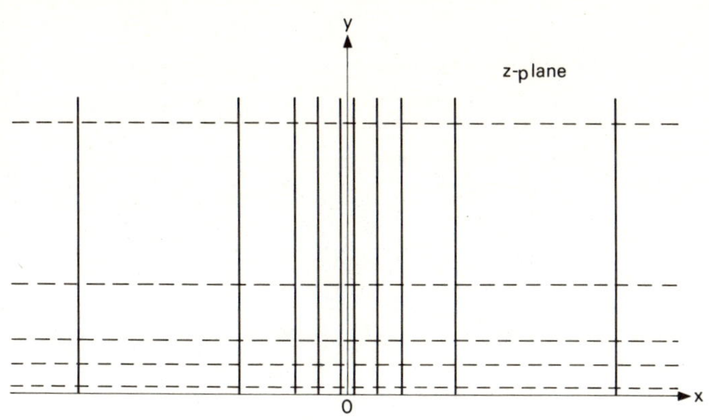

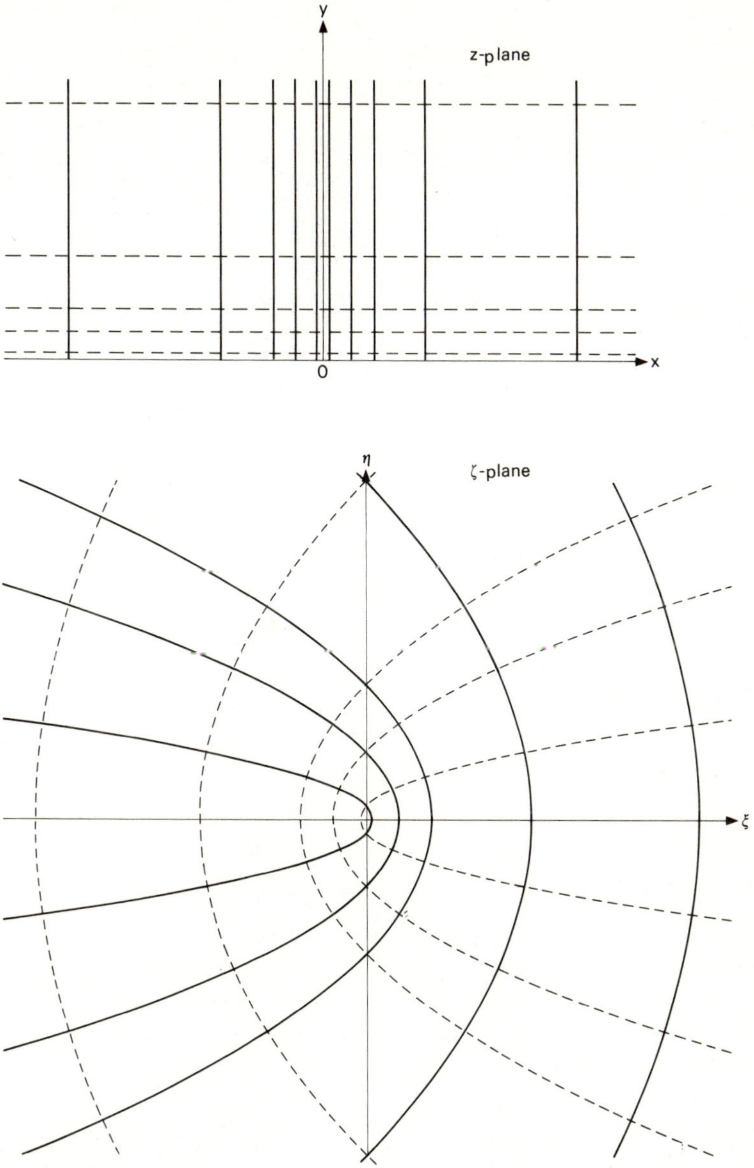

FIGURE 9.3

$(a^2 - b^2, 2ab).)$ Hence if we draw any straight line
through O in the z-plane such as POQ in Figure 9.4,
then there is a one-to-one correspondence between every
point on one side of POQ and every point of the

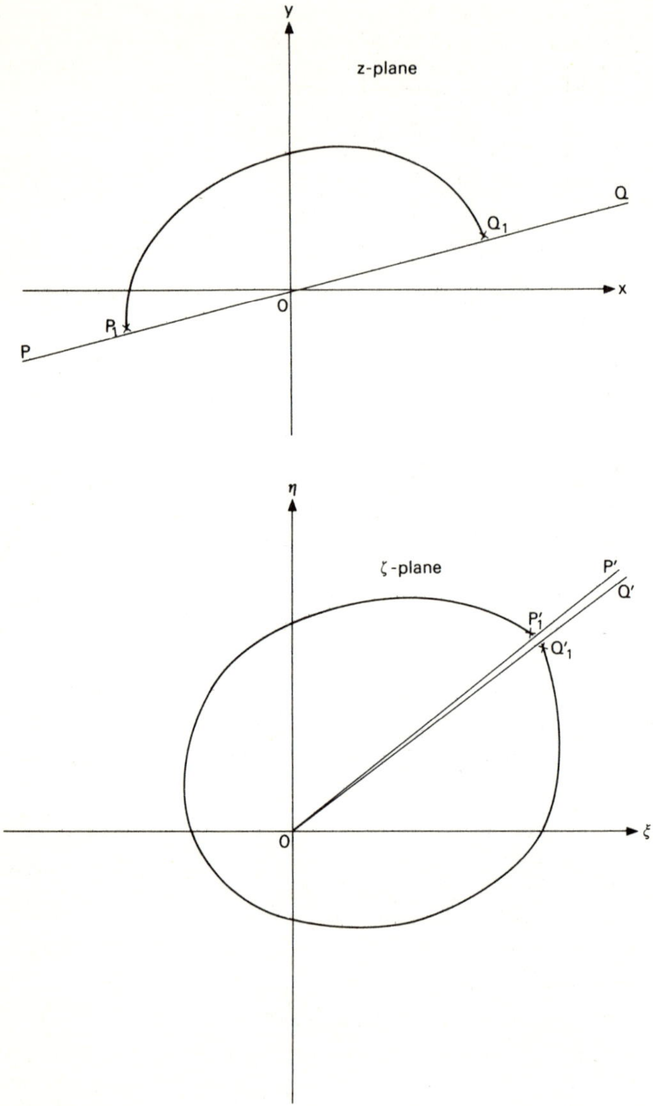

FIGURE 9.4

ζ-plane   (if, for the moment, we exclude the line   POQ
itself) - and also a one-to-one correspondence between
every point on the other side of   POQ   and every point
of the ζ-plane.   The lines   OP   and   OQ   map into the
same line as shown approximately in the figure.   Hence
an infinite ditch (or wall)   POQ   in the z-plane

corresponds to the semi-infinite ditch (or wall) OP' in
the ζ-plane - this is technically called a *cut*. It is
clear that points which are near to one another but are
on opposite sides of the cut in the ζ-plane do not
necessarily correspond to points which are near to one
another in the z-plane. In fact, a continuous curve
joining the points P_1, Q_1 in the z-plane transforms

into an apparently closed continuous curve joining the
points P_1', Q_1' in the ζ-plane, which are points having

the same coordinates but are on opposite sides of the
cut. We shall not concern ourselves here with the
interesting theoretical problems which arise because of
this; our problems will involve only cases in which we
are concerned with some domain in the z-plane which has
a one-to-one correspondence with some domain in the
ζ-plane. Thus we might be concerned only with the
domain to one side or the other of the line POQ in the
z-plane - but we shall never be interested in both sides
at once when there is a cut in the ζ-plane. The only
thing we have to do is to remember that the cut is there
and determine which side of POQ in the z-plane is the
relevant one.
    Usually it is convenient to take the line POQ in
this example to be the real axis in the z-plane (so
that OP' is the positive half of the real axis in the
ζ-plane). The region in the z-plane for which $y > 0$
is often called the upper half plane, and we shall use
this terminology without further explanation. In this
example, therefore, we say that the upper half z-plane
maps into the whole of the ζ-plane with a cut along the
positive part of the real axis in the ζ-plane. In
any mapping it is, of course, necessary to know whether
the interior of the simple closed curve C in the
z-plane maps into the interior or exterior of the
corresponding simple closed curve C' in the ζ-plane.*
To do this we could evaluate the value of ζ for some
point z which is in the interior of C. However, a
simpler method (and one which is easy to justify using
the conformal property of the mapping) is to imagine one
is walking along C in the z-plane from P to Q, say,
with the interior of C on the left: then the
corresponding part of the ζ-plane is that on the left
of C' when one is walking along C' from P' to Q'.

---

* Or, if the curve C or C' or both extend to
  infinity, which side of C corresponds to which
  side of C'.

9.3 CORRESPONDING FLOWS

Now we shall see how we can apply this theory to the
problem of determining the complex potential of a flow
in the z-plane where certain streamlines and
singularities of the flow are given. To do this we look
for a mapping of the z-plane into a ζ-plane where the
specified streamlines become simple curves like circles
or straight lines. We then find the complex potential
of a flow in the ζ-plane having these curves as streamlines (and having appropriate singularities at points
corresponding to the original ones in the z-plane).

First we define a flow in the ζ-plane by the complex
potential $W(\zeta) = \Phi(\xi,\eta) + i\Psi(\xi,\eta)$ where the numerical
value of W (that is, of Φ and Ψ) at the point P'
in the ζ-plane is the same as that of the flow defined
by $w(z)$ at the point P to which it corresponds in
the z-plane; that is $w(z) = W[\zeta(z)]$. Now the curve
C in the z-plane is a streamline in the flow in that
plane - that is, ψ is constant along C. It follows
that Ψ is constant along C', and so C' is a
streamline in the flow in the ζ-plane.

Since ζ is an analytic function of z, there are
no singularities in the flow in the ζ-plane except at
points in the ζ-plane corresponding to points in the
z-plane where there are singularities; thus, if there
is a singularity of $w(z)$ at A in the z-plane, there
will also be one at A' in the ζ-plane. Further, if
C_0 is any curve in the z-plane in whose interior $w(z)$
is analytic except at the single singularity A, then
the corresponding curve C_0' encloses a domain in the
ζ-plane in which $W(\zeta)$ is analytic everywhere except
at the point A'. For example, if there is a source of
strength m at A, then the change in ψ when we go
once round C_0 is $2\pi m$; but the values of Ψ at the
points of C_0' are (by definition) the same as the
values of ψ at the corresponding points of C_0, and
so the change in Ψ when we go once round C_0' is
also $2\pi m$. It follows that the singularity at A' is
also a source of strength m. Similarly, if A is a
vortex of strength κ, then the change in ϕ going
once round C_0 is κ : and, since the values of ϕ
and Φ at corresponding points are the same, the change
in Φ in going once round C_0' is also κ. Hence the
singularity at A' is a vortex of strength κ.

The same result holds for the circulation κ round any closed curve C in the flow in the z-plane - it corresponds in the ζ-plane flow to a circulation κ round the curve C'.

If there are doublets in the flow in the z-plane, the situation is slightly less simple. This is because, if there is a source of strength m at P and a sink of strength m at Q in the z-plane, and we write $\mu = m|PQ|$, then there will be a source of strength m at P' and a sink of strength m at Q', the corresponding points in the ζ-plane, and we can write $\mu' = m|P'Q'|$. If now $Q \to P$ (so that $Q' \to P'$ as well), the two systems reduce to doublets of strength μ and μ' respectively. But

$$\mu' = \mu \frac{|P'Q'|}{|PQ|} = \mu \left|\frac{d\zeta}{dz}\right|$$

using the result of equation (178).

We must also consider the situation when $|z| \to \infty$. In many cases we choose a mapping such that $\zeta(z) \sim z$ as $|z| \to \infty$, and in this situation it is clear that if $w(z) \sim w_\infty(z)$ as $|z| \to \infty$, then $W(\zeta) \sim w_\infty(\zeta)$

as $|\zeta| \to \infty$. If the mapping into the ζ-plane does not have this simple form, we must make sure that nothing sinister happens there: it is easier to do this for particular problems than to try to formulate a criterion for the general case.

If, then, we can find a suitable mapping into a ζ-plane where the corresponding flow $W(\zeta)$ can be determined, we can immediately write down the complex potential $w(z)$ of the required flow in the z-plane. We illustrate this by means of some simple examples.

9.4 APPLICATION TO PARTICULAR PROBLEMS

EXAMPLE 9.4.1   A line source of strength m through the point $z = ib$ (where b is real) in a flow having the plane $y = 0$ as a boundary.
We use the mapping $\zeta = z^2$ which we discussed in Section 9.2. The point $z = ib$ maps into the point $\zeta = -b^2$ in the ζ-plane. Now the inviscid flow in the ζ-plane due to a source of strength m at $\zeta = -b^2$ in the presence of a wall along the positive part of the real axis is

$$W(\zeta) = m \log(\zeta + b^2),$$

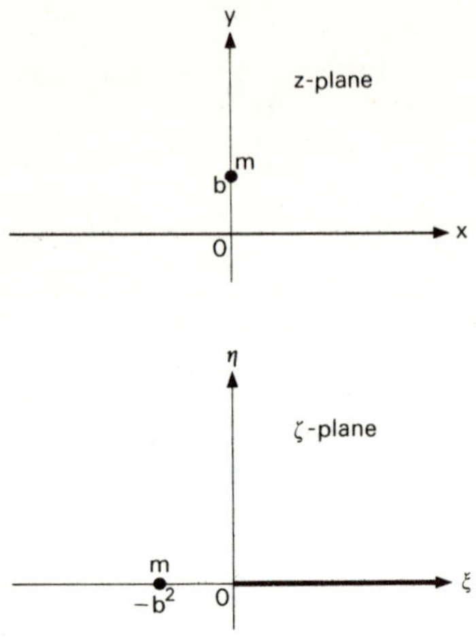

FIGURE 9.5

since the real axis is a streamline of this flow.
Hence we have immediately that the required flow in
the z-plane is given by

$$w(z) = m \log(z^2 + b^2). \tag{181}$$

This is the same complex potential as would be
obtained with a source m at z = ib together with
a source m at z = -ib. This second source is
often referred to as the *image* of the given source
in the plane y = 0.

EXAMPLE 9.4.2   A line source of strength m through
(b,0) outside the cylinder $|z|$ = a.
We use the mapping* $\zeta = z + \dfrac{a^2}{z}$.   Then

$$\xi + i\eta = \left(r + \frac{a^2}{r}\right)\cos\theta + i\left(r - \frac{a^2}{r}\right)\sin\theta .$$

Hence the cylinder r = a maps into the flat plate
η = 0, $|\xi| \le 2a$; further θ = 0 maps into η = 0,
ξ ≥ 2a and θ = π into η = 0, ξ ≤ -2a.   Finally,
as $|z| \to \infty$, ζ ~ z.   This is shown in Figure 9.6.
The flow in the ζ-plane corresponding to the required
flow is that due to a line source of strength m
through $b + \dfrac{a^2}{b}$, 0   and is given by

$$W(\zeta) = m \log\left(\zeta - b - \frac{a^2}{b}\right).$$

---

* This mapping is suggested by the Example 8.4.6.

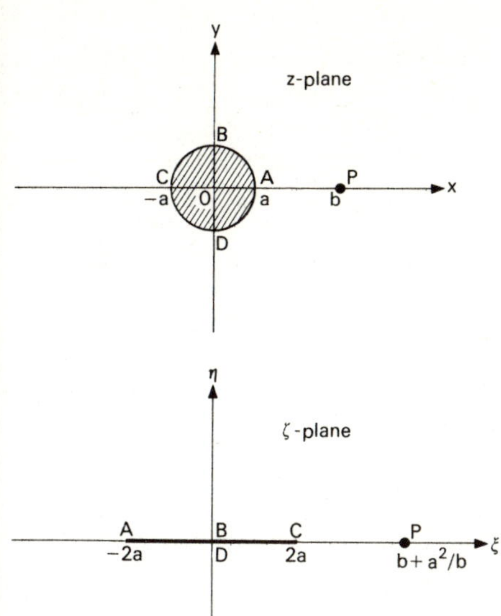

FIGURE 9.6

Hence the complex potential for the required flow
in the z-plane is

$$w(z) = m \log\left(z + \frac{a^2}{z} - b - \frac{a^2}{b}\right)$$

$$= m \log(z - b) + m \log\left(z - \frac{a^2}{b}\right) - m \log z. \quad (182)$$

This is the same complex potential as would be
obtained by the given source through $(b,0)$ together
with a source of strength m at $z = \frac{a^2}{b}$, the
inverse point of P in the circle $|z| = a$, and a
sink of strength m through $z = 0$. This source
and sink form an *image system* for the given source
in the cylinder.

Note that in Figure 9.6 we have labelled corresponding
points in the z-plane and the ζ-plane by the same letter,
having dropped the prime we have used till now for points
in the ζ-plane. This is common practice and causes no
confusion as long as we always make it clear to which
plane we are referring.

EXAMPLE 9.4.3 A line source of strength m through
(a,b) within the region bounded by plane walls
along $y = 0, y = c$ and $x = 0$ (see Figure 9.7).

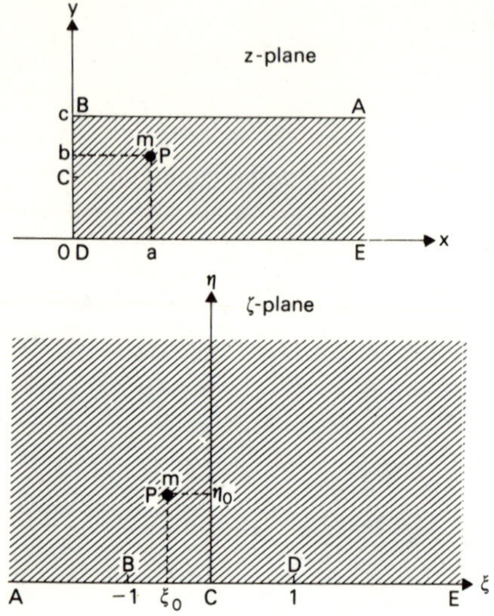

FIGURE 9.7

We use the mapping $\zeta = \cosh \dfrac{\pi z}{c}$. Then

$$\xi + i\eta = \cosh \frac{\pi x}{c} \cos \frac{\pi y}{c} + i \sinh \frac{\pi x}{c} \sin \frac{\pi y}{c}.$$

Hence $y = 0$, $x \geq 0$ maps into $\eta = 0$, $\xi \geq 1$; $y = c$, $x \geq 0$ into $\eta = 0$, $\xi \leq -1$; and $x = 0$, $0 \leq y \leq c$ into $\eta = 0$, $1 \geq \xi \geq -1$. The semi-infinite strip of fluid in the z-plane maps into the whole of the upper half ζ-plane.

The flow in the ζ-plane corresponding to the given flow in the z-plane is that due to a line source of strength m through the point ζ_0 in the ζ-plane, where $\zeta_0 = \cosh \dfrac{\pi z_0}{c}$ with $z_0 = a + ib$. This is given by the complex potential

$$W(\zeta) = m \log(\zeta^2 - \zeta_0^2).$$

Hence the appropriate complex potential in the z-plane is

$$w(z) = m \log\left(\cosh^2 \frac{\pi z}{c} - \cosh^2 \frac{\pi z_0}{c}\right).$$

EXAMPLE 9.4.4   The flow past the flat plate $y = 0$,
$|x| \leq 2a$ which is uniform far from the plate with
velocity U parallel to Ox, having circulation
κ round the plate.

We use the mapping $\zeta = \frac{1}{2}\left[z + (z^2 - 4a^2)^{\frac{1}{2}}\right]$; this is

the inverse of the mapping $z = \zeta + \dfrac{a^2}{\zeta}$ (compare

Example 9.4.2) to the outside of the cylinder
$|\zeta| = a$ in the ζ-plane from the outside of the given
plate in the z-plane (see Figure 9.8).

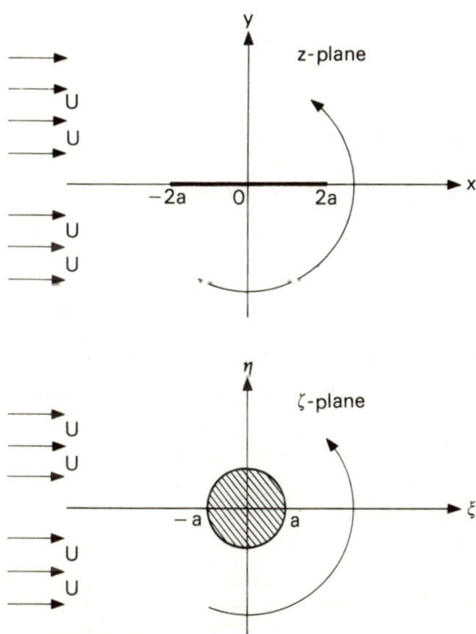

FIGURE 9.8

The corresponding flow in the ζ-plane is the flow
past the cylinder $|\zeta| = a$, which is uniform far
from the cylinder with velocity U parallel to O
with circulation κ round the cylinder. The complex
potential due to this flow is given by a form like
that of equation (174) (which was derived for a flow
in the z-plane). So we take

$$W(\zeta) = U\left(\zeta + \frac{a^2}{\zeta}\right) - \frac{i\kappa}{2\pi} \log \zeta .$$

Hence the required flow in the z-plane is given by
the complex potential

$$w(z) = Uz - \frac{i\kappa}{2\pi} \log \tfrac{1}{2}\left[z + (z^2 - 4a^2)^{\frac{1}{2}}\right]$$

$$= Uz - \frac{i\kappa}{2\pi} \log\left[\frac{z}{2a} + \left(\frac{z^2}{4a^2} - 1\right)^{\frac{1}{2}}\right] + \text{constant.}$$

But $\log\left[\dfrac{z}{2a} + \left(\dfrac{z^2}{4a^2} - 1\right)^{\frac{1}{2}}\right] = \cosh^{-1}\left(\dfrac{z}{2a}\right)$, and we can ignore the constant. Hence the required complex potential is

$$w(z) = Uz - \frac{i\kappa}{2\pi} \cosh^{-1}\left(\frac{z}{2a}\right). \tag{183}$$

It is of interest to note that the velocity is given by

$$u - iv = \frac{dw}{dz} = U - \frac{i\kappa}{2\pi} \frac{1}{(z^2 - 4a^2)^{\frac{1}{2}}}. \tag{184}$$

Hence the flow defined by (183) gives, for general κ, infinite velocities at the ends of the plate, which is unrealistic. This is not surprising because the effect of the circulation on the uniform stream must be (on the present theory) to force the stream-lines to turn through 180° at each end of the plate, as indicated in Figure 9.9(b).

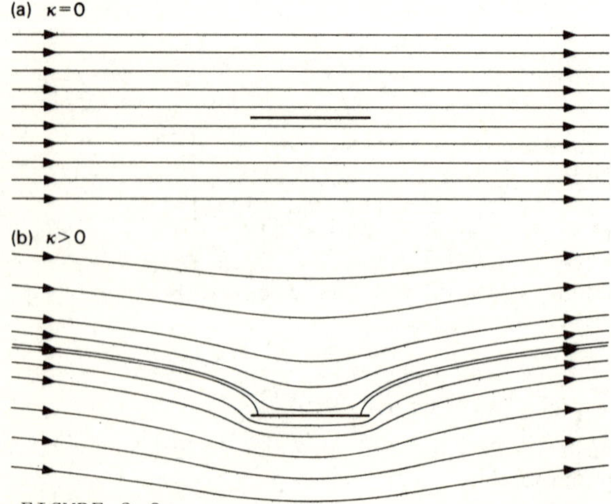

FIGURE 9.9

There is nothing we can do about this except to say that a flow like this is impossible in a real

fluid. In the next example, however, we shall be
able to get over part of a similar difficulty by
choosing a particular value of κ.

EXAMPLE 9.4.5 The flow past the flat plate of
length 4a which passes through $z = 0$ and which
makes an angle α with Ox, the flow far from the
plate being uniform with velocity U parallel to
Ox and having a circulation κ round the plate
(see Figure 9.10).

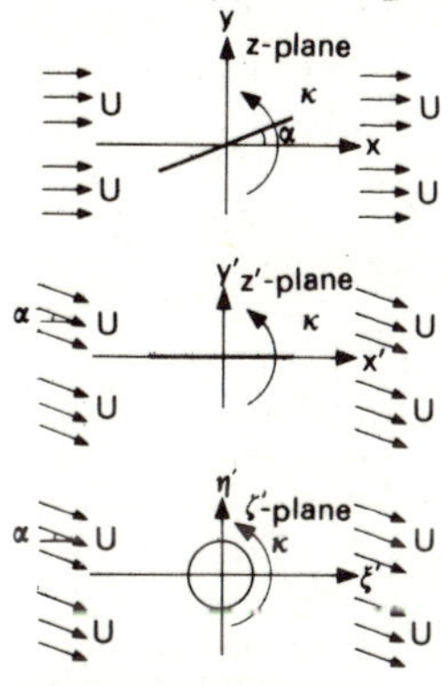

FIGURE 9.10

Here we use the intermediate mapping $z' = z\,e^{-i\alpha}$,
so that the whole system is turned through an angle
α. Then we use the mapping $\zeta' = \tfrac{1}{2}\left\{z' + (z'^2 + 4a^2)^{\frac{1}{2}}\right\}$
which maps the plate into the cylinder $|\zeta'| = a$.

Finally, we take the mapping $\zeta = \zeta'e^{i\alpha}$ which turns
the system in the opposite direction through the
same angle α. These mappings ensure that $\zeta \sim z$
as $|z| \to \infty$; in fact we have

$$\zeta = \zeta'e^{i\alpha} = \tfrac{1}{2}\,e^{i\alpha}\left[z' + (z'^2 - 4a^2)^{\frac{1}{2}}\right]$$

$$= \tfrac{1}{2}\,e^{i\alpha}\left[ze^{-i\alpha} + (z^2e^{-2i\alpha} - 4a^2)^{\frac{1}{2}}\right]$$

$$= \tfrac{1}{2}\left[z + (z^2 - 4a^2e^{2i\alpha})^{\frac{1}{2}}\right].$$

The flow in the ζ-plane which corresponds to the
given flow in the z-plane is again of the form given
in equation (174) and we have

$$W(\zeta) = U\left(\zeta + \frac{a^2}{\zeta}\right) - \frac{i\kappa}{2\pi}\,\log\zeta.$$

Hence the corresponding flow in the original z-plane is given by

$$w(z) = U\left[\frac{z + (z^2 - 4a^2e^{2i\alpha})^{\frac{1}{2}}}{2} + \frac{2a^2}{z + (z^2 - 4a^2e^{2i\alpha})^{\frac{1}{2}}}\right]$$

$$- \frac{i\kappa}{2\pi} \log\left[z + (z^2 - 4a^2e^{2i\alpha})^{\frac{1}{2}}\right]$$

$$= U\left[ze^{-i\alpha}\cos\alpha + i(z^2 - 4a^2e^{2i\alpha})^{\frac{1}{2}}e^{-i\alpha}\sin\alpha\right]$$

$$- \frac{i\kappa}{2\pi} \cosh^{-1}\left(\frac{ze^{-i\alpha}}{2a}\right) + \text{constant}. \quad (185)$$

This gives the complex velocity

$$\frac{dw}{dz} = U\left[e^{-i\alpha}\cos\alpha + \frac{ize^{-i\alpha}\sin\alpha}{(z^2 - 4a^2e^{2i\alpha})^{\frac{1}{2}}}\right]$$

$$- \frac{i\kappa}{2\pi}\frac{1}{(z^2 - 4a^2e^{2i\alpha})^{\frac{1}{2}}}. \quad (186)$$

In general we see that this solution gives infinite velocities at $z = \pm 2ae^{i\alpha}$ - that is, at both edges of the plate. However, if we can arrange that $\kappa = 2\pi U(2ae^{i\alpha})e^{-i\alpha}\sin\alpha = 4\pi Ua \sin\alpha$, equation (186) becomes

$$\frac{dw}{dz} = Ue^{-i\alpha}\left[\cos\alpha + i\left(\frac{z - 2ae^{i\alpha}}{z + 2ae^{i\alpha}}\right)^{\frac{1}{2}}\sin\alpha\right]. \quad (187)$$

For this value of κ, then, the velocity at the trailing edge, $z = 2ae^{i\alpha}$, is not infinite.

We can do nothing about the infinite velocity at the leading edge, $z = -2ae^{i\alpha}$, within the framework of the present theory, without re-introducing the infinite velocity at the trailing edge. However, in a real fluid, the boundary layer which must be present will have a 'rounding-off' effect so that the effective leading edge is not sharp. The expression (185) will therefore probably be a reasonably good approximation to the flow which really does occur when $\kappa = 4\pi Ua \sin\alpha$, as long as boundary-layer separation does not occur. In practice, of course, separation is likely to occur unless α is very small.

None of the flows discussed in this section is of direct practical application to real fluids. They do illustrate the type of problem which arises and, in particular, the way in which we can (theoretically) adjust the circulation to avoid an infinite velocity at the trailing edge of a body. That this circulation can be achieved in practice for a suitably shaped body in a real fluid is dicussed in Section 9.6. The condition for the circulation to adjust to this value is usually called the Joukowski condition.

9.5 JOUKOWSKI AEROFOILS

Here we take the flow in the ζ-plane to be the flow past a circular cylinder whose axis passes through the point ζ_0 and whose radius is c. There is a uniform flow having velocity $(U \cos \alpha, U \sin \alpha)$ far from the cylinder and a circulation κ round the cylinder. This is shown in Figure 9.11. If $\zeta_0 = 0$ and $\alpha = 0$

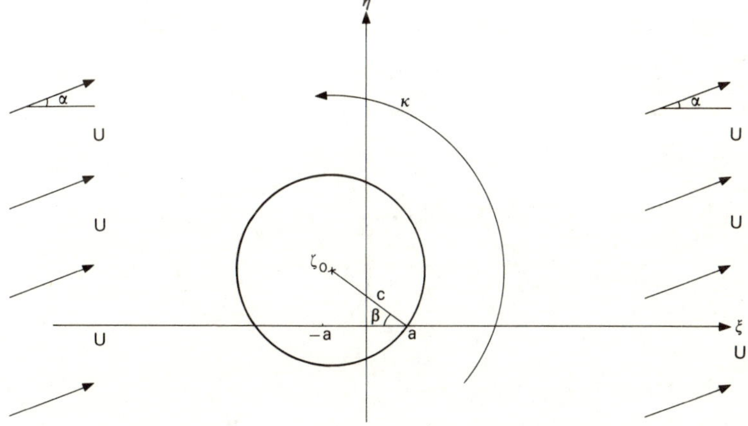

FIGURE 9.11

the flow is given by a complex potential of the form given in equation (174) with a replaced by c, and z by ζ. The displacement of the origin changes ζ in that expression to $\zeta - \zeta_0$ and the subsequent rotation through an angle α changes it to $(\zeta - \zeta_0)e^{-i\alpha}$. Hence the required flow in the ζ-plane is given by

$$W(\zeta) = U\left[(\zeta - \zeta_0)e^{-i\alpha} + \frac{c^2 e^{i\alpha}}{\zeta - \zeta_0}\right] - \frac{i\kappa}{2\pi}\log(\zeta - \zeta_0), \quad (188)$$

where we have dropped the unnecessary constant $-\frac{i\kappa}{2\pi}\log(e^{-i\alpha})$.

Joukowski suggested a series of shapes which are obtained from this flow by using the mapping

$$z = \zeta + \frac{a^2}{\zeta}, \quad (189)$$

where $\zeta = a$ is one of the points in which the circle $|\zeta - \zeta_0| = c$ intersects Ox, and where c is so large that the point $\zeta = -a$ is within the circle $|\zeta - \zeta_0| = c$. Figure 9.12 shows the form of the new

(a)   c=4a/3, β=0°

(b)   c=8a/3, β=0°

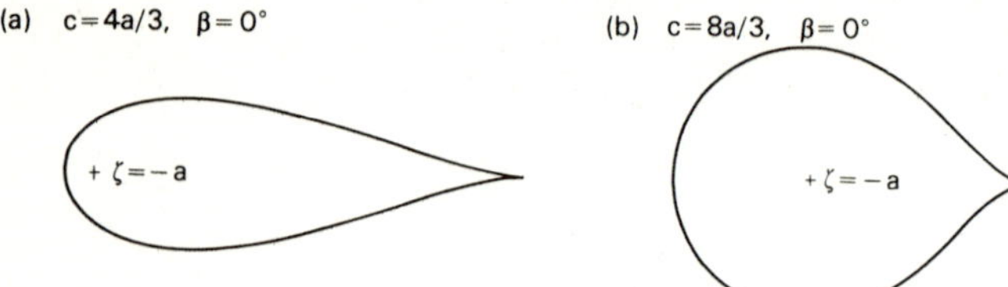

(c)   c=4a/3, β=30°

(d)   c=8a/3, β=30°

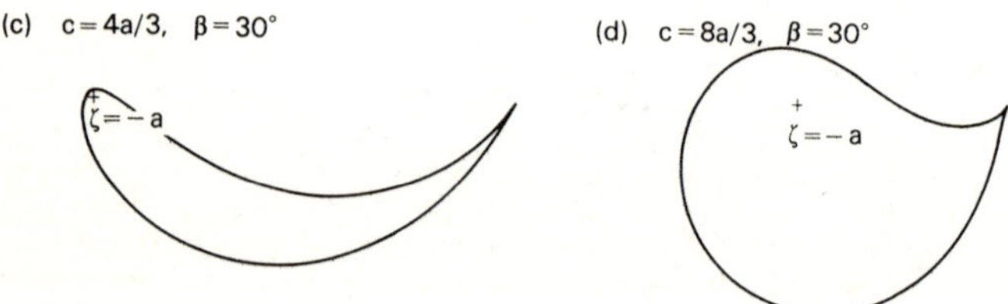

FIGURE 9.12

cylinder cross-section in several special cases (in practice, more slender and less curved shapes would be used, of course). The point in the z-plane corresponding to the point $\zeta = -a$ is inside the cylinder and therefore not in the fluid. Now, in the flow in the z-plane the complex potential is obtained by putting

$$\zeta = \tfrac{1}{2}\left[z + (z^2 - 4a^2)^{\frac{1}{2}}\right] \quad (190)$$

(which is the appropriate inverse* of equation (189))
in the expression (188). The complex velocity is

$$\frac{dw}{dz} = \frac{dW}{d\zeta}\frac{d\zeta}{dz}$$

$$= \tfrac{1}{2}\left[U\left(e^{-i\alpha} - \frac{c^2 e^{i\alpha}}{(\zeta - \zeta_0)^2}\right) - \frac{i\kappa}{2\pi(\zeta - \zeta_0)}\right] \times$$

$$\times \left[1 + \frac{z}{(z^2 - 4a^2)^{\frac{1}{2}}}\right]. \qquad (191)$$

The only point in the fluid at which this can become
infinite is where $z = 2a$ (since neither the point
corresponding to $\zeta = \zeta_0$ nor the point $z = -2a$ is

in the fluid), and this corresponds to the point $\zeta = a$
in the ζ-plane. If κ can be adjusted so that
$\frac{dw}{d\zeta} = 0$ when $\zeta = a$ (that is when $z = 2a$),   the problem
disappears and the velocity is no longer infinite at the
trailing edge. The required value of κ is
$-4\pi Uc\,\sin(\alpha + \beta)$,   where β is the angle made by the
line $a - \zeta_0$ in the ζ-plane with the negative direction
of the ξ-axis (that is, $\zeta_0 = a - c\,e^{-i\beta}$),   and in this

case the velocity is finite everywhere in the flow. We
shall discuss in the next section how such an adjustment
of κ might take place in a real fluid.
     We saw in Section 8.6 that the force per unit length
on a cylindrical obstacle is $-\rho\kappa U$ normal to the
stream. For a Joukowski aerofoil, this force takes the
value $4\pi\rho U^2 c\,\sin(\alpha + \beta)$,   and when α and β are
small (and positive), this is positive. It follows
that there is a positive lift on the aerofoil when the
Joukowski condition is satisfied, and this (fortunately)
is what is required to enable the aerofoil to fly.
     It must be remembered that this flow as described by
equations (188) and (190) is that corresponding to
Step 1 of Section 6.7. It is of interest to compare the
results with experimental values, as given by Preston
for a symmetric Joukowski aerofoil with zero lift (that
is, with $\alpha = \beta = 0$). This is shown in Figure 9.13.

---

* The other inverse, $\zeta = \tfrac{1}{2}\left[z - (z^2 - 4a^2)^{\frac{1}{2}}\right]$,   maps
  the interior of the aerofoil on to the exterior of
  the circular cylinder.

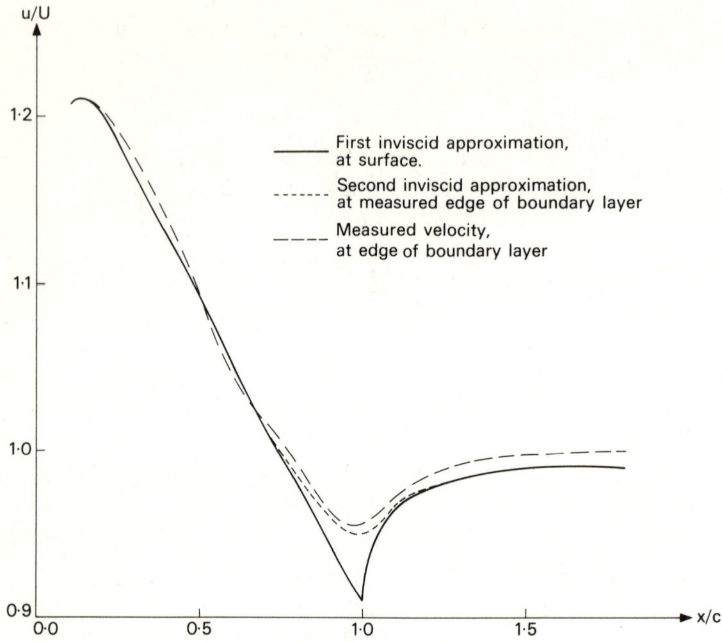

FIGURE 9.13

The curve corresponding to the second inviscid approximation is obtained by going as far as Step 3 of
Section 6.7, and it can be seen that this gives a very
good approximation to the observed flow.

9.6 GENERAL AEROFOIL THEORY

So far in this chapter we have discussed how to find
the flow past a cylinder of given shape, but in the
theory of aerofoil design we have to look for a shape
which gives a flow with suitable characteristics. We
choose, for simplicity, the x-axis to be such that
$\alpha = 0$ (at least, at the time under discussion), and
neglect variations of U with time. Usually we need
to have a shape which gives a high lift and comparatively
small drag. One of the important non-dimensional
parameters is, therefore, the *lift coefficient*

$$C_L = \frac{L}{\frac{1}{2}\rho U^2 A} ,$$

(192)

where L is the component in the positive direction of
the y-axis of the force per unit length on the cylinder
and A is a convenient reference area (often the plan
area of the wing). Note that L depends on α, the
angle of incidence of the stream on the aerofoil, and
it will also probably depend on the Reynolds number of
the flow. If D is the component in the positive
direction of the x-axis of the force per unit length on
the cylinder, then we define the *drag coefficient* as

$$C_D = \frac{D}{\frac{1}{2}\rho U^2 A},\qquad(193)$$

and this also depends on α and on the Reynolds number
of the flow. Another important quantity is the
lift-drag ratio

$$\frac{L}{D} = \frac{C_L}{C_D},\qquad(194)$$

and this is a measure of the aerodynamic efficiency.
The problem, of course, in designing the aerofoil is that
any design which increases C_L is likely to increase C_D
as well and so increase the resistance to the motion.
As α is increased it usually happens that C_L
increases until a certain critical value of α is
reached; at this point the position at which the
boundary layer separates is crucial. For values of α
smaller than the critical value, the boundary layer is
attached almost everywhere on the aerofoil: separation
may occur, but only near the trailing edge, as shown in
Figure 9.14(a). Near the critical value of α, however,

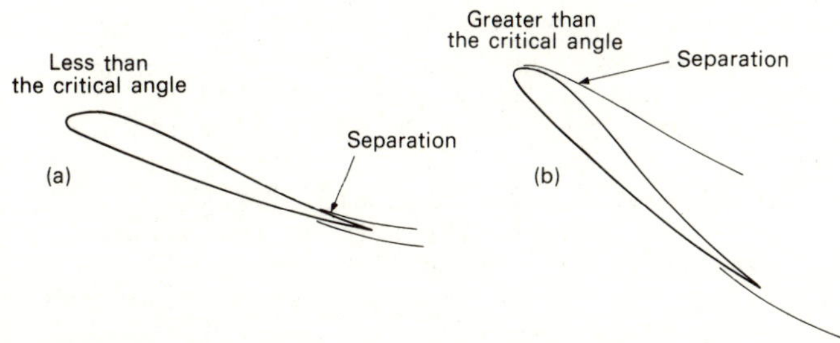

FIGURE 9.14

the point of separation suddenly moves near to the
leading edge (as in Figure 9.14(b)), and the whole flow
regime changes. This is indicated in Figure 9.15(a)
where x is the distance from the leading edge of the

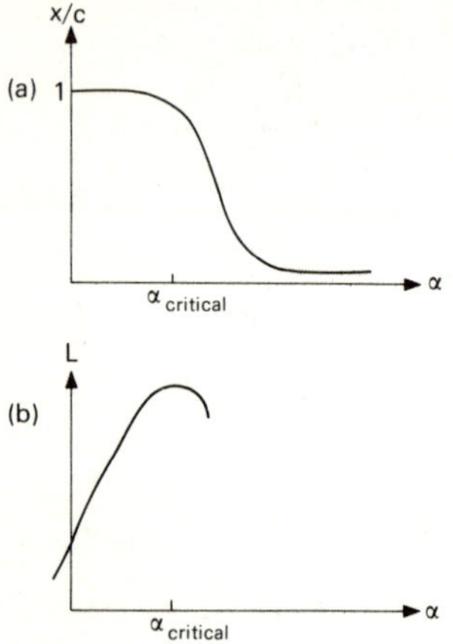

FIGURE 9.15

aerofoil of the point where separation occurs, and c
is the value of x at zero incidence (that is, when
$\alpha = 0$). The forces exerted on the cylinder therefore
change very suddenly (for example, a typical graph
showing the lift L as a function of α is shown in
Figure 9.15(b) and, if α reaches its critical value
in flight, loss of control of the aircraft may be
caused. This phenomenon is called *stalling*.
    When a good aerofoil shape is found, it always has
one and only one point at which (for arbitrary
circulation) the velocity becomes infinite: we saw
this is the particular case of the Joukowski aerofoil.
The Joukowski condition states that the circulation
round the aerofoil will adjust itself so that the
velocity everywhere becomes finite. The mechanism by
which this happens in a real fluid is illustrated in
Figure 9.16. We suppose the motion to start from rest:
initially, therefore, the motion will be irrotational
everywhere (since it requires a finite time for the
vorticity due to the presence of the surface to diffuse
into the fluid) and there will be no circulation.
There will, therefore, be an infinite velocity at the
trailing edge as the fluid from below the aerofoil
turns to go back along the upper surface to the
stagnation point on that surface. For a real fluid,
viscous forces in the neighbourhood of this singularity

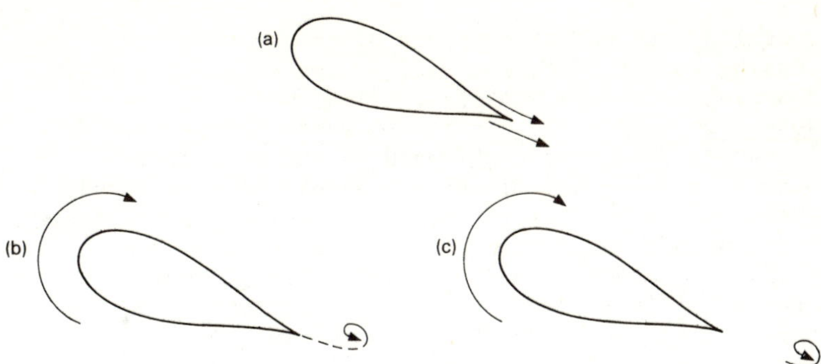

FIGURE 9.16

will be large, the boundary layer from the lower surface
will separate as it tries to turn on to the upper surface
and a vortex will be formed at the trailing edge in the
sense shown in Figure 9.16(b). This vortex will be
carried along by the fluid in the irrotational flow
outside the boundary layer and will be swept along with
it and away from the aerofoil, as shown in Figure 9.16(c).
Since there is no mechanism to produce circulation round
a closed curve which encloses both the aerofoil and the
vortex, and there is a positive circulation round the
vortex, it follows that a negative circulation must exist
round the cylinder and this will have the effect of
moving the stagnation point of the irrotational flow
nearer to the trailing edge. If the circulation is of
the right magnitude to bring the stagnation point
exactly to the trailing edge (so that the velocity
singularity disappears and the Joukowski condition is
satisfied), the motion becomes steady. If the magnitude
of the circulation is too small, another vortex must
form and be cast off to improve the situation further.
If, on the other hand, it is too large, so that the
irrotational flow (now with some circulation) has to
turn round the trailing edge from the upper to the lower
surface, a vortex of the opposite sense is formed and
cast off, and this results in the magnitude of the
circulation decreasing again. So, in the end, the
circulation round the aerofoil will satisfy the
Joukowski condition. (It should perhaps be noted that
this physical argument is a modern one and was not
available to Joukowski, who put forward his condition
as a plausible hypothesis confirmed by experiment.)
The cast-off vortices are, in a real fluid, ultimately
dissipated by viscous forces.
   Nowadays the design of aerofoils involves far more
considerations than these. When aircraft fly at speeds

approaching that of sound, wings of the type we have been discussing are not the most useful, and swept-back wings have advantages. For supersonic flow, delta wings must be considered, as well. In these cases, furthermore, the fluid may no longer be considered as incompressible throughout most of the flight (it should be remembered, however, that *all* aircraft have to fly slowly sometimes - when taking off and landing). The theory described in this chapter cannot, therefore, be applied directly to the design of fast aircraft - but it sometimes happens that it can be used as part of the theory (see, for example, question 4 at the end of this chapter).

9.7 FREE-STREAMLINE THEORY

We have seen in Chapter 6 that, in the flow of a real fluid past an obstacle, a boundary layer forms on the surface of the obstacle when the Reynolds number is sufficiently high, and that in many cases this boundary layer separates to form a wake behind the obstacle. In Chapter 7 we saw that we can sometimes idealize such a situation by replacing a layer of high velocity gradient by a vortex sheet. For the unseparated boundary layer, it is clear that the position of this vortex sheet is on the surface (at least to the first approximation), but when separation has occurred the position of the vortex sheet has to be determined. It is, of course, a surface of discontinuity and there is no flow across it and we have therefore a free-streamline flow.

It should be emphasized that this model is not always a good model of the real flow; it is essentially a steady flow and it is well known that the wake behind a bluff body is unsteady. For the flow behind a circular cylinder, for example, there is a Kármán vortex street (Figure 7.12). However, it is a considerable improvement on the elementary theory in which the flow does not separate at all: it is the equivalent of Step 3 of Section 6.7. For example, we might hope that the drag on the obstacle (which is predicted to be zero by the elementary theory) may be predicted at least to within an order of magnitude.

We shall consider here only problems of flow past cylindrical obstacles with sharp edges; we can then assume that separation occurs at these edges and that the free streamlines (or vortex sheets) are tangential to the surface at the point of separation. We have, for a given geometrical configuration, to find a relationship between the complex potential w of the

flow and the position   z.   It turns out that it is
necessary first to establish a relationship between   w
and its derivative $\frac{dw}{dz}$: this can then be integrated to
give the required relationship between   w   and   z.
Usually, of course, we also wish to establish the
position of the vortex sheets.

   This is most easily seen by investigating some
particular problems, but first we introduce a new
variable Ω .   We have already seen in equation (161)
that we can write

$$\frac{dw}{dz} = q\, e^{-i\theta},$$

where   q   is the speed of the fluid at a point and θ
is the angle made by the direction of the flow there with
the position direction of the x-axis.   It is convenient
to introduce the new variable

$$\Omega = \log\left(U \Big/ \frac{dw}{dz}\right) = \log\left(\frac{U}{q}\right) + i\theta \ , \tag{195}$$

where   U   is a reference speed appropriate to the flow.
We look for a mapping in the first place from the
Ω-plane to the w-plane.

   In the examples below, we assume that there is no
motion* inside the wake region behind the body: it
follows that the dynamic pressure there is constant
(in fact it is zero) and so the dynamic pressure on the
free streamline itself is also constant.   However, we
can use Bernoulli's equation in the form (21) in which
the external forces are zero, to see that $p_d + \frac{1}{2}q^2$

is constant along the free streamline.   It follows at
once that, along the free streamline separating the main
flow from the stationary wake, the speed   q   is constant.
On the other hand, for a streamline along a given rigid
surface, the angle θ is known.   In terms of the
variable Ω defined in equation (195), we note that the
real part is constant along a free streamline and the
imaginary part is given along a fixed streamline.

   EXAMPLE 9.7.1 The impact of a uniform stream on an
   infinitely long flat plate which is of width   2a
   and normal to the stream (see Figure 9.17(a)).

---

* In practice, there will be some reverse flow there.

(a)
z-plane
ψ>0
q=U
ψ=0
D
φ=+∞
θ=0
C φ=φ₀, y=a
A ψ=0, θ=0
B φ=0, q=0
φ=−∞
q=U
C'
D'
y
x

(b)
w-plane
ψ
φ₀
A
B
C
D
φ

(c)
θ
Ω-plane
π/2 C
B
D,A
B
log (U/q)

(d)
η
ζ-plane
B
C
D A
B
ξ

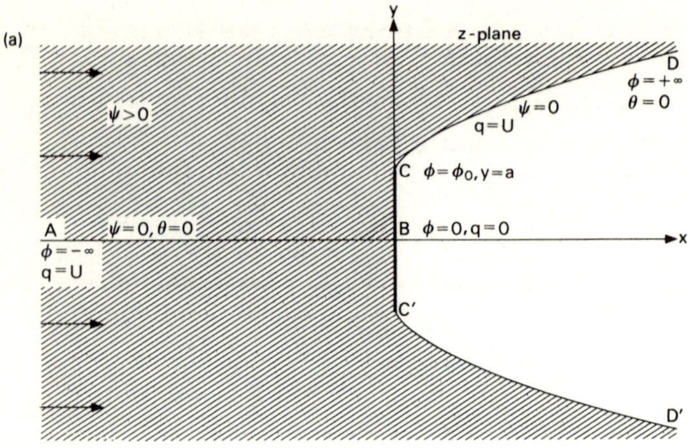

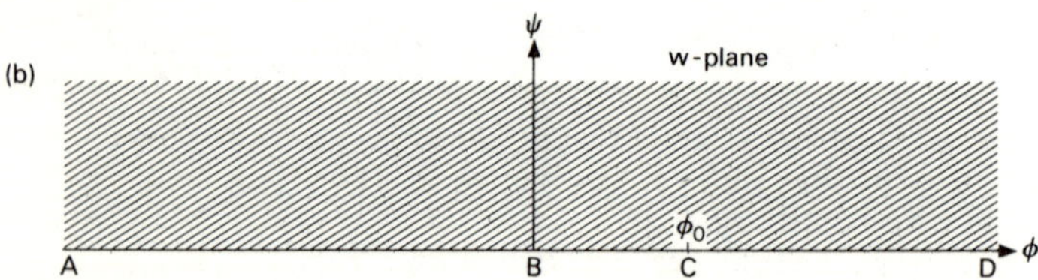

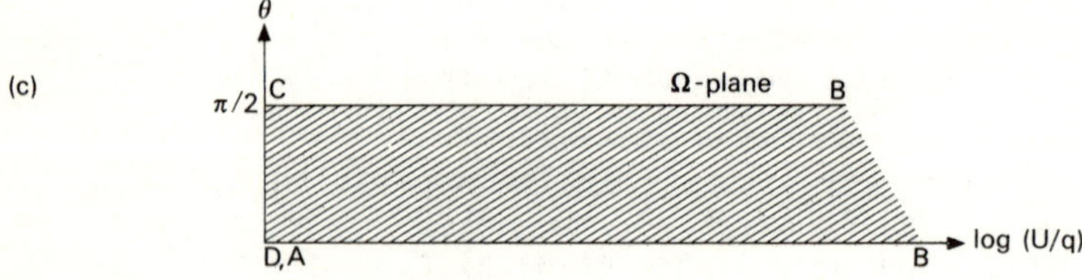

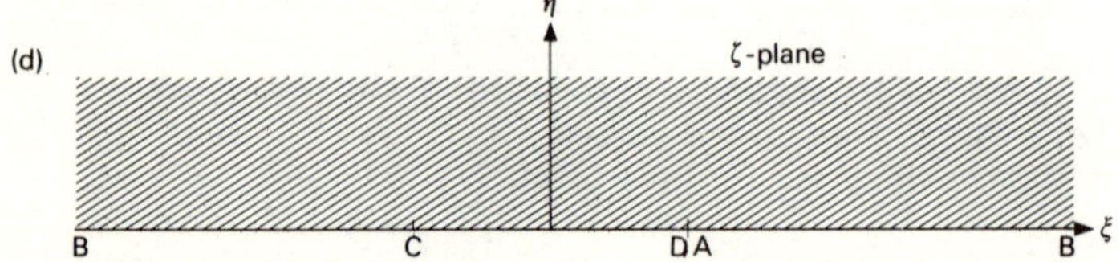

FIGURE 9.17

The velocity far upstream of the plate has components
(U,0) and the plate itself is given by x = 0,
$|y| \leq a$. From symmetry, the negative part of the
real axis is a streamline and we shall, for
simplicity,* take this to be $\psi = 0$; this part of
the streamline starts at A $(-\infty,0)$ and proceeds to
B $(0,0)$ where it splits into two halves; the upper
half is along the flat plate to C $(0,a)$ and then
along the free streamline CD. (The system is
symmetrical about the x-axis, and so we do not need
to define the other half of the streamline $\psi = 0$
in more detail.) The x-component of velocity is
everywhere non-negative, and so $\dfrac{\partial\psi}{\partial y} = \dfrac{\partial\phi}{\partial x} \geq 0$, with
equality only along CBC'. For simplicity we choose
the origin for ϕ to be at B, and we note that
$\dfrac{\partial\phi}{\partial s}$, the rate of change of ϕ along the streamline
ABCD, is always positive (except at B, where it is
zero). It follows that ϕ increases from $-\infty$ at A
to zero at B and then to $+\infty$ at D. We suppose
that at C (and from symmetry also at C') the value
of the velocity potential is ϕ_0; this is a positive

number which is to be determined. For every point in
the fluid which has y > 0, ψ is necessarily
positive but ϕ may have any value; for every point
which has y < 0, on the other hand, ψ is
necessarily negative but ϕ again may have any value.
There must, therefore, be a mapping from the region
in the z-plane containing fluid into the whole
w-plane, this plane having a cut (corresponding to
DCBC'D') along the positive part of the real axis.
It is our purpose to determine this mapping, and it
is convenient to use the symmetry of the problem and
to consider only the upper half plane.
    We now consider the variable Ω defined in
equation (195). First we consider the streamline
ABCD; along AB, we have $\theta = 0$ everywhere and
q decreasing from U at A to zero at B. So
AB is the real axis in the Ω-plane as shown in
Figure 9.17(c). On BC we have $\theta = \pi/2$ everywhere
and q increases from zero at B to U at C, so
that the real part of Ω decreases from infinity at
B to zero at C. On the free streamline CD the

---

* Both ϕ and ψ are defined only within an arbitrary
additive constant, and we may therefore take them to
be zero at an arbitrary point.

speed q is constant and equal to U, so we have
that the real part of Ω is zero everywhere and
θ decreases from $\pi/2$ at C to zero at D. (We
note that the points at infinity A,D in the z-plane
map into the same point in the Ω-plane: it is a
result of complex variable theory that, if a point
at infinity maps into a point in the finite part of
another plane, then all points at infinity in the
first plane must map into the same point in the
second plane.) It is easy to verify (by 'walking'
along the streamline ABCD) that the domain
containing the fluid in the z-plane corresponds to
the domain inside the semi-infinite strip in the
Ω-plane.

All we have to do now is to find a mapping between
the shaded domains of the w-plane and the Ω-plane,
as shown in Figure 9.17. This is comparatively easy,
because the boundaries of both domains are composed
of straight lines and there are theorems in the
theory of functions of a complex variable which
enable such mappings to be determined. We shall
proceed here without these theorems as we already
have experience of the semi-infinite strip and so
we can easily guess a suitable mapping. It is worth
noting at this stage that the rest of the work,
though long and somewhat tedious, is purely
'arithmetic': the fluid dynamics of the problem is
now complete.

It is usually convenient when trying to find the
mapping between the w-plane and the Ω-plane to map
each into an upper half auxiliary plane: with
experience it is often possible to map them both
into the same plane straight away, but here we shall
choose the simplest mapping we can think of and then
find a mapping between the two auxiliary planes.

In fact we do not need a mapping into an auxiliary
plane for the w-plane, since this is already an upper
half plane. For the Ω-plane, we note that we want
the boundary BCDAB to map into the real axis of the
auxiliary ζ-plane, and that this will happen if the

imaginary part of ζ is $\sinh 2\left[\log\left(\dfrac{U}{q}\right)\right] \sin 2\theta.$

This corresponds to the mapping $\zeta = \pm \cosh 2\Omega.$ It
is easy to verify that the shaded portion of the
Ω-plane maps into the upper half of the ζ-plane as
shown in Figure 9.17(d) if we take the positive sign,
so that

$$\zeta = \cosh 2\Omega \,.$$

In particular, at C we have $\zeta = -1$ and at D (and A) we have $\zeta = +1$.

We see by inspection* of the w-plane and the ζ-plane that

$$\zeta = \frac{w - 2\phi_0}{w} .$$

We therefore have

$$\cosh 2\Omega = \frac{w - 2\phi_0}{w} ,$$

and so

$$\sinh \Omega = \left[\tfrac{1}{2}(\cosh 2\Omega - 1) \right]^{\frac{1}{2}} = -\left(\frac{\phi_0}{w} \right)^{\frac{1}{2}} ,$$

$$\cosh \Omega = \left[\tfrac{1}{2}(\cosh 2\Omega + 1) \right]^{\frac{1}{2}} = \left(1 - \frac{\phi_0}{w} \right)^{\frac{1}{2}} ,$$

where both these quantities have positive (or zero) real and imaginary parts. It follows that

$$U\frac{dz}{dw} = e^{\Omega} = \cosh \Omega + \sinh \Omega = \left(1 - \frac{\phi_0}{w} \right)^{\frac{1}{2}} + \left(- \frac{\phi_0}{w} \right)^{\frac{1}{2}} .$$

Note that, when $|\phi| \to \infty$, the right-hand side of this expression approaches unity: it follows that $\frac{dw}{dz} \to U$ everywhere far enough from the plate (not just on the free streamline CD), and this confirms the hypothesis that far from the plate the velocity of the fluid is always U in the positive direction of the x-axis.

This relation between dz/dw and w can be integrated formally to give a mapping between the z-plane and the w-plane. Here we shall merely evaluate the constant ϕ_0 and determine the force on the plate. We note that on BC we have $x = 0$, $0 \le y \le a$ and $\psi = 0$, $0 \le \phi \le \phi_0$. Hence

---

* Note that, even if we cannot do this immediately by inspection, we can always use a mapping of the form $\zeta_2 = (a\zeta_1 + b)/(c\zeta_1 + d)$ and determine the constants by mapping three given points in one half plane to three given points in the other. The constants will all be real.

$$w = \phi(y), \quad \frac{dz}{dw} = -\frac{1}{iv} = i\frac{dy}{d\phi},$$

$$\left(-\frac{\phi_0}{w}\right)^{\frac{1}{2}} = i\left(\frac{\phi_0}{\phi}\right)^{\frac{1}{2}}, \quad \left(1 - \frac{\phi_0}{w}\right)^{\frac{1}{2}} = i\left(\frac{\phi_0}{\phi} - 1\right)^{\frac{1}{2}}.$$

Hence

$$U\frac{dy}{d\phi} = \left(\frac{\phi_0}{\phi}\right)^{\frac{1}{2}} + \left(\frac{\phi_0}{\phi} - 1\right)^{\frac{1}{2}},$$

where both square roots are real and positive. We integrate this expression (the second term is most easily integrated by first making the substitution $\phi = \phi_0 \cos\gamma$, $0 \le \gamma \le \pi/2$ and by then integrating with respect to γ), and obtain

$$Uy = 2(\phi_0\phi)^{\frac{1}{2}} + \phi_0 \sin^{-1}\left[(\phi/\phi_0)^{\frac{1}{2}}\right] + \left[\phi(\phi_0 - \phi)\right]^{\frac{1}{2}}.$$

In particular, at C we have $y = a$ and $\phi = \phi_0$, and it follows that

$$Ua = 2\phi_0 + \tfrac{1}{2}\pi\phi_0$$

and so

$$\phi_0 = \frac{2Ua}{\pi + 4}.$$

Using this value of ϕ_0, we can determine the mapping from the z-plane into the w-plane and also the shape of the free streamlines: it is found that they are asymptotic to a parabola.

To determine the force D per unit length on the plate we note that if p_∞ is the pressure behind the plate, and in particular on the free streamlines, then using Bernoulli's equation we have

$$p + \tfrac{1}{2}\rho q^2 = p_\infty + \tfrac{1}{2}\rho U^2.$$

Hence we have, from symmetry,

$$D = 2\int_{BC} (p - p_\infty)\,dy = \rho\int_{BC} (U^2 - q^2)\,dy$$

$$= \rho\int_0^a \left[U^2 - \left(\frac{d\phi}{dy}\right)^2\right]dy$$

since, on BC, $\phi = \phi(y)$ and $q = d\phi/dy$. We may write this in the form

$$D = \rho U \int_0^{\phi_0} \left(U \frac{dy}{d\phi} - \frac{1}{U} \frac{d\phi}{dy} \right) d\phi.$$

But, on $`BC$, we have seen that

$$U \frac{dy}{d\phi} = \left(\frac{\phi_0}{\phi} \right)^{\frac{1}{2}} + \left(\frac{\phi_0}{\phi} - 1 \right)^{\frac{1}{2}},$$

and so

$$\frac{1}{U} \frac{d\phi}{dy} = \left(\frac{\phi_0}{\phi} \right)^{\frac{1}{2}} - \left(\frac{\phi_0}{\phi} - 1 \right)^{\frac{1}{2}}.$$

Hence

$$D = 2\rho U \int_0^{\phi_0} \left(\frac{\phi_0}{\phi} - 1 \right)^{\frac{1}{2}} d\phi$$

$$= \pi \rho U \phi_0 = \frac{2\pi \rho U^2 a}{\pi + 4},$$

since $\phi_0 = 2Ua/(\pi + 4)$. It follows that the drag coefficient is

$$C_D = \frac{D}{\rho U^2 \times 2a} = \frac{2\pi}{\pi + 4} = 0.88.$$

This is to be compared with the experimental value of approximately 2. It is, therefore, somewhat low (as is to be expected because we know that the wake is unsteady and so, in a real fluid, there will also be a wave drag present), but this is a considerable improvement on the value of zero we obtained with the more primitive hypothesis.

EXAMPLE 9.7.2 The efflux of a fluid at a constant rate through an infinite slit of width $2a$ (see Figure 9.18(a)).
We take the axis of symmetry ED as the streamline $\psi = 0$ and the streamline ABC (consisting partly of the rigid wall AB and partly of the free stream-line BC) as $\psi = \psi_0$, where ψ_0 is a constant to be determined. We suppose that the fluid speed on the free streamline is U, that the width of the issuing jet is $2b$ for large values of x and we choose the origin for ϕ to be at B. It is easy

(a)
z-plane
φ=−∞
A ψ=ψ₀
θ=−π/2
B
φ=0
y=a
ψ=ψ₀, q=U
C y=b
φ=+∞
E
φ=−∞
D x
ψ=0, θ=0
C'
B'
A'

(b)
ψ
w-plane
ψ₀ B
A
C
E
D
φ

(c)
θ
Ω-plane
C,D
E
log (U/q)
−π/2
B
A

(d)
η₁
ζ₁-plane
B
C
−1
A,E
D
ξ₁

(e)
η₂
ζ₂-plane
D,C
B
E
−1
1
A
ξ₂

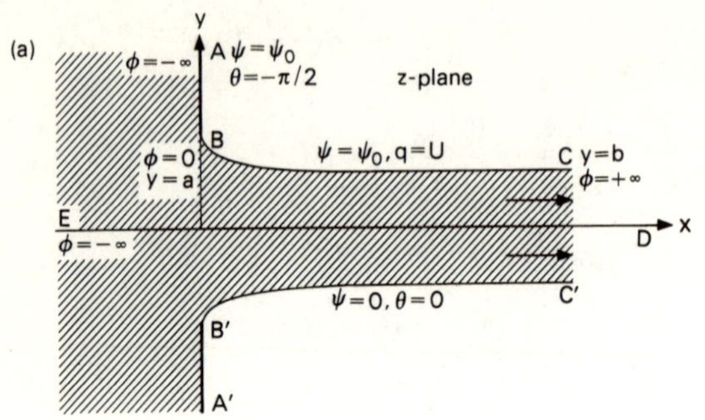

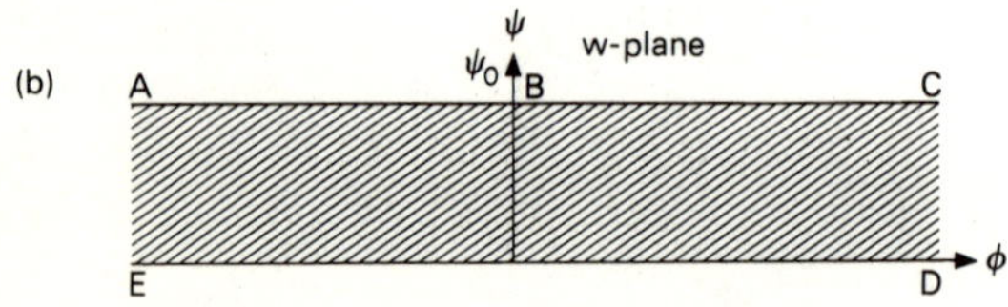

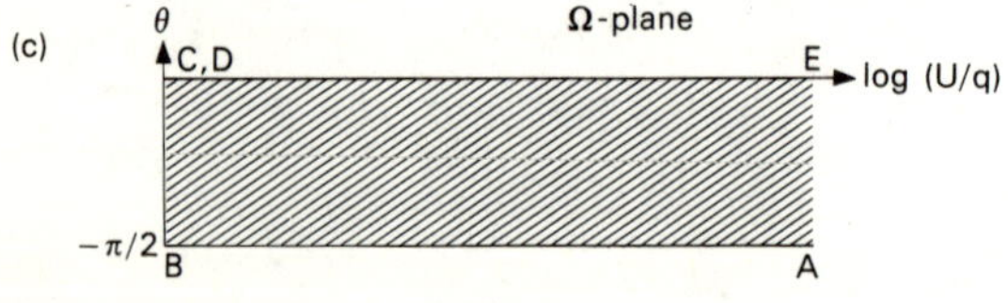

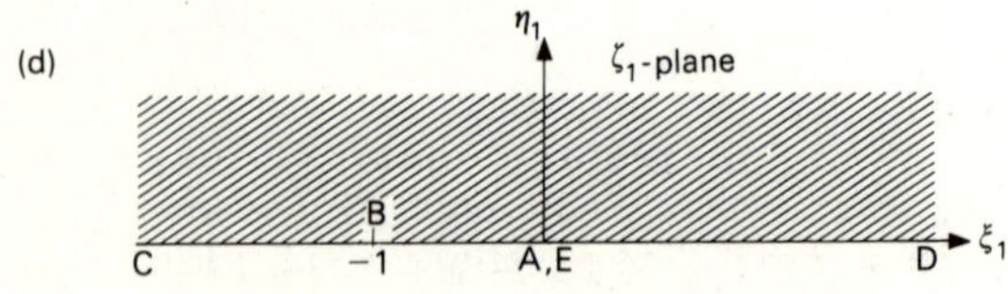

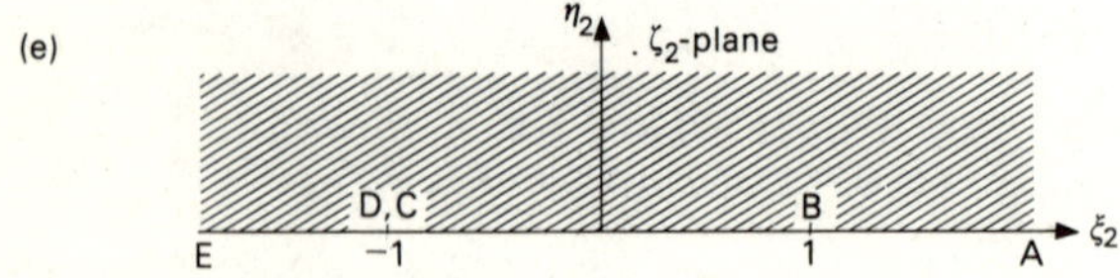

FIGURE 9.18

to verify that the upper half of the domain in the
z-plane which contains the fluid must map into an
infinite strip in the w-plane as shown in Figure
9.18(b), and into the semi-infinite strip in the
Ω-plane as shown in Figure 9.18(c). We emphasize
again that we now need only to find a mapping between
the w-plane and the Ω-plane and that this is mere
(though tedious) 'arithmetic': the fluid dynamics of
the problem is now complete. This time we need two
auxiliary half planes. We consider first the mapping

$$\zeta_1 = \exp\left(\frac{\pi w}{\psi_0}\right)$$

which maps the strip in the w-plane into the upper
half ζ_1-plane as shown in Figure 9.18(d). Next we
consider the mapping

$$\zeta_2 = - \cosh 2\Omega$$

which maps the strip in the Ω-plane into the upper
half ζ_2-plane as shown in Figure 9.18(e). By
inspection we see that

$$\zeta_2 = \frac{- 2 - \zeta_1}{\zeta_1}$$

and so

$$\cosh 2\Omega = 1 + 2 \exp (- \pi w/\psi_0).$$

Hence

$$\sinh \Omega = \left[\tfrac{1}{2}(\cosh 2\Omega - 1)\right]^{\frac{1}{2}} = \exp(- \tfrac{1}{2} \pi w/\psi_0),$$

where we have taken the root having positive real
part (since ψ ψ_0 everywhere) and

$$\cosh \Omega = \left[\tfrac{1}{2}(\cosh 2\Omega + 1)\right]^{\frac{1}{2}} = \left[1 + \exp(- \pi w/\psi_0)\right]^{\frac{1}{2}},$$

where we take the branch of the square root which
has positive (or zero) real part (and negative, or
zero, imaginary part). It follows that

$$U \frac{dz}{dw} = e^{\Omega} = \cosh \Omega + \sinh \Omega$$

$$= \exp(- \pi w/2\psi_0) + \left[1 + \exp(- \pi w/\psi_0)\right]^{\frac{1}{2}} .$$

We note immediately that, as $\phi \to \infty$, the right-hand side approaches unity and so $\frac{dw}{dz} \to U$ there; thus the velocity, as might be expected, is uniform across the issuing jet far enough from the orifice. Hence we can write $\psi_0 = Ub$ and so

$$U \frac{dz}{dw} = \exp\left(- \frac{\pi w}{2Ub}\right) + \left[1 + \exp\left(- \frac{\pi w}{Ub}\right)\right]^{\frac{1}{2}}.$$

As is usually the case in such problems this can be integrated formally to find the appropriate mapping between the z-plane and the w-plane. Here we shall merely integrate along the free streamline BC, where we have $w = \phi + iUb$ and so

$$U \frac{d(x + iy)}{d\phi} = - i \exp\left(- \frac{\pi \phi}{2Ub}\right) + \left[1 + \exp\left(- \frac{\pi \phi}{Ub}\right)\right]^{\frac{1}{2}}.$$

Equating the imaginary parts we find that, on BC,

$$U \frac{dy}{d\phi} = - \exp\left(- \frac{\pi \phi}{2Ub}\right).$$

Since $y = a$ when $\phi = 0$, this integrates to give

$$U(y - a) = \frac{2Ub}{\pi} \left[\exp\left(- \frac{\pi \phi}{2Ub}\right) - 1\right]$$

and, if we now let $y \to b$ as $\phi \to \infty$, we obtain

$$\pi(b - a) = - 2b,$$

so that

$$b = \frac{\pi a}{\pi + 2} = 0.61\, a.$$

If we suppose that the volume flux V through the slit is given, then we have

$$U = \frac{V}{2b} = \frac{V}{1.22\, a} = 0.82\, \frac{V}{a}.$$

To determine the shape of the free streamline we notice that, on BC,

$$U \frac{dx}{d\phi} = \left[1 + \exp\left(- \frac{\pi \phi}{Ub}\right)\right]^{\frac{1}{2}},$$

and that $x = 0$ when $\phi = 0$. We can integrate this (most easily by using the substitution

$$s = \left[1 + \exp\left(-\frac{\pi\phi}{Ub}\right)\right]^{\frac{1}{2}} \quad) \quad \text{to obtain}$$

$$x = \frac{\phi}{U} + \frac{2b}{\pi} \log\left\{1 + \left[1 + \exp\left(-\frac{\pi\phi}{Ub}\right)\right]^{\frac{1}{2}}\right\}$$

$$- \frac{2b}{\pi}\left[1 + \exp\left(-\frac{\pi\phi}{Ub}\right)\right]^{\frac{1}{2}} - \frac{2b}{\pi}\log(1 + \sqrt{2}) + \frac{2b}{\pi}\sqrt{2}.$$

We have already found that

$$y = a + \frac{2b}{\pi}\left[\exp\left(-\frac{\pi\phi}{2Ub}\right) - 1\right],$$

and so we have parametric equations for the curve BC.

EXERCISES

1 An incompressible fluid is contained between two fixed planes inclined at an angle π/n and is in irrotational motion due to the presence of a line source which is parallel to the line of intersection of the planes. Find the equations (in polar coordinates) for the streamlines of the flow.

2 An incompressible fluid is contained between two fixed planes inclined at an angle π/n and is in irrotational motion due to the presence of a line vortex which is parallel to the line of intersection of the planes. Find the equation (in polar coordinates) of the path of the vortex.

3 The two-dimensional irrotational flow of an incompressible fluid in a region with no rigid boundaries is described by the complex potential $w(z)$, and this function $w(z)$ is analytic at every point within the circle $|z| = a$. The flow is disturbed by the introduction of a circular cylinder whose boundary is $|z| = a$. Show that the complex potential of the resulting flow is $w(z) + \bar{w}(a^2/z)$.

4 By superposing a uniform velocity parallel to the generators of a suitable cylindrical aerofoil in a two-dimensional irrotational flow, show how to obtain the motion due to a swept-back aerofoil of given cross-

section.

5   Sketch the shape of the aerofoil generated by applying the conformal transformation

$$\zeta = z + b + \frac{(a - b)^2}{z + b}, \quad 0 < b < a,$$

to the circle $|z| = a$. If b/a is small, show that the aerofoil has the appropriate parametric equations

$$\xi = 2a \cos \theta + 2b(1 - \cos \theta - \cos^2\theta),$$

$$\eta = 2b \sin \theta \ (1 + \cos \theta),$$

where $\zeta = \xi + i\eta$.

6   An aerofoil is generated by applying the conformal transformation $\zeta = z + a^2/z$ to the circle whose centre is $z = b\,e^{i\pi/4}$, where $b \ll a$, and which passes through the point $z = -a$. Show that the aerofoil has the approximate parametric equations

$$\xi = 2a \cos \theta, \quad \eta = \sqrt{2}b \sin \theta \ (1 + \sin \theta + \cos \theta),$$

where $\zeta = \xi + i\eta$. Draw a sketch of the aerofoil.
   The aerofoil is placed in a uniform stream whose velocity is U in the negative direction of the ξ-axis. Find the value of the circulation necessary to avoid singularities in the flow.

7   An elliptic cylinder is inclined at an angle so that the major axis of its cross-section is inclined at an angle α to a uniform stream having velocity U in a direction normal to the generators of the cylinder. There is a circulation κ round the cylinder. Find the complex potential of the flow.

8   Cisotti's paradox is stated as follows:
   Consider a flat plate placed in a uniform stream at and angle α to the plate, the circulation round the plate being κ. Using Blasius's formula for the force on the plate, we find that the force on the plate is normal to the stream. On the other hand, the fluid is inviscid and so the pressure forces on the plate are normal to the plate: hence the resultant force is normal to the plate. This is in contradiction to the earlier result.
How can this paradox be resolved?

9 Find the parametric equations for the free streamlines in a flow due to the impact of a uniform stream
on an infinite flat plate of width 2a normal to the
stream.

10 Fluid escapes, under pressure, from a large vessel
through a two-dimensional 'tube' $x \geq 0, -c \leq y \leq c$
inserted into the vessel whose walls are assumed to be
very far from the origin (Borda's mouthpiece). The
flow is steady, two-dimensional and irrotational. If
$w(z)$ is the complex potential of the motion, find a
relation between w and dz/dw.
   Show that the width of the stream in the tube far
downstream of the origin is c .

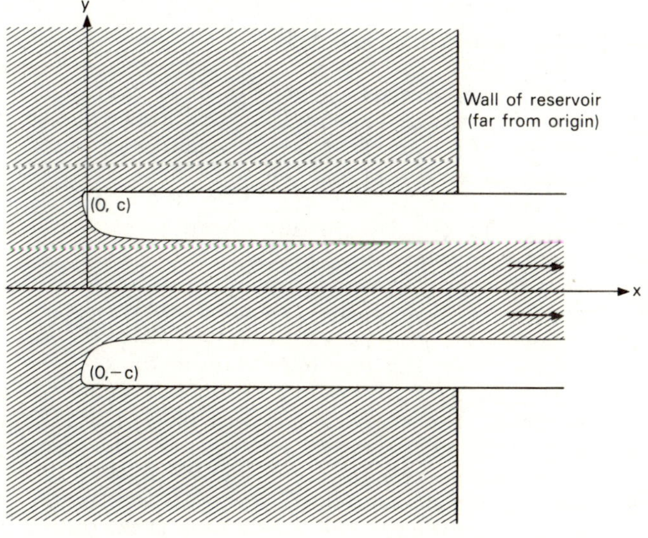

FIGURE 9.19

WAVES ON THE SURFACE OF A LIQUID

10.1 BOUNDARY CONDITIONS AT A FREE SURFACE

Most of the flows we have discussed so far have been
either of fluids having rigid boundaries, of fluids which
are in uniform motion far away from all boundaries, or of
fluids in steady two-dimensional motion with free streamlines. (The exception to this was the case of shallow
water waves discussed in Chapter 3, but there it was
assumed that the surface was essentially horizontal.)
The inviscid boundary conditions have been reasonably
easy to apply because we have either known exactly where
to apply them or been able to use complex variable theory
to solve the problem. For the unsteady motion of a
liquid having a free surface, however, we do not know
in advance the shape of the surface nor, in many cases,
can we apply complex variable theory.

We shall restrict ourselves to flows in which the
only external body force in the liquid is that due to
the gravitational attraction of the earth, so that in
equilibrium the free surface is a horizontal flat plane
which we shall take as the plane $z = 0$. We shall
assume further that the liquid is bounded below by the
rigid horizontal surface $z = -h$. Viscous forces will
be neglected and we shall assume that the flows are
irrotational. Finally we shall consider only cases in
which the free surface is in contact with a gas at
constant pressure Π.

At time t, the height of the free surface above its
equilibrium position at given values of x and y will
be written $z = \zeta(x,y,t)$, and we shall consider only
cases in which ζ is very small compared with the
horizontal scale of the motion and in which $\partial\zeta/\partial x$ and
$\partial\zeta/\partial y$ are very small compared with unity. We shall
apply two boundary conditions at this surface: the first

is that the normal component of velocity of the surface
as it changes shape is equal to the normal component of
velocity of the fluid at the surface,* and the second
concerns the normal component of force on the surface.
We shall now consider each of these in more detail.

For simplicity we shall restrict ourselves to the
case in which the shape of the surface is given by
$z = \zeta(x,t)$ and the component of the liquid velocity
parallel to Oy is zero. We shall suppose that, at
time t, the surface at the point P whose coordinates
are $(x,0,\zeta(x,t))$ makes an angle θ with the
horizontal; then the assumption that $\partial\zeta/\partial x$ is small
means that θ is small. If in addition $(x + \delta x, 0,$
$\zeta(x+\delta x,t+\tau))$ is a point Q which is on the surface
$z = \zeta(x,t+\tau)$ where τ is small, and PQ is normal

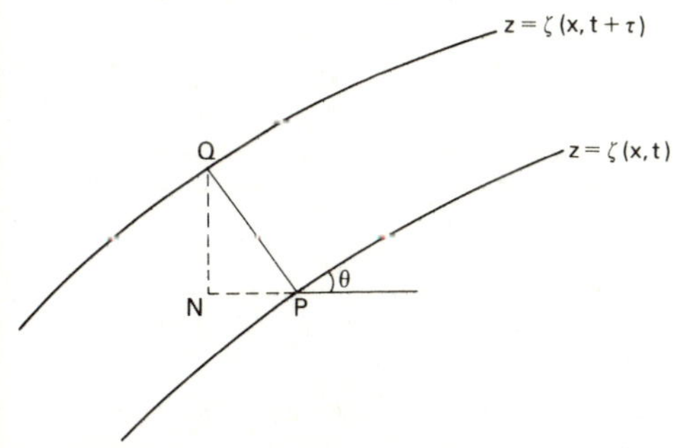

FIGURE 10.1

to the surface $z = \zeta(x,t)$ (that is, PQ makes an
angle θ with the vertical), then the normal component
of velocity of the surface is

$$\lim_{\tau\to 0}\left(\frac{PQ}{\tau}\right) = \lim_{\tau\to 0}\frac{\zeta(x+\delta x,t+\tau) - \zeta(x,t)}{\tau}.$$

Now, to the first order in δx and τ, we have

$$\zeta(x+\delta x,t+\tau) - \zeta(x,t) = \frac{\partial\zeta}{\partial x}\delta x + \frac{\partial\zeta}{\partial t}\tau$$

$$= \frac{\partial\zeta}{\partial x} PQ \sin\theta + \frac{\partial\zeta}{\partial t}\tau.$$

---

* This means that an element of fluid which is once in
  the surface remains in the surface.

Hence

$$\lim_{\tau \to 0}\left(\frac{PQ}{\tau}\right) = \frac{\partial \zeta}{\partial t} + \frac{\partial \zeta}{\partial x} \sin \theta \; \lim_{\tau \to 0}\left(\frac{PQ}{\tau}\right)$$

$$= \frac{\partial \zeta}{\partial t} \left(1 - \frac{\partial \zeta}{\partial x} \sin \theta\right)^{-1}.$$

But θ is small and $\partial \zeta / \partial x = \tan \theta$, so we have that
the required normal component of velocity of the surface
is $\frac{\partial \zeta}{\partial t}(1 + 0(\theta^2))$. Further, the component of the liquid
velocity in the same direction is

$$w \cos \theta - u \sin \theta = w(1 + 0(\theta)).$$

Hence, for sufficiently small θ, the first condition
on the surface is that

$$w = \frac{\partial \zeta}{\partial t} \quad \text{when} \quad z = \zeta. \tag{196}$$

We can simplify this even further because ζ is small,
and note that

$$w(x,\zeta) = w(x,0) + \zeta\left(\frac{\partial w}{\partial z}\right)_{z=0} + 0(\zeta^2),$$

using Taylor's theorem. So, to the first order in the
small quantity ζ, the condition (196) becomes

$$w = \frac{\partial \zeta}{\partial t} \quad \text{when} \quad z = 0, \tag{197}$$

and this is one boundary condition for the free surface
when ζ and its spatial derivatives are small. Although
we have limited our discussion to the case where everything is independent of y, it is easy to show that
equation (197) holds in the more general case when ζ
is a function of y as well as of x and t, and when
v is not zero.
    The other boundary condition at the free surface is,
in most of the cases we shall consider, that the pressure
in the liquid at the surface is equal to that in the gas
with which the liquid is in contact. This means that

$$p = \Pi \quad \text{when} \quad z = \zeta. \tag{198}$$

There are cases, however, in which we have to take
account of surface-tension effects: this happens if the

curvature of the surface is sufficiently large, and this case will be discussed in Section 10.3. There are also occasions when it is necessary to take into account the tangential component of the force on the surface. This, like the normal component p, must be continuous across the surface: however, in most cases (and in all those which we shall discuss here) the tangential components of force at the surface are negligible anyway. (We do not have the same problems at a free surface as we have at a rigid one where the relative velocity must be reduced to zero in a real fluid, thus necessitating a large velocity gradient.)

The boundary condition at the bottom of the liquid is, of course,

$$w = 0 \quad \text{when} \quad z = -h. \tag{199}$$

We shall start by considering the case of a liquid of infinite depth (that is, we let $h \to \infty$) and in this case the condition (199) becomes

$$w \to 0 \quad \text{when} \quad z \to -\infty. \tag{200}$$

The theory applies to any inviscid liquid, although it is conventional to label the types of wave which can occur as if the liquid were water.

10.2 UNI-DIRECTIONAL WAVES - GRAVITY WAVES IN DEEP WATER

Here we shall consider the case in which $\zeta = \zeta(x,t)$ and $v = 0$, so that the motion of the liquid is two-dimensional and waves are propagated parallel to the x-axis. Such motions can occur in a horizontal channel of uniform cross-section.*

First of all we find an alternative form for the boundary condition at the free surface. Since the motion is irrotational we can use the form of Bernoulli's equation given in equation (141), which is

$$\frac{\partial \phi}{\partial t} + \tfrac{1}{2} \underset{\sim}{u} \cdot \underset{\sim}{u} + \Omega + \frac{p}{\rho} = \text{constant},$$

and this is true throughout the liquid. In this case $\Omega = gz$ and, if we assume that the velocity is

---

* There will, of course, be boundary layers on the sides and bottom of the channel, but the effect of these is negligible.

everywhere so small that $\underset{\sim}{u}.\underset{\sim}{u}$ is negligible, we have

$$\frac{\partial \phi}{\partial t} + gz + \frac{p}{\rho} = \text{constant.}$$

We can add an arbitrary function of t to ϕ without affecting the motion, and we choose such a function to make the constant on the right-hand side equal to Π/ρ. Then we have that everywhere in the liquid

$$\frac{\partial \phi}{\partial t} + gz + \frac{p - \Pi}{\rho} = 0. \tag{201}$$

Since we are neglecting the effect of surface tension in this section, (198) and (201) combine to give

$$\frac{\partial \phi}{\partial t} + g\zeta = 0 \quad \text{when} \quad z = \zeta.$$

But

$$\left(\frac{\partial \phi}{\partial t}\right)_{z=\zeta} = \left(\frac{\partial \phi}{\partial t}\right)_{z=0} + \zeta \left(\frac{\partial^2 \phi}{\partial z \partial t}\right)_{z=0} + 0(\zeta^2),$$

and so, to the first order in ζ, we have

$$\left(\frac{\partial \phi}{\partial t}\right)_{z=0} + g\zeta = 0. \tag{202}$$

We can combine this condition with the condition (197) and obtain, since $w = \partial \phi/\partial z$,

$$g \frac{\partial \phi}{\partial z} + \frac{\partial^2 \phi}{\partial t^2} = 0 \quad \text{when} \quad z = 0. \tag{203}$$

We note again that this is valid for disturbances in which the surface displacement and its spatial derivatives are small.

Now the motion of the liquid is incompressible and irrotational; hence

$$\frac{\partial u}{\partial x} + \frac{\partial w}{\partial z} = 0 \, ,$$

where $u = \partial \phi/\partial x$ and $w = \partial \phi/\partial z$ everywhere. Hence

$$\frac{\partial^2 \phi}{\partial x^2} + \frac{\partial^2 \phi}{\partial z^2} = 0 \tag{204}$$

everywhere in the liquid.

We have, therefore, to look for a solution to equation (204) which satisfies the boundary conditions (200) and (203). Now the solution to equation (204) can be written in the form

$$\phi(x,z,t) = \mathcal{R}\left\{\sum_k \left[\phi_{0k}(t)e^{kz}e^{ikx} + \phi_{1k}(t)e^{-kz}e^{ikx}\right]\right\}. \qquad (205)$$

Since the motion is small everywhere (including very large values of $|x|$), each value of k must be real, and there is no loss of generality in taking each k to be non-negative.* We next apply the boundary condition (200), and it follows at once that $\phi_{1k}(t) = 0$ for each k. Finally we apply the boundary condition (203) and obtain

$$\sum_k \left[\frac{d^2\phi_{0k}}{dt^2} + gk\phi_{0k}\right]e^{ikx} = 0$$

for all x and t. This is possible only if

$$\frac{d^2\phi_{0k}}{dt^2} + gk\phi_{0k} = 0$$

and we know already that k is real and positive. (If $k = 0$, then ϕ_{0k} increases or decreases indefinitely with t, and this contradicts the assumption that the motion is small everywhere: so we must exclude this value of k from the summation.) It follows that

$$\phi_{0k}(t) = A_k e^{i\omega t} + B_k e^{-i\omega t},$$

where A_k, B_k are constants and $\omega = \sqrt{(gk)}$. So we have the required solution for ϕ which is

$$\phi(x,z,t) = \mathcal{R}\left\{\sum_{k>0} e^{kz}\left[A_k e^{i(kx+\omega t)} + B_k e^{i(kx-\omega t)}\right]\right\},$$

$$\omega = (gk)^{\frac{1}{2}}. \qquad (206)$$

---

* The spectrum of the constants k may be discrete (as is suggested by the form for ϕ given in equation (205)), or it may be continuous. In the second case, the summation in equation (205) must be replaced by an integral.

Thus the motion can be regarded as the sum of a number of
sinusoidal waves. The speed c of propagation of
sinusoidal waves whose wavelength λ is $2\pi/k$ is given
by

$$c = \frac{\omega}{k} = \left(\frac{g}{k}\right)^{\frac{1}{2}} = \left(\frac{g\lambda}{2\pi}\right)^{\frac{1}{2}}. \tag{207}$$

Since this speed depends on the wavelength, the different
waves making up the motion of equation (206) travel at
different speeds and the medium is said to be
dispersive. We shall discuss what happens to the
general disturbance in Section 10.5 .
   The velocity of the liquid has components

$$u = \frac{\partial\phi}{\partial x} = \mathcal{R}\left\{\sum_k ike^{kz}\left[A_k e^{i(kx+\omega t)} + B_k e^{i(kx-\omega t)}\right]\right\}$$

$$w = \frac{\partial\phi}{\partial z} = \mathcal{R}\left\{\sum_k ke^{kz}\left[A_k e^{i(kx+\omega t)} + B_k e^{i(kx-\omega t)}\right]\right\}.$$

In particular, when the disturbance is sinusoidal with
wavelength $\lambda = 2\pi/k$ and is travelling in the positive
direction of the x-axis (that is, $A_k = 0$) we have,
with a suitable origin of time so that B_k is real,

$$\phi = -\frac{U}{k} e^{kz} \cos(kx - \omega t),$$

where we have written $B_k = -U/k$. In this case

$$u = U e^{kz}\sin(kx - \omega t), \quad w = -U e^{kz}\cos(kx - \omega t)$$

and this is the velocity distribution discussed in
Example 2.2.5 (except that there the motion was parallel
to the x-y plane, whereas here it is parallel to the
x-z plane). We can therefore use the results obtained
there to see that the elements of the liquid move in
circles whose radii decrease exponentially with depth.

10.3 THE EFFECTS OF SURFACE TENSION AND FINITE DEPTH

When we include the effect of surface curvature we have
to modify the condition (198) at the free surface. It
is replaced, as shown in Appendix 6, by the condition

$$p = \Pi - \sigma \frac{\partial^2 \zeta}{\partial x^2} \quad \text{when } z = 0,$$

where σ is the surface tension. The value of σ for various interfaces is shown in Table 10.1. The only

TABLE 10.1 Surface tension σ, of various liquids in contact with air, with their own vapour, or with other liquids, at various temperatures.

Interface	T(°C)	$\sigma\,(kg/s^2)$
Acetone/vapour	20	0.0237
	50	0.0200
Benzene/air	20	0.0289
Carbon tetrachloride/vapour	20	0.0270
Mercury/vapour	20	0.472
Water/olive oil	20	0.020
Water/paraffin oil	20	0.048
Water/air	0	0.0757
	10	0.0742
	20	0.07275
	30	0.0712
	80	0.0626
	100	0.0588

effect of this on the calculation of the previous section is to replace equation (202) by

$$\left(\frac{\partial \phi}{\partial t}\right)_{z=0} + g\zeta - \frac{\sigma}{\rho}\left(\frac{\partial^2 \zeta}{\partial x^2}\right)_{z=0} = 0. \tag{208}$$

We can combine this with equation (197) to give

$$\frac{\partial^2 \phi}{\partial t^2} + g \frac{\partial \phi}{\partial z} - \frac{\sigma}{\rho}\frac{\partial^3 \phi}{\partial x^2 \partial t} = 0 \quad \text{when } z = 0, \tag{209}$$

and this replaces equation (203).

We now have to solve equation (204) subject to the conditions (208) and, when the depth is finite, (199) instead of (200). The solution to equation (204) is, as before, given in (205) and, since the motion is small everywhere, we must have k real, and there is no loss of generality in taking k to be non-negative. Using equation (199), we have

$$\phi_{0k}(t) \, e^{-kh} - \phi_{1k}(t) \, e^{kh} = 0.$$

So, if we write $\Phi_{0k}(t) = \tfrac{1}{2}\phi_{0k}(t) \, e^{-kh}$, we have $\phi_{1k}(t) = \tfrac{1}{2}\Phi_{0k}(t) \, e^{-kh}$ and equation (205) becomes

$$\phi(x,z,t) = \mathcal{R}\left\{\sum_k \Phi_{0k}(t) \cosh[k(z+h)]e^{ikx}\right\}.$$

Substituting this into equation (209), we have

$$\mathcal{R}\left\{\sum_{k>0} e^{ikx}\left[\frac{d^2\Phi_{0k}}{dt^2}\cosh kh + (gk \sinh kh \right.\right.$$
$$\left.\left. + \frac{\sigma k^3}{\rho}\sinh kh) \, \Phi_{0k}\right]\right\} = 0$$

for all x and t. This means that

$$\frac{d^2\Phi_{0k}}{dt^2} + \omega^2\Phi_{0k} = 0$$

where
$$\omega^2 = (gk + \frac{\sigma k^3}{\rho}) \tanh kh, \qquad\qquad (210)$$

and the solution for ϕ is then given by

$$\phi(x,z,t) = \mathcal{R}\left\{\sum_{k>0}\left[A_k e^{i(kx+\omega t)} + \right.\right.$$
$$\left.\left. + B_k e^{i(kx-\omega t)}\right]\cosh[k(z+h)]\right\}, \quad (211)$$

where ω is given by equation (210). Hence the speed c of propagation of the waves whose wave-length λ is $2\pi/k$ is given by

$$c = \frac{\omega}{k} = \left[\left(\frac{g}{k} + \frac{\sigma k}{\rho}\right)\tanh kh\right]^{\frac{1}{2}} = \left[\left(\frac{g\lambda}{2\pi} + \frac{2\pi\sigma}{\rho\lambda}\right)\tanh\left(\frac{2\pi h}{\lambda}\right)\right]^{\frac{1}{2}}. \quad (212)$$

We note that when kh is large, then $\tanh kh \simeq 1$, and the dependence of ω on k does not involve h : we have the case of a deep liquid. In fact $\tanh kh$ differs from unity by less than 0.4 per cent when

$h > \frac{1}{2}\lambda$. So, to all intents and purposes, we can treat the liquid as infinitely deep as long as the depth is greater than half the length of the waves we are considering.

Next we note that when λ^2 is very much greater than $4\pi^2\sigma/g\rho$, the term in the relation (212) containing σ is negligible.* When (as often happens) $h^2 >> 4\pi^2\sigma/g\rho$, we have three distinct types of waves.

(i) $\lambda << 2\pi(\sigma/g\rho)^{\frac{1}{2}} << h$: *capillary waves.*
In this case we can neglect the term containing g in equation (212) and also put $\tanh(2\pi h/\lambda) = 1$. Then

$$c = \left(\frac{k\sigma}{\rho}\right)^{\frac{1}{2}} = \left(\frac{2\pi\sigma}{\rho\lambda}\right)^{\frac{1}{2}}. \tag{213}$$

(ii) $2\pi(\sigma/g\rho)^{\frac{1}{2}} << \lambda << h$: *deep-water gravity waves.*
In this case we can neglect the term containing σ in equation (212) and also put $\tanh(2\pi h/\lambda) = 1$. Then

$$c = \left(\frac{g}{k}\right)^{\frac{1}{2}} = \left(\frac{g\lambda}{2\pi}\right)^{\frac{1}{2}}, \tag{214}$$

and this is the case that we discussed in the previous section.

(iii) $\lambda >> h$: *shallow water waves.*
Here we can neglect the term containing σ in equation (212) and also put $\tanh(2\pi h/\lambda) = 2\pi h/\lambda$. Then

$$c = (gh)^{\frac{1}{2}}, \tag{215}$$

and this is the case which we treated using quite different methods in Section 4.4.

It is of interest to plot on a graph the quantity c as a function of λ. This can be sketched readily as in Figure 10.2 using the expressions in equations (213), (214) and (215) for different ranges of λ and fairing in the curve joining up the pieces. We note that as long as $h >> 2\pi(\sigma/g\rho)^{\frac{1}{2}}$, c has a minimum value and that this occurs when $\lambda = 2\pi(\sigma/g\rho)^{\frac{1}{2}}$, where $\frac{dc}{d\lambda}$ as calculated from equation (212) vanishes. For a water-air surface the minimum value of c is 23 cm/s and this is appropriate to waves of length 1.7 cm.

---

* For a water-air surface, this means that λ must be much greater than 1.7 cm.

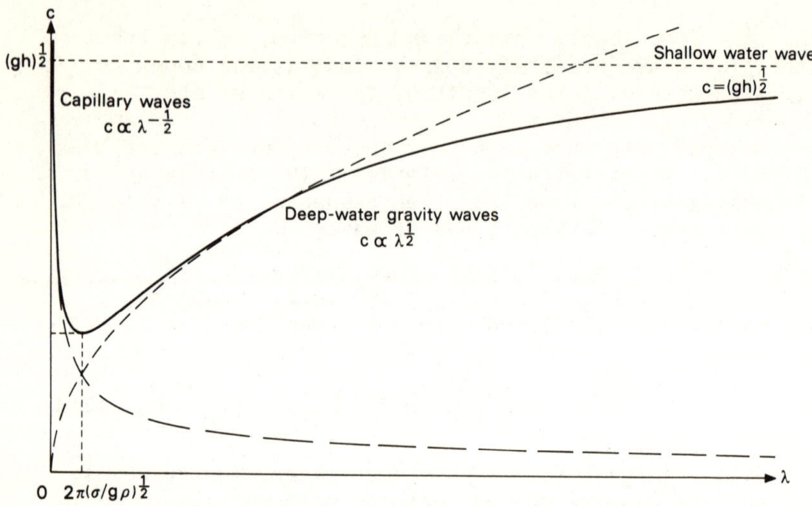

FIGURE 10.2

The relation (210) is called the *frequency equation* or the *dispersion relation* for the surface waves under discussion. The speed c is called the *wave velocity* or the *phase velocity*.

The path of an element of the liquid is, for capillary waves, as well as for deep-water gravity waves, a circle; the radius of such a circle decreases with the depth of the element in the liquid. It can be shown, by methods similar to those of Example 2.2.5, that when we have to take the depth of the liquid into account (that is, when $\lambda > h$) each element moves in an ellipse whose minor axis is vertical; the length of the minor axis decreases with the depth of the mean position of the liquid element. In the limit as $\lambda/h \to \infty$,   these ellipses degenerate into horizontal lines and we have the one-dimensional motion discussed in Section 4.4.

10.4 THE EFFECT OF FINITE AMPLITUDE

The form of the equation of the free surface can be derived from equations (202) and (206) or, in the more general case, from equation (208) and (211). It will clearly be of the form

$$\zeta = \mathcal{R}\left\{\sum_{k>0}\left[a_k e^{i(kx+\omega t)} + b_k e^{i(kx-\omega t)}\right]\right\}, \qquad (216)$$

and we have assumed throughout that the ka_k, kb_k are
all small compared with unity.

It is beyond the scope of this book to discuss in
detail what happens when this is not so. It is worth
while, however, noting that we have discussed a particular
case of a wave of finite amplitude in Sections 3.4 and 3.5:
the bore.

The case in which the amplitude, a, of the wave is
small, but not so small that a/λ can be neglected
compared with unity (as assumed in the treatment given
in this chapter) was discussed by Stokes; in general,
the shape of waves varies with time. There is a special
case in which a/λ is small, but not very small, when
the waves are propagated without change of shape; these
are *cnoidal waves*. Another special case is that of a
solitary wave, where there is a single disturbance
(that is, where there is no periodicity with respect
to the horizontal) propagated. All these are discussed
in some detail by Lamb in his classic book on
hydrodynamics.

Figure 10.3 shows diagrammatically the various regimes
of flow in terms of the quantities a/λ and λ/h which

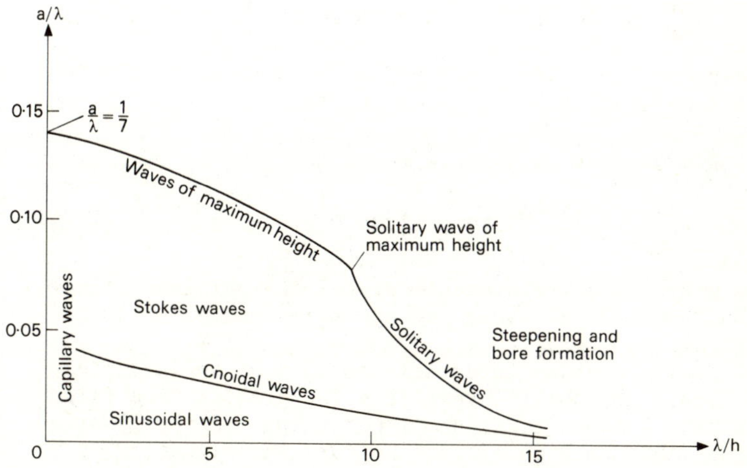

FIGURE 10.3

are the most suitable parameters. For sufficiently
large values of a/λ, laminar flow does not occur and
some turbulence is present (for example, white horses
at sea). The discussion in the earlier part of this

chapter was restricted to the regions near the axis.

10.5 WAVES BEHIND A TWO-DIMENSIONAL OBSTACLE: WAVE
RESISTANCE

It is well known to anyone who watches a river or stream
that, if an obstacle is placed in the flow, a system of
stationary surface waves is formed downstream of the
obstacle. The wavelength of these waves depends only
on the speed of the water and, possibly, on its depth.
(The latter observation, of course, requires a more
detailed study than the casual observer normally makes.)
As suggested in Section 8.7, such a system of waves
provides a mechanism for energy dissipation, and we
would expect that the presence of such waves would
involve a positive wave drag on the obstacle.
    We consider here a fixed, infinitely long obstacle
placed across the stream. If the stream is moving with
constant velocity U in the positive direction of the
x-axis, we could use a system of axes moving with the
liquid. If we do this, the coordinate x in the actual
flow (referred to axes attached to the obstacle) is
replaced by $x + Ut$:   that is, we would have waves
travelling in the negative direction of the x-axis with
speed U in a liquid with zero mean velocity. If
surface-tension effects are negligible, this corresponds
to waves whose length is $2\pi/k$ where

$$U^2 k = g \text{ for an infinitely deep liquid,} \qquad (217a)$$

$$U^2 k = g \tanh kh \text{ for a liquid of depth } h, \qquad (217b)$$

so we would expect the wavelength of the original steady
motion to be $2\pi/k$ where k satisfies one or other of
the equations (217).
    We shall consider in detail the case of an infinitely
deep liquid and shall investigate the problem from first
principles - that is, we shall not assume in the first
instance that equation (217a) holds. The situation is
represented schematically in Figure 10.4. The station
$x = x_1$ is far enough upstream of the obstacle for the

stream to be indistinguishable from a uniform stream
whose velocity is U in the direction shown. The
station $x = x_2$ is far enough downstream for the stream
to be indistinguishable from a similar uniform stream on

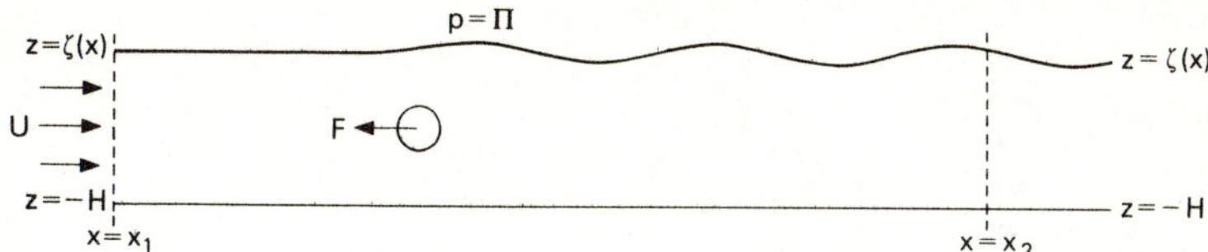

FIGURE 10.4

which is superposed a perturbation whose wavelength (in
the x-direction) is $2\pi/k$; it is supposed that this
perturbation is sinusoidal in x . We also choose a
horizontal surface z = - H which is so deep that the
flow is indistinguishable from a uniform stream. The
pressure at the surface is taken to have the constant
value Π and we suppose that F is the force on unit
length of the body due to the motion of the liquid, as
shown. The surface of the liquid is at $z = \zeta(x)$ and
we have $\zeta(x_1) = 0$.

In the neighbourhood of $x = x_1$ we therefore have
$\phi(x,z) = Ux$, $u = U$ and $w = 0$; and in the neighbourhood of $x = x_2$ we take $\phi(x,z) = Ux + Z(x)X(x)$.

Since $\nabla^2\phi = 0$, we have $X(x) \propto \sin kx$ for some k
(where we have chosen a suitable origin for x, so
that the $\cos kx$ term vanishes) and $Z(z) \propto e^{kz}$
since $\partial\phi/\partial z \to 0$ as $z \to -\infty$. We can therefore write
the downstream velocity potential in the form

$$\phi(x,z) = Ux - Ua\, e^{kz} \sin kx , \qquad (218)$$

where a is a constant of proportionality having the
dimensions of length. Hence in the neighbourhood of
$x = x_2$ we have

$$u = U(1 - ka\, e^{kz} \cos kx), \qquad w = - Uka\, e^{kz} \sin kx.$$

Now we consider the rate of mass flow in the domain $x_1 < x < x_2$, $-H < z < \zeta(x)$; since the motion is steady, we have

$$\int_{-H}^{0} (\rho u)_{x=x_1}\, dz + \int_{x_1}^{x_2} (\rho w)_{z=-H}\, dx$$

$$= \int_{-H}^{\zeta(x_2)} (\rho u)_{x=x_2}\, dz$$

$$= \rho U \int_{-H}^{\zeta(x_2)} (1 - ka\, e^{kz} \cos kx)\, dz;$$

that is,

$$\rho U H + I(H) = \rho U \left[z - a\, e^{kz} \cos dx \right]_{-H}^{\zeta(x_2)}$$

$$= \rho U\, (H + \zeta(x_2) - ae^{k\zeta(x_2)} \cos kx$$

$$+ ae^{-kH} \cos kx),$$

where the integrand in the integral $I(H)$ is either zero (upstream) or proportional to e^{-kH} (downstream). If we let $H \to \infty$, we obtain in the neighbourhood of $x = x_2$

$$\zeta = a\, e^{k\zeta} \cos kx. \tag{219}$$

Note that, for small values of ka, this approximates to $\zeta = a \cos kx$; but the result (219) applies whether ka is small or not. (It corresponds exactly to the condition that $w_{z=\zeta} = \dfrac{D\zeta}{Dt}$, which is automatically satisfied when (219) is true.)

Next we consider the momentum balance in the domain, noting that the force on the liquid due to the body is F in the opposite direction to that shown in Figure 10.4. We clearly have

$$\int_{-H}^{0} (p + \rho u^2)_{x=x_1}\, dz + \int_{x_1}^{x_2} (\rho uw)_{z=-H}\, dx + \int_{x_1}^{x_2} \Pi \frac{d\zeta}{dx}\, dx - F$$

$$= \int_{-H}^{\zeta(x_2)} (p + \rho u^2)_{x=x_2}\, dz. \tag{220}$$

Now we have, from Bernoulli's equation for steady irrotational motion, that

$$\frac{p}{\rho} + \tfrac{1}{2}(u^2 + w^2) + gz = \text{constant throughout the flow}$$

$$= \frac{\Pi}{\rho} + \tfrac{1}{2}U^2,$$

its value where $x = x_1$, $z = 0$. Considering separately the terms of equation (220), we see that

$$\int_{-H}^{0} (p + \rho u^2)_{x=x_1} \, dz = \int_{-H}^{0} (\Pi - g\rho z + \rho U^2) \, dz$$

$$= \Pi H + \tfrac{1}{2}g\rho H^2 + \rho U^2 H,$$

$$\int_{x_1}^{x_2} (\rho uw)_{z=-H} \, dx = J(H),$$

and this is of order e^{-kH} or less,

$$\int_{x_1}^{x_2} \Pi \frac{d\zeta}{dx} \, dx \quad - \quad \Pi\zeta(x_2) \quad \text{and}$$

$$\int_{-H}^{\zeta(x_2)} (p + \rho u^2)_{x=x_2} \, dz =$$

$$= \int_{-H}^{\zeta(x_2)} \left[\Pi - g\rho z + \tfrac{1}{2}\rho U^2 + \tfrac{1}{2}\rho(u^2 - w^2)\right]_{x=x_2} dz$$

$$= \Pi H + \Pi\zeta(x_2) + \tfrac{1}{2}g\rho H^2 - \tfrac{1}{2}g\rho\zeta^2(x_2) + \tfrac{1}{2}\rho U^2(\zeta(x_2) + H)$$

$$+ \rho U^2 \int_{-H}^{\zeta(x_2)} \left[\tfrac{1}{2} - kae^{kz}\cos kx \right.$$

$$\left. + \tfrac{1}{2}k^2 a^2 e^{2kz}(\cos^2 kx - \sin^2 dx)\right] dz$$

$$= \Pi H + \Pi\zeta(x_2) + \tfrac{1}{2}g\rho H^2 - \tfrac{1}{2}g\rho\zeta^2(x_2)$$

$$+ \rho U^2\zeta(x_2) + \rho U^2 H - \rho U^2 a(e^{k\zeta(x_2)} - e^{-kH})\cos kx_2$$

$$+ \tfrac{1}{4}\rho U^2 ka^2 (e^{2k\zeta(x_2)} - e^{-2kH})(2\cos^2 kx_2 - 1).$$

Hence the right-hand side of equation (220) is

$$\Pi H + \Pi\zeta(x_2) + \tfrac{1}{2}g\rho H^2 + \rho U^2 H$$

$$+ \rho U^2 a e^{-kH}\cos kx_2 - \tfrac{1}{2}\rho U^2 ka^2 e^{-2kH}\cos^2 kx_2$$

$$- \tfrac{1}{2}\rho(g - U^2 k)a^2 e^{2k\zeta(x_2)}\cos^2 kx_2$$

$$- \tfrac{1}{4}\rho U^2 ka^2(e^{2k\zeta(x_2)} - e^{-2kH}),$$

where we have used equation (219) which certainly holds when $x = x_2$.

Substituting for these integrals in equation (220), we obtain

$$F = J(H) + \tfrac{1}{2}\rho(g - U^2 k)a^2 e^{2k\zeta(x_2)}\cos^2 kx_2$$

$$+ \tfrac{1}{4}\rho U^2 ka^2 e^{2k\zeta(x_2)} + \text{other terms}$$

$$\text{containing } e^{-kH} \text{ or } e^{-2kH} \text{ as a factor.}$$

If we now let $H \to \infty$, we have that $J(H) \to 0$ and so do the other terms in e^{-kH} and e^{-2kH}. It follows that

$$F = \tfrac{1}{2}\rho(g - U^2 k)a^2 e^{2k\zeta(x_2)}\cos^2 kx_2 + \tfrac{1}{4}\rho U^2 ka^2 e^{2k\zeta(x_2)}$$

$$= \tfrac{1}{2}\rho(g - U^2 k_0)a^2\cos^2 k_0 x_2 + \tfrac{1}{4}\rho U^2 k_0 a^2$$

$$+ \text{terms of order } k_0^3 a^3,$$

where k_0 is the first approximation (for small a) to the value of k.

Now the exact position of x_2 was arbitrary (subject to the condition that it was far downstream of the obstacle), so the value of F cannot depend on the particular value of x_2 which is chosen. It follows that the coefficient of $\cos^2 k_0 x_2$ in the last expression must vanish and we have

$$k_0 = g/U^2 \tag{221}$$

(in fact, $k = g[1 + 0(ga/U^2)]/U^2$) and this is the value we expected for the wave number of the downstream perturbation. We therefore have for the force on the obstacle

$$F = \tfrac{1}{4}\rho U^2 k_0 a^2 = \tfrac{1}{4}\rho g a^2, \qquad (222)$$

to the smallest order of the small quantity $k_0 a = ga/U^2$.

The corresponding result for a liquid of finite depth h is slightly easier to obtain because we replace the surface $z = -H$ by the base $z = -h$ of the liquid, and we do not have to let $h \to \infty$. Also, of course, $w_{z=-h} = 0$ and so the integrals corresponding to I(H) and J(H) in the above argument are identically zero. We then find that $k = k_0[1 + 0(k_0 a^2)]$, where

$$k_0 \coth k_0 h = g/U^2, \qquad (223)$$

as we might have expected, and that, to the smallest order of the small quantity $k_0 a$,

$$F = \tfrac{1}{4}\rho g a^2 (1 - 2k_0 h \operatorname{cosech} 2k_0 h). \qquad (224)$$

It is left as an exercise to the reader to verify this result and also to verify that, in the limit as $k_0 h \to \infty$ (that is, for a liquid of infinite depth), the value of F given in equation (224) reduces to that of equation (222).

Note that, if $gh/U^2 < 1$, equation (223) shows that no waves can form downstream of the obstacle (at least as long as surface-tension effects are negligible), and that when gh/U^2 is unity or just over, $k_0 h$ must be small, the length of the waves very large and the wave resistance F will be very small. So we see that, in a channel of given depth, the force required to tow an obstacle is substantially reduced if the speed at which it is made to move is greater than $\sqrt{(gh)}$, since then gravity waves cannot form and dissipate energy. In any case we see from equation (224) that $0 \leq F \leq \tfrac{1}{4}\rho g a^2$.

In a real liquid, the waves will be damped out by viscous forces when they are far enough away from the obstacle. In practice it is found that capillary waves are damped out within a few centimetres whereas gravity waves whose wavelength is about a metre are not damped out for distances of the order of about 20 metres. This means that capillary waves are not of great importance in this kind of problem.

10.6 WAVE FRONTS OF GENERAL SHAPE

So far we have considered only uni-directional waves which, of necessity, have plane wave fronts normal to the direction of propagation. For a sinusoidal wave, whose wave number is k and whose frequency is ω, the equation of a given wave front is $kx + \omega t = N$, a constant, and this travels with phase velocity $\underset{\sim}{c}$ whose magnitude is $|\omega|/|k|$ and whose direction is the negative direction of the x-axis (if k and ω are of the same sign).

    We now suppose that the plane sinusoidal wave is not normal to the x-axis, but is normal to the direction defined by a horizontal vector $\underset{\sim}{k}$, where $|\underset{\sim}{k}|$ is the wave number of the flow. We choose new axes OX, OY

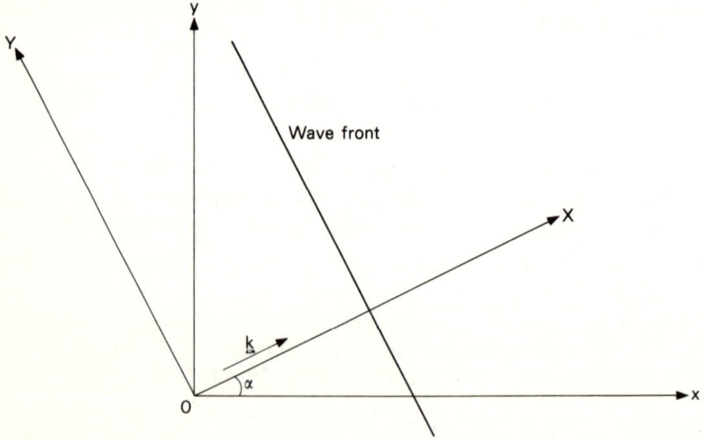

FIGURE 10.5

so that OX is in the direction of $\underset{\sim}{k}$ and suppose that the angle between Ox and OX is α, as shown in Figure 10.5. Then the equation of the wave front referred to the new axes is clearly

$$|\underset{\sim}{k}|X + \omega t = N .$$

But $X = x \cos \alpha + y \sin \alpha$, and so

$$|\underset{\sim}{k}|X = x|\underset{\sim}{k}|\cos \alpha + y|\underset{\sim}{k}|\sin \alpha$$

$$= \underset{\sim}{k}.\underset{\sim}{r} ,$$

where $\underset{\sim}{r}$ is the horizontal vector (x,y). Hence, referred to the original axes, the equation of a plane wave front normal to the direction $\underset{\sim}{k}$ is $\underset{\sim}{k}.\underset{\sim}{r} + \omega t = N$ and the phase velocity is $\underset{\sim}{c}$ whose magnitude is $|\omega|/|\underset{\sim}{k}|$ and whose direction is in the negative direction of the X-axis when ω is positive. So we may write

$$\underset{\sim}{c} = - \frac{\omega}{|\underset{\sim}{k}|}\,\underset{\sim}{k} = - \frac{\omega}{|\underset{\sim}{k}|^2}\,\underset{\sim}{k}.$$

The vector $\underset{\sim}{k}$ is called the *wave number vector* for the wave.

More generally we consider the situation in which the equation of a given wave front is

$$\chi(\underset{\sim}{r},t) = N, \tag{225}$$

where $\underset{\sim}{r} = (x,y)$ and N is a constant.* We define the quantities

$$\underset{\sim}{k} = \text{grad } \chi, \quad \omega = \frac{\partial \chi}{\partial t} \tag{226}$$

and note that, in the neighbourhood of the point $\underset{\sim}{r}_0$ near the time t_0, where $\chi(\underset{\sim}{r}_0,t_0) = N$, we can write

$$\chi(\underset{\sim}{r},t) = \chi(\underset{\sim}{r}_0,t_0) + \underset{\sim}{k}.(\underset{\sim}{r} - \underset{\sim}{r}_0) + \omega(t - t_0) \tag{227}$$

to the first order in the small quantities $|\underset{\sim}{r} - \underset{\sim}{r}_0|$, $|t - t_0|$, where $\underset{\sim}{k}_0$, ω_0 are evaluated at $\underset{\sim}{r}_0$, t_0. Now $\underset{\sim}{r},t$ and $\underset{\sim}{r}_0,t_0$ both satisfy equation (225) and so $\chi(\underset{\sim}{r},t) = \chi(\underset{\sim}{r}_0,t_0)$ and the equation reduces to

$$\underset{\sim}{k}.\underset{\sim}{r} + \omega t = \underset{\sim}{k}.\underset{\sim}{r}_0 + \omega t_0 = \text{constant.} \tag{228}$$

It follows that $\underset{\sim}{k}$ and ω correspond locally to the wave-number vector and the frequency respectively as we have defined them for plane waves, and we can regard the tangent to the wave front at $\underset{\sim}{r}_0$, t_0 as a plane wave front given by equation (228).

---

* For waves on the surface of a liquid, $\underset{\sim}{r}$ is a horizontal coordinate, as given here, but the subsequent theory can be applied directly to the case in which $\underset{\sim}{r}$ is a three-dimensional space vector.

It is not immediately apparent, however, that the point $\underset{\sim}{r}_0$ at time t_0 of the actual wave front (225) will move in exactly the same way as the same point of the plane wave front (228). To show that it does we

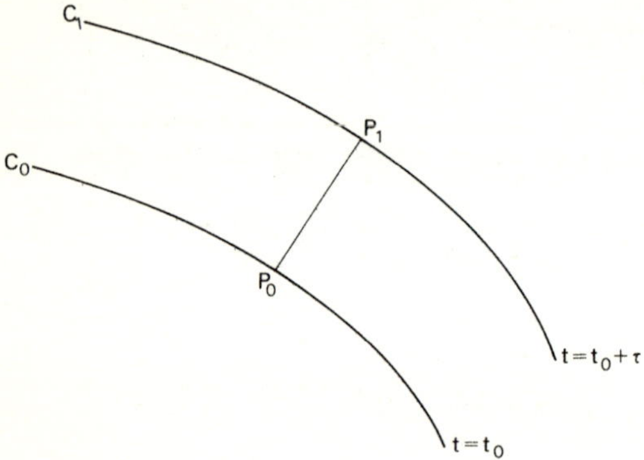

FIGURE 10.6

sketch, as in Figure 10.6, the curves

$$C_0: \chi(\underset{\sim}{r},t_0) = N, \quad C_1: \chi(\underset{\sim}{r},t_0+\tau) = N.$$

Then if C_0 is the position of a wave front at time t, it has moved to the position C_1 at time $t_0 + \tau$. We suppose that P_0 is the point $\underset{\sim}{r}_0$ on C_0, so that $\chi(\underset{\sim}{r}_0,t_0) = N$ and that P_1 is the point $\underset{\sim}{r}_1$ on C_1, so that $\chi(\underset{\sim}{r}_1,t_0+\tau) = N$, and where P_0P_1 is normal to the curve C_0. It follows that

$$\underset{\sim}{r}_1 - \underset{\sim}{r}_0 = \overrightarrow{P_0P_1} = \sigma\underset{\sim}{k}_0,$$

where σ is a constant of proportionality and where $\underset{\sim}{k}_0$ is the value of $\underset{\sim}{k}$ evaluated at $\underset{\sim}{r}_0$. It follows that

$$N = \chi(\underset{\sim}{r}_1,t_0+\tau) = \chi(\underset{\sim}{r}_0 + \sigma\underset{\sim}{k}_0,t_0+\tau)$$

$$= \chi(\underset{\sim}{r}_0,t_0) + \sigma\underset{\sim}{k}_0\cdot(\text{grad }\chi)_0 + \tau\left(\frac{\partial\chi}{\partial t}\right)_0$$

to the first order in σ,

$$= N + \sigma|\underset{\sim}{k}_0|^2 + \tau\omega_0,$$

using the definitions (226) for $\underset{\sim}{k}$ and ω. Hence

$$\sigma \, |\underset{\sim}{k}_0|^2 + \tau\omega_0 = 0$$

to the first order in σ and τ. But the local phase velocity of the wave front at P_0 is

$$\underset{\sim}{c} = \lim_{\tau \to 0} \frac{\overrightarrow{P_0 P_1}}{\tau} = \lim_{\tau \to 0} \frac{\sigma \underset{\sim}{k}_0}{\tau} = - \frac{\omega_0}{|\underset{\sim}{k}_0|} \hat{\underset{\sim}{k}}_0 \, .$$

This is precisely the phase velocity of the plane wave given by equation (228).

It is usually convenient to think of a general disturbance as a superposition of plane waves,* and we generalize equations (206), (211) to the form

$$\phi(r,z,t) = \mathcal{R}\left\{ \sum_n \phi_n(z) e^{i(\underset{\sim}{k}_n \cdot \underset{\sim}{r} - \omega_n t)} \right\},$$

where we have dropped the terms with a plus sign in the exponent, since ω_n may be positive or negative. Since $\nabla^2 \phi = 0$, we must have

$$\phi_n''(z) = |\underset{\sim}{k}_n|^2 \, \phi_n(z),$$

so

$$\phi_n(z) \propto e^{|\underset{\sim}{k}_n| z} \quad \text{for a liquid of infinite depth,}$$

and

$$\phi_n(z) \propto \cosh[|\underset{\sim}{k}_n|(z + h)] \quad \text{for a liquid of depth } h.$$

The surface will then be given by $z = \zeta(r,t)$, and we have the boundary conditions

$$\left(\frac{\partial \phi}{\partial z}\right)_{z=\zeta} = \frac{D\zeta}{Dt}$$

and

$$g\zeta + \left(\frac{\partial \phi}{\partial t}\right)_{z=\zeta} + \tfrac{1}{2}\left[\left(\frac{\partial \phi}{\partial x}\right)^2 + \left(\frac{\partial \phi}{\partial y}\right)^2 + \left(\frac{\partial \phi}{\partial z}\right)^2\right]_{z=\zeta} = 0.$$

---

* This merely means that we are splitting it up into Fourier components.

(These may be compared with equation (196) and that leading to equation (202).) In their linearized form, these become

$$\left(\frac{\partial \phi}{\partial z}\right)_{z=0} = \frac{\partial \zeta}{\partial t}, \quad g\zeta + \left(\frac{\partial \phi}{\partial t}\right)_{z=0} = 0,$$

and so equation (203) still holds and we have

$$g\,\frac{\partial \phi}{\partial z} + \frac{\partial^2 \phi}{\partial t^2} = 0 \quad \text{when } z = 0.$$

It follows that $g\left|\underset{\sim}{k}_n\right| = \omega_n^2$ for waves in deep water.

10.7 WAVES BEHIND A SMALL OBSTACLE IN A UNIFORM STREAM: BOW WAVES

We idealize the situation and suppose that the obstacle is concentrated at the origin and that far upstream of the obstacle the flow has uniform velocity U in the positive direction of the x-axis and that waves form behind the obstacle. This is an idealized description of the disturbance behind a ship travelling with speed U in the negative direction of the x-axis, as observed from the ship itself (see Plate 4). We consider first the case of a liquid of infinite depth.

In the region where waves are present we write

$$\phi = Ux + \phi'(\underset{\sim}{r}, z, t) \tag{229}$$

where

$$\phi'(\underset{\sim}{r}, z, t) = \mathcal{R}\left\{\sum_n A_n e^{\left|\underset{\sim}{k}_n\right| z} e^{i(\underset{\sim}{k}_n \cdot \underset{\sim}{r} - \omega_n t)}\right\}, \tag{230}$$

where we expect the ω_n to be positive, since we expect the waves to travel in the positive direction of the x-axis. Although we expect the motion to be steady, we have included the possibility of unsteady flow for reasons which not be apparent until the next chapter. We use the boundary conditions

$$\left(\frac{\partial \phi}{\partial z}\right)_{z=\zeta} = \frac{D\zeta}{Dt}, \quad g\zeta + \left(\frac{\partial \phi}{\partial t}\right)_{z=\zeta} + \tfrac{1}{2}(\text{grad }\phi . \text{grad }\phi)_{z=\zeta} = \tfrac{1}{2}U^2$$

in their linearized form

$$\left(\frac{\partial\phi'}{\partial z}\right)_{z=0} = \frac{\partial\zeta}{\partial t} + U\frac{\partial\zeta}{\partial x}, \quad g\zeta + \left(\frac{\partial\phi'}{\partial t}\right)_{z=0} + U\left(\frac{\partial\phi'}{\partial x}\right)_{z=0} = 0.$$

These may be combined to give

$$g\frac{\partial\phi'}{\partial z} + \left(\frac{\partial}{\partial t} + U\frac{\partial}{\partial x}\right)^2\phi' = 0 \quad \text{when} \quad z = 0, \qquad (231)$$

which may be compared with equation (203) in which U was zero. Substituting from equation (230) into equation (231), we obtain

$$g|\underset{\sim}{k}_n| = (\omega_n - U\lambda_n)^2, \qquad (232)$$

where $\underset{\sim}{k}_n = (\lambda_n, \mu_n)$.

It is clear that, for a given value of U, the steady wave motion $(\omega = 0)$ behind the disturbance no longer has a unique value of $\underset{\sim}{k}$. The determination of the way in which these waves are propagated will be left to the next chapter.

For a liquid having finite depth h, the expression for the perturbation velocity potential ϕ' is no longer given by equation (230), but instead by

$$\phi'(\underset{\sim}{r},z,t) = \mathcal{R}\left\{\sum_n A_n \cosh[|\underset{\sim}{k}_n|(z+h)]e^{i(\underset{\sim}{k}_n\cdot\underset{\sim}{r} - \omega_n t)}\right\},$$

$$(233)$$

since $\dfrac{\partial\phi'}{\partial z} = 0$ when $z = -h$. The surface boundary condition is

$$g|\underset{\sim}{k}_n|\tanh(|\underset{\sim}{k}_n|h) = (\omega_n - U\lambda_n)^2 \qquad (234)$$

instead of equation (232).

EXERCISES

1 Small amplitude irrotational plane waves of length λ are propagated in water of infinite depth. Show that, to the first order of small quantities, the pressure experienced by an element of water is constant and equal to its value in the equilibrium position.

2   Sinusoidal waves of infinitesimal amplitude occur in a liquid of depth h, the velocity potential of the flow being given by

$$\phi = \phi_0 \cosh[k(z + h)]\cos(kx - \omega t).$$

Show that the 'paths of the particles' are ellipses which flatten to a straight line when $z = -h$.

   Discuss the limiting forms of these curves (a) when $kh \ll 1$, and (b) when $kh \gg 1$.

3   Liquid flows irrotationally between the rigid horizontal plane $z = -h$ and the free surface $z = \zeta(x)$ which is given approximately by $\zeta = a \cos kx$ with $ka \ll 1$. The motion is steady and its mean velocity is U in the positive direction of the x-axis. Show that the velocity potential of the flow is

$$\phi = Ux - aU \operatorname{cosech}(kh)\sin(kx)\cosh[k(z + h)];$$

and that this steady motion can exist only if $kU^2 = g \tanh(kh)$.

   If these waves are caused by the presence of a cylinder whose generators are parallel to the y-axis and which is in the neighbourhood of the origin, show that the force per unit length required to hold the cylinder in position is $\frac{1}{4} g\rho a^2(1 - 2kh \operatorname{cosech} 2kh)$.

4   The bed of a water channel of uniform width is of the form $z = a \sin kx$ where x is measured along the channel, z is measured vertically upwards and a, k are given constants such that $ka \ll 1$. Water whose mean depth is h $(h \gg a)$ flows steadily along the channel with mean speed U. Show that the equation of the free surface, when $kU^2 \neq g \tanh kh$, is

$$z = h + \frac{U^2 ka \sin kx}{kU^2\cosh kh - g \sinh kh}.$$

What happens when $kU^2 = g \tanh kh$?

5   The free surface of a liquid of unbounded depth has the form $z = \zeta(x)$ where $\zeta = ae^{k\zeta}\cos kx$. The liquid is in steady irrotational motion and, for $z \ll 0$, its velocity is U in the positive direction of the x-axis. Show that the quantity

$$- g\zeta + U^2 kae^{k\zeta}\cos kx - \tfrac{1}{2} U^2 k^2 a^2 e^{2k\zeta}$$

is constant if such a motion can be sustained with uniform pressure on the free surface.

Show that, when powers of ka greater than the second can be neglected compared with unity, $kU^2 = g(1 + k^2a^2)$ and

$$\zeta = \tfrac{1}{2}ka^2 + a(1 + \tfrac{9}{8}k^2a^2)\cos kx + \tfrac{1}{2}ka^2\cos 2kx +$$

$$+ \frac{3}{8}k^2a^3\cos 3kx.$$

6 A liquid of density ρ_2 and unbounded height flows horizontally with velocity V_2 parallel to the x-axis over a liquid of density ρ_1 $(\rho_1 > \rho_2)$ and unbounded depth which flows with constant velocity V_1 in the same direction. Show that a stationary wave of small amplitude can exist such that the equation of the interface is $z = a \cos kx$ as long as

$$k(\rho_1 V_1^2 + \rho_2 V_2^2) = g(\rho_1 - \rho_2).$$

Show that a travelling wave with the same wavelength and wave velocity c can also exist, where

$$c = 2(\rho_1 V_1 + \rho_2 V_2)/(\rho_1 + \rho_2).$$

7 A layer of liquid of uniform thickness h is at rest on a horizontal plane rigid boundary and air blows over it with uniform velocity U in a horizontal direction. The density of the liquid is ρ_1, that of the air is ρ_2 and σ is the surface tension at the interface.

Show that stable waves of small amplitude whose wavelength is $2\pi/k$ cannot occur if

$$\rho_1\rho_2 U^2 > (\rho_1 + \rho_2\tanh kh)[\sigma k + (\rho_1 - \rho_2)g/k].$$

8 An incompressible fluid is in equilibrium under the action of gravity, its density ρ_0 and its pressure p_0 being known functions of height. Small disturbances are such that the velocity is $\underset{\sim}{u}$, the density is $\rho_0 + \rho'$ and the pressure is $p_0 + p'$; the incompressibility of the fluid ensures that the rate of change of the density following the motion is zero. Show that,

to the first order of small quantities,

$$\frac{\partial \rho'}{\partial t} + \underset{\sim}{u}.\text{grad } \rho_0 = 0,$$

$$\text{div } \underset{\sim}{u} = 0,$$

$$\rho_0 \frac{\partial \underset{\sim}{u}}{\partial t} = g\rho' - \text{grad } p'.$$

If the disturbance consists of plane waves whose frequency is ω and whose vector wave-number is $\underset{\sim}{k}$, where $|\underset{\sim}{k}| \gg -\frac{1}{\rho_0}\frac{d\rho_0}{dz}$ and $\underset{\sim}{k}$ makes an angle α with the vertical, show that

$$\omega^2 = -\frac{g}{\rho_0}\frac{d\rho_0}{dz}\sin^2\alpha.$$

Chapter 11

GROUP VELOCITY: ENERGY PROPAGATION

11.1 DISPERSION OF PLANE WAVES

We consider a general disturbance which may be expressed in the form*

$$\phi(x,z,t) = \mathcal{R}\left\{ \sum_k \phi_k(z) \, e^{i(kx - \omega t)} \right\}, \qquad (235)$$

where

$$\omega = \omega(k) \qquad (236)$$

is a given function which we call the *frequency equation* (or the *dispersion relation*). The function ϕ in equation (235) may be a velocity potential, and we shall consider only problems in which this is the case; the ensuing theory, however, is appropriate for any quantity which may be expressed in the form (235).

At any given instant, t_0, equation (235) serves to express the function ϕ as a Fourier series (or integral). When $\omega(k) = ck$ and c is a constant (that is, when the phase velocity is independent of wave number, as for the sound waves of Section 4.2 and for the shallow water waves of Section 4.4), it is easy to see that the value of ϕ is unchanged at time t_1 as long as we replace x by $x + c(t_1 - t_0)$. It follows that the disturbance is propagated without change of form and the system is said to be *non-dispersive*. If, on the other hand, the phase velocity $c = \omega/k$ does depend on k, this is no longer true. The separate

---

* In some cases the sum may be replaced by an integral with respect to k.

sinusoidal components of the right-hand side of equation (235) travel with different speeds and, when they are added up, no longer give the same value for ϕ : the system is *dispersive*. (It is implicitly assumed here that the sign of the ratio ω/k is the same for all k in the summation of equation (235), so that the disturbance is a travelling wave.)

We are commonly interested in the situation in which at some instant $(t = 0,$ say) the value of ϕ is negligible for all values of x except in some small finite range. For example, we may suppose that the disturbance is caused by a stone dropped into a large lake or by a ship on the ocean. It turns out that in such a case the values of the coefficients ϕ_k are small except for values of k near some particular value k_0 (usually $2\pi/k_0$ is approximately the horizontal extent of the disturbance). An example of this is given in question 1 at the end of the chapter. When $|k - k_0|$ is small, we can write equation (236) in the form

$$\omega = \omega(k_0) + (k - k_0) \frac{d\omega}{dk}\bigg|_0 \tag{237}$$

to the first order in the small quantity $|k - k_0|$. We can therefore write equation (235) in the form

$$\phi(x,z,t) = \mathcal{R}\left\{A\ e^{i(k_0 x - \omega_0 t)}\right\},$$

where $\omega_0 = \omega(k_0)$ and

$$A = \sum_k \phi_k\ e^{i[(k - k_0)x - (\omega - \omega_0)t]}$$

can be regarded as the slowly varying amplitude of a plane sinusoidal wave with wave number k_0. We can use the equation (237) in this expression and obtain

$$A = \sum_k \phi_k\ e^{i(k - k_0)[x - (d\omega/dk)_0 t]} \tag{238}$$

to a first approximation. So A itself can be regarded as the superposition of a number of plane waves each having phase velocity $v = (d\omega/dk)_0$.

Physically, the interpretation of all this is that we have a group of waves (whose wavelengths differ little from $2\pi/k_0$)   forming a disturbance of finite extent. The maximum amplitude of this disturbance is given by the expression for A in equation (238), and this travels with the constant velocity whose magnitude is

$$v = \frac{d\omega}{dk} \tag{239}$$

evaluated at the appropriate value of k. This is called the *group velocity* and is a measure of the way in which the disturbance as a whole travels. The wave crests still travel with the phase velocity $c = \omega/k$ and so they will be propagated relative to the group (unless $c = v$, which happens if and only if ω is directly proportional to k).

In the special cases which we have already discussed of waves on the surface of a liquid, we have the following results:

(i) Gravity waves on deep water: $\omega = (gk)^{\frac{1}{2}}$. Here

$$c = (g/k)^{\frac{1}{2}}, \quad v = \tfrac{1}{2}(g/k)^{\frac{1}{2}} = \tfrac{1}{2}c. \tag{240}$$

(ii) Shallow water waves: $\omega = (gh)^{\frac{1}{2}}k$. Here

$$c = (gh)^{\frac{1}{2}}, \quad v = (gh)^{\frac{1}{2}} = c. \tag{241}$$

(iii) Capillary waves: $\omega = (\sigma k^3/\rho)^{\frac{1}{2}}$. Here

$$c = (\sigma k/\rho)^{\frac{1}{2}}, \quad v = \tfrac{3}{2}(\sigma k/\rho)^{\frac{1}{2}} = \tfrac{3}{2}c. \tag{242}$$

11.2 WAVE PACKETS

It is useful to consider what happens when we have two sinusoidal waves of equal amplitude superposed. We take

$$\phi = a\,\cos(k_1 x - \omega_1 t) + a\,\cos(k_2 x - \omega_2 t)$$

$$= 2a\,\cos\left\{\tfrac{1}{2}\left[(k_2 - k_1)x - (\omega_2 - \omega_1)t\right]\right\} \times$$

$$\times \cos\left\{\tfrac{1}{2}\left[(k_1 + k_2)x - (\omega_1 + \omega_2)t\right]\right\}$$

$$= A\,\cos(\bar{k}x - \bar{\omega}t)$$

where $\bar{k}$ and $\bar{\omega}$ are the mean values of the wave numbers and the frequencies respectively and where

$$A = 2a \cos\left\{\tfrac{1}{2}\left[(k_2 - k_1)x - (\omega_2 - \omega_1)t\right]\right\}$$
$$= 2a \cos\left\{\tfrac{1}{2}\left[(\Delta k)x - (\Delta\omega)t\right]\right\},$$

where we have written $\Delta k = k_2 - k_1$ and $\Delta\omega = \omega_2 - \omega_1$.

So A can be regarded as a slowly varying amplitude (as long as $\Delta k \ll \bar{k}$ and $\Delta\omega \ll \bar{\omega}$) travelling with speed

$$v = \frac{\Delta\omega}{\Delta k} . \tag{243}$$

(An example of this is shown in Figure 11.1, where $k_1 = 5$, $k_2 = 6$, $\omega_1 = \tfrac{1}{2}$, $\omega_2 = 1$; in Figure (a) t = 0

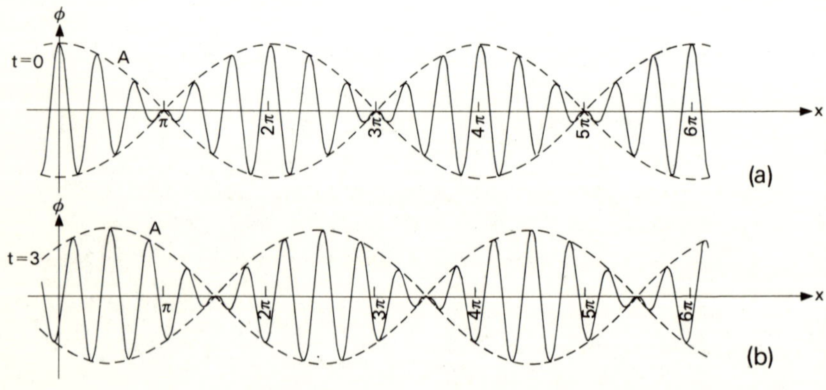

FIGURE 11.1

and in Figure (b) t = 3.) It is evident that the disturbance can be regarded as consisting of a series of wave packets, each of which moves with a speed v given by equation (243). This is exactly analagous to the situation in the previous section where the sum of a number of sinusoidal waves satisfying a frequency equation $\omega = \omega(k)$ formed a single disturbance and this travelled with the group velocity $v = d\omega/dk$.

Other situations in which this sort of thing applies
are those in which the waves originate from a small
region during a small interval of time. As usual, we
assume that the disturbance can be regarded as the sum
of sinusoidal waves of different wavelength and that all
wave numbers k are present. As time goes on those
waves having larger phase velocity will have travelled
farther than the others and the general impression will
be that of a wave-like disturbance whose wavelength
varies slowly with distance. An example is shown in
Figure 11.2, where it is assumed that phase velocity

FIGURE 11.2

increases with wavelength (as for deep-water gravity
waves). A particular wave crest travels with the speed
corresponding to the local value of the wave number,
but as it does so, the distance between neighbouring
crests changes. In fact the wave number corresponding
to a given crest changes with time. If we choose new
axes which move in such a way that the wave number at
the origin is independent of time, then the crests will
not be stationary with respect to these axes: suppose
that such a set of axes moves with speed v parallel
to the original x-axis. Referred to the new axes, the
new coordinate X is given by $x = X + vt$ and so a
wave proportional to $e^{i(kx-\omega t)}$ is, referred to the new
axes, $e^{i\chi}$ where $\chi = kX + (kv - \omega)t$. At the origin,
$\frac{\partial \chi}{\partial k}$ must be independent of t and so $(v - \frac{d\omega}{dk})t$ is
independent of t, and this means that $v = \frac{d\omega}{dk}$,
the group velocity as defined in equation (239). An
observer at the new origin (and so moving with speed
$v = \frac{d\omega}{dk}$) will then have wave crests passing him; each
crest as it passes him has the same local wavelength
$2\pi/k$, although its wavelength changes as it moves away
from the origin.

11.3 ENERGY AND ENERGY TRANSPORT FOR PLANE TRAVELLING SINUSOIDAL GRAVITY WAVES IN DEEP WATER

We consider first of all the total amount of wave energy (per unit width in the y-direction) contained between two planes $x = x_1$ and $x = x_2$ for the motion described by

$$\phi(x,z,t) = -\frac{\omega a}{k} e^{kz} \cos(kx - \omega t) \qquad (244)$$

subject to the conditions (202) and (207), so that the equation of the surface is $z = \zeta(x,t)$ where

$$\zeta(x,t) = a \sin(kx - \omega t). \qquad (245)$$

This energy is partly kinetic and partly potential. The kinetic energy is

$$\int_{x=x_1}^{x=x_2} \int_{z=-\infty}^{z=\zeta} \tfrac{1}{2}\rho(\phi_x^2 + \phi_z^2)\,dzdx$$

$$= \int_{x=x_1}^{x=x_2} \int_{z=-\infty}^{z=\zeta} \left(\frac{\omega a}{k}\right)^2 k^2 e^{2kz}\,dzdx$$

$$= \tfrac{1}{4}\rho\left(\frac{ga}{\omega}\right)^2 k \int_{x=x_1}^{x=x_2} e^{2k\zeta}dx$$

$$= \tfrac{1}{4}\rho\,\frac{g^2 a^2}{\omega^2} k \int_{x=x_1}^{x=x_2} [1 + 0(ka)]dx$$

$$= \tfrac{1}{4}\rho g a^2 (x_2 - x_1)$$

to the lowest order in the small quantity, since $\omega^2 = gk$.

The excess potential energy in the same region due to the wave motion is

$$\lim_{H\to\infty} \int_{x=x_1}^{x=x_2} \left[\int_{z=-H}^{z=\zeta} \rho gz\,dz - \int_{z=-H}^{z=0} \rho gz\,dz\right] dx =$$

$$= \int_{x=x_1}^{x=x_2} \int_{z=0}^{z=\zeta} \rho gz\,dzdx$$

$$= \tfrac{1}{2}\rho g \int_{x=x_1}^{x=x_2} \zeta^2 dx$$

$$= \tfrac{1}{4}\rho g a^2 \int_{x=x_1}^{x=x_2} [1 - \cos 2(kx - \omega t)]dx$$

$$= \tfrac{1}{4}\rho g a^2 \left[x_2 - x_1 - \frac{\sin 2(kx_2 - \omega t) - \sin 2(kx_1 - \omega t)}{2k} \right].$$

Hence the total energy E per unit length of the wave
train is the sum of these two quantities divided by
$(x_2 - x_1)$ and we have, to the lowest order in the small
quantity ka,

$$E = \tfrac{1}{2}\rho g a^2 + \frac{\sin 2(kx_2 - \omega t) - \sin 2(kx_1 - \omega t)}{8\,k(x_2 - x_1)}\,\rho g a^2.$$

By taking a suitable average, we can simplify this
expression, for if we suppose that $k(x_2 - x_1)$ is large
(that is, that we are averaging over a distance which is
large compared with a wavelength), the sinusoidal term
is negligibly small. Alternatively we can suppose that
$k(x_2 - x_1)$ is an integer multiple of 2π (that is, that
we are averaging over an integral number of wavelengths);
the sinusoidal term vanishes identically. Another
possibility is to take the average with respect to time
(either over a time which is long compared with a period,
or over an integral number of periods). Whichever method
of averaging we use, we get the same answer, that the
mean wave energy per unit length is

$$E = \tfrac{1}{2}\rho g a^2, \tag{246}$$

and we do not need to specify how the average is
calculated.
   Now, due to the wave motion there is energy propagated
across the plane $x = x_1$. The rate of transfer of energy
(per unit width in the y-direction) can be written as
vE, and we can regard v as the speed of propagation
of the energy. The energy transfer is partly due to the
energy which is carried across the plane by the fluid
itself; the rate at which this happens is

$\int_{-\infty}^{\zeta} [\rho gz + \frac{1}{2}\rho(\phi_x^2 + \phi_z^2)]\phi_x dz$ (per unit width in the y-direction).   In addition there is an effective transfer of energy due to the pressure acting on the plane $x = x_1$; the rate at which this works is $\int_{-\infty}^{\zeta} p\phi_x dz$ per unit width in the y-direction).   It follows that

$$vE = \int_{-\infty}^{\zeta} [p + g\rho z + \tfrac{1}{2}\rho(\phi_x^2 + \phi_z^2)]\phi_x dz$$

$$= \int_{-\infty}^{\zeta} (C - \rho\phi_t)\phi_x dz,$$

using Bernoulli's equation for irrotational flow in the form of equation (141).   Now C is a constant and $\int_{-\infty}^{\zeta} Cdz = 0$ in the mean.   It follows that

$$vE = -\rho \int_{-\infty}^{\zeta} \phi_t \phi_x dz$$

$$= -\rho \int_{-\infty}^{\zeta} \left(\frac{ga}{\omega}\right)^2 \omega k\ e^{2kz} \sin^2(kx_1 - \omega t)dz,$$

$$\text{using equation (244)}$$

$$= -\frac{\rho g^2 a^2}{2\omega}\ e^{k\zeta} \sin^2(kx_1 - \omega t)$$

$$= -\frac{\rho g^2 a^2}{4\omega}$$

in the mean to the required order in a^2.   Using equation (246), we see at once that

$$v = \frac{\omega}{2k}, \tag{247}$$

and this is the group velocity for waves of this type.
   When we have a number of such waves forming a group like that discussed in Section 11.1, this result is not surprising, since it is clear that the energy travels with the speed of the group.   Similarly, if we have a system of wave packets like that discussed in Section 11.2, we would expect each packet to conserve its own energy and this would mean that the energy travels with the group velocity.   We have now shown that (in some sense) energy is transported with the group velocity, not only for groups of waves and wave packets, but also

for a simple sinusoidal wave.

11.4 ENERGY PROPAGATION FOR A MORE GENERAL PLANE DISTURBANCE

We consider a given disturbance which is travelling parallel to the x-axis and whose amplitude is a. We suppose that the mean kinetic energy per unit length is T and the mean potential energy per unit length is V. Next we imagine that superposed on this, at time $t = t_0$, is a disturbance of the same magnitude at every point, but that at every point the velocity is exactly reversed. It, too, will have mean kinetic energy T and mean potential energy V. The composite disturbance, however, has amplitude 2a and zero velocity everywhere. Its mean energy per unit length is therefore entirely potential and exactly double the total energy of the given disturbance: that is, its energy is 2T + 2V. But potential energy is proportional to the square of the amplitude, so the mean potential energy of the composite disturbance is 4V. We have already seen that this is the total energy for the disturbance, and so 4V = 2T + 2V. Hence the mean potential energy and the mean kinetic energy of the given disturbance are equal. We found that this was true for the special case of sinusoidal gravity waves in the previous section.

To discuss the speed of propagation of this energy, we can use a method originally due to Lord Rayleigh. We imagine that the system consists of elementary 'particles', a typical one of which has mass m and is under the action of an oscillatory exciting force F in the direction of the x-axis, together with a small dissipative force proportional to the momentum of the 'particle'. If the displacement of the 'particle' from its mean position x is ξ at time t, we have

$$F - \epsilon m \frac{d\xi}{dt} = m \frac{d^2\xi}{dt^2} , \qquad (248)$$

where ϵ is the small constant of proportionality. We suppose that the exciting force is proportional to $e^{i\omega t}$ and that the displacement ξ of the 'particle' is $\xi = a\, e^{i(k'x - \omega't)}$. When there is no dissipative force, $\omega' = \omega$ and the displacement is $a\, e^{i(kx - \omega t)}$. So we have, for the same exciting force F,

$$- \omega^2 - i\omega'\epsilon = - \omega'^2.$$

It follows that $\omega^2 = \omega'^2 - i\omega'\epsilon = (\omega' - \tfrac{1}{2}i\epsilon)^2 + (\tfrac{1}{2}\epsilon)^2.$ Hence

$$\omega = \omega' - \tfrac{1}{2}i\epsilon + \text{terms of order } \epsilon^2.$$

Further, to the first order in the small quantity   ,

$$k' = k + \frac{dk}{d\omega} (\omega' - \omega) = k + \tfrac{1}{2} i\epsilon \frac{dk}{d\omega}.$$

Hence the displacement of the 'particle' is

$$\xi = a \; e^{i(kx-\omega t) \; - \; \tfrac{1}{2}\epsilon(dk/d\omega)x \; + \; \tfrac{1}{2}\epsilon t} = a' \; e^{i(kx-\omega t)},$$

where

$$a' = a \; e^{-\tfrac{1}{2}\epsilon(dk/d\omega)x \; + \; \tfrac{1}{2} \; t}$$

$$= a \; e^{-\tfrac{1}{2}\epsilon(dk/d\omega)x} \quad \text{when } t = 0.$$

Now the energy of the 'particle' at time $t = 0$ is proportional to a'^2 and this is $e^{-\epsilon(dk/d\omega)x}$ times the value it would have it there were no dissipation (that is, if $\epsilon = 0$). This is true for each 'particle' whose mean position is at x, and so, if E is the mean energy of all the 'particles' in the region $x > 0$, then

$$E = Ka^2 \int_0^{\infty} e^{-\epsilon(dk/d\omega)x} \; dx,$$

where Ka^2 is the corresponding mean energy for the undissipated flow. Performing the integration, we see that

$$E = \frac{1}{\epsilon} \frac{d\omega}{dk} Ka^2. \tag{249}$$

Now the rate of working of the dissipative force on the typical 'particle' is $(\epsilon m \frac{d\xi}{dt}) \frac{d\xi}{dt} = 2\epsilon$ times its kinetic energy. So the mean rate of working, W, of the dissipative force on the fluid in the region $x > 0$ is 2ϵ times the mean kinetic energy there; and this is ϵ times the mean total energy (since, as we have already seen, the kinetic energy is half the total energy for a travelling wave), and using equation (249), we have

$$W = \epsilon E = \frac{d\omega}{dk} Ka^2.$$

We note that this is independent of ϵ, and that the energy must be supplied by the inflow of energy across the surface $x = 0$. We suppose that this energy which, as we have seen, is Ka^2 per unit length when there is no dissipation, crosses the surface $x = 0$ with speed v. Then we must have $Ka^2v = W = (d\omega/dk)Ka$. It follows at once that $v = (d\omega/dk)$, as we have found for some special cases in the last section.

11.5 WAVE FRONTS OF GENERAL SHAPE

We can extend the idea of group velocity to more general disturbances like those we discussed in Section 10.6. There we found it convenient to think of a general disturbance as a superposition of plane waves and to write

$$\phi(\underset{\sim}{r},z,t) = \mathcal{R}\left\{ \sum_{\underset{\sim}{k}} \phi_{\underset{\sim}{k}}(z) \ e^{i(\underset{\sim}{k}\cdot\underset{\sim}{r} - \omega t)} \right\}, \qquad (250)$$

where $\underset{\sim}{r}$ is a horizontal vector. Usually there is a frequency equation of the form

$$\omega = \omega(\underset{\sim}{k}). \qquad (251)$$

These two equations correspond exactly to equations (235) and (236) at the beginning of this chapter. We can follow the argument of that section and consider a disturbance where each $\underset{\sim}{k}$ differs only slightly from some given value $\underset{\sim}{k_0}$. Then equation (237) is replaced by

$$\omega = \omega(\underset{\sim}{k_0}) + (\underset{\sim}{k} - \underset{\sim}{k_0})\cdot(\nabla'\omega)_0$$

to the first order in the small quantity $|\underset{\sim}{k} - \underset{\sim}{k_0}|$, where $\nabla' = \frac{\partial}{\partial\lambda}, \frac{\partial}{\partial\mu}$ is the gradient in wave number space, where $\underset{\sim}{k} = (\lambda, \mu)$. It follows that we can write

$$\phi(\underset{\sim}{r},z,t) = \mathcal{R}\left\{ A \ e^{i(\underset{\sim}{k_0}\cdot\underset{\sim}{r} - \omega_0 t)} \right\}$$

where $\omega_0 = \omega(\underset{\sim}{k}_0)$ and (after suitable manipulation in the manner of Section 11.1)

$$A = \sum_{\underset{\sim}{k}} \phi_{\underset{\sim}{k}}(z) \; e^{i(\underset{\sim}{k} - \underset{\sim}{k}_0) \cdot [\underset{\sim}{r} - t(\nabla'\omega)_0]}$$

to the first order in $|\underset{\sim}{k} - \underset{\sim}{k}_0|$. So the slowly varying amplitude A is propagated with group velocity

$$\underset{\sim}{v} = \nabla'\omega = \left(\frac{\partial\omega}{\partial\lambda} , \frac{\partial\omega}{\partial\mu} \right) \tag{252}$$

evaluated for $\underset{\sim}{k} = \underset{\sim}{k}_0$. We have already seen in Section 10.6 that the phase velocity is

$$\underset{\sim}{c} = \frac{\omega}{|k|} \, \underset{\sim}{k} = \left(\frac{\omega\lambda}{\lambda^2 + \mu^2} , \frac{\omega\mu}{\lambda^2 + \mu^2} \right), \tag{253}$$

where we have changed the minus sign of Section 10.5 to a plus because, in equation (250), we changed the sign of ω.

We notice at once that, in general, $\underset{\sim}{v}$ and $\underset{\sim}{c}$ are in different directions: in fact it is easily verified that the directions are the same if and only if equation (251) is of the form

$$\omega = \omega(\underset{\sim}{k}) = \omega\left[(\lambda^2 + \mu^2)^{\frac{1}{2}} \right],$$

so that the frequency depends only on the magnitude of the wave number, not on its direction. This may seem surprising at first sight, but in fact it is not difficult to imagine an isolated group of almost plane waves travelling in a different direction from that of the normal to the wave crests: such a situation is suggested in Figure 11.3.

As a simple example of this type of disturbance, we consider the bow waves on deep water caused by an obstacle at the origin in a uniform stream U parallel to the x-axis. This is the situation discussed in Section 10.7 where we found, in equation (232), that

$$g|\underset{\sim}{k}| = (\omega - U\lambda)^2, \tag{254}$$

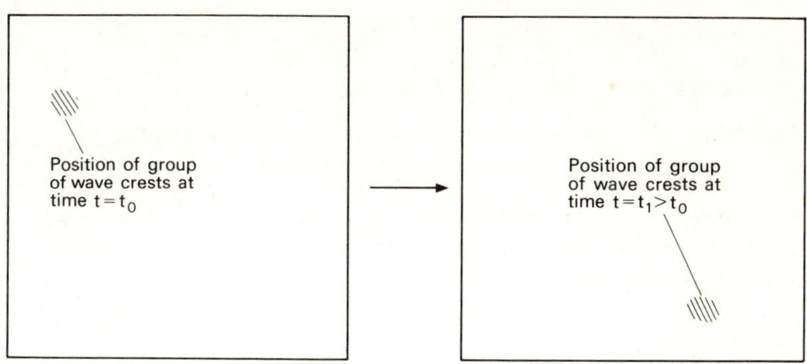

FIGURE 11.3

and this may be written as

$$\omega = U\lambda \pm g^{\frac{1}{2}}(\lambda^2 + \mu^2)^{\frac{1}{4}}. \qquad (255)$$

In this case we have

$$\underset{\sim}{v} = \left[U \pm \frac{1}{2}\,\frac{g^{\frac{1}{2}}\lambda}{(\lambda^2 + \mu^2)^{\frac{3}{4}}}\,,\; \pm\;\frac{g^{\frac{1}{2}}\mu}{(\lambda^2 + \mu^2)^{\frac{3}{4}}} \right]$$

$$= \underset{\sim}{U} + \underset{\sim}{v}_0\,,$$

where

$$\underset{\sim}{v}_0 = \frac{g^{\frac{1}{2}}\underset{\sim}{k}}{2|\underset{\sim}{k}|^{\frac{3}{2}}} = \frac{1}{2}\left(\frac{g}{|\underset{\sim}{k}|}\right)^{\frac{1}{2}} \hat{\underset{\sim}{k}}$$

is the group velocity for these waves in a fluid at rest and is precisely that found for one-dimensional waves in equation (240).

The arguments of Section 11.2 can also be extended to the more general case discussed here, and so can the energy argument for waves with sufficiently small amplitude. It should be noted, however, that the energy of waves with finite amplitude is not necessarily propagated with the group velocity as defined in equation (252).

11.6 WAVE-NUMBER SPACE

We consider a particular frequency equation like that of equation (251) and draw axes $O\lambda$, $O\mu$ in the same directions as Ox, Oy respectively. Then, in the wave-number space formed by these axes, we draw the curves* $\omega(\underset{\sim}{k})$ = constant. It is clear that, for a particular wave number $\underset{\sim}{k}$ = (λ,μ), a point P is defined in this space such that $\overline{OP}$ = $\underset{\sim}{k}$; further, the normal at P to the surface of constant $\omega(\underset{\sim}{k})$ which passes through P is in the direction of $\underset{\sim}{v}$ = $\nabla'\omega$. The appropriate direction of the normal will, of course, be in the sense of increasing ω, and it is convenient to put small arrows on the curve $\omega(\underset{\sim}{k})$ = ω_0, where ω_0 is a given constant pointing towards a neighbouring curve $\omega(\underset{\sim}{k})$ = $\omega_0' > \omega_0$.

The usefulness of this procedure is most easily seen by considering an example and we choose the problem of bow waves for this purpose. We have already seen that, in this case, the equation (254) holds. Then the curve $\omega(\underset{\sim}{k})$ = ω_0 is given by

$$g^2(\lambda^2 + \mu^2) = (\omega_0 - U\lambda)^4. \qquad (256)$$

The form of these curves can easily be found and details of the derivation in the range $0 \leq \omega_0 < g/4U$ are given in Appendix 7. In practice we are, as we saw in Section 10.7, interested primarily in the case ω_0 = 0 : these are the waves which are stationary with respect to the body causing them. (Waves with $\omega_0 \neq 0$ may be present, but they travel away and tend to cancel each other out.) In this case we have

$$g^2(\lambda^2 + \mu^2) = U^4\lambda^4, \qquad (257)$$

and the graph of this curve in wave-number space is as shown in Figure 11.4. The arrows show the direction in which the curves move as ω_0 becomes positive. We note that there are waves with $\underset{\sim}{k}$ = 0, and these have arrows pointing in all directions; however, we are not interested in these as they depend neither on position nor on time.

---

* If $\underset{\sim}{k}$ = (λ,μ,ν) is a three-dimensional wave number, then $\omega(\underset{\sim}{k})$ = constant represents a surface in wave-number space.

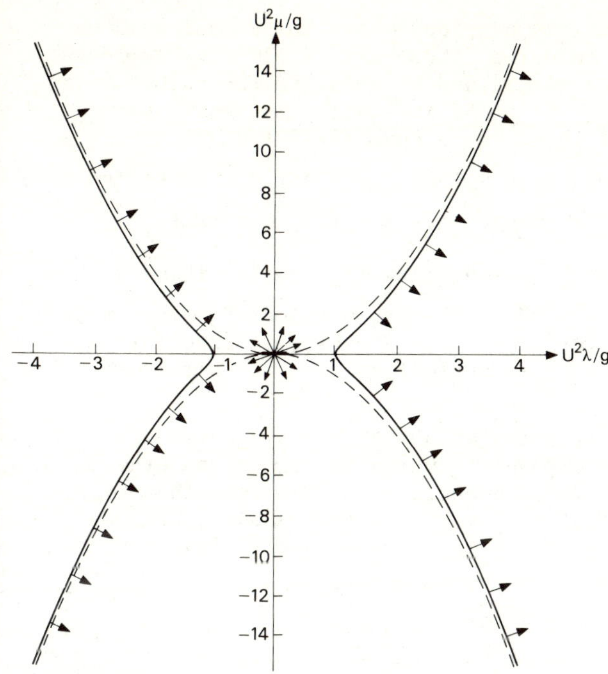

FIGURE 11.4

For other possible values of $\underset{\sim}{k}$ (for example, those represented by P, Q, S in Figure 11.5) the possible

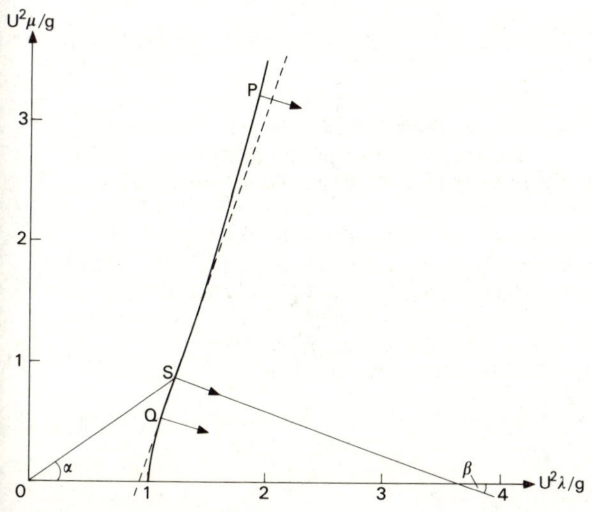

FIGURE 11.5

directions of the group velocity are those indicated by
the arrows. We note that these arrows all make angles
with the axis $O\lambda$ less than or equal to the value β at
S, where S is a point of inflexion, as shown in
Figure 11.5. It is shown in Appendix 8 that
$\tan \beta = \dfrac{1}{2\sqrt{2}}$ so that $\beta = 19°$, and that OS makes an
angle α with $O\lambda$ where $\tan \alpha = \sqrt{2}$, so that
$\alpha = 54\tfrac{1}{2}°$.

For values of $\underset{\sim}{k}$ making an angle greater than α
with $O\lambda$ (such as $\overrightarrow{OP}$), the group velocity makes an
angle with $O\lambda$ less than β. The same is true for
values of $\underset{\sim}{k}$ making an angle less than α with $O\lambda$
(such as $\overrightarrow{OQ}$).

Since energy is propagated with the group velocity,
the waves cannot transport energy in directions other
than those which make an angle with Ox less than β.
So we have a situation similar to that shown in Figure
11.6, where two pairs of wave crests are shown, each

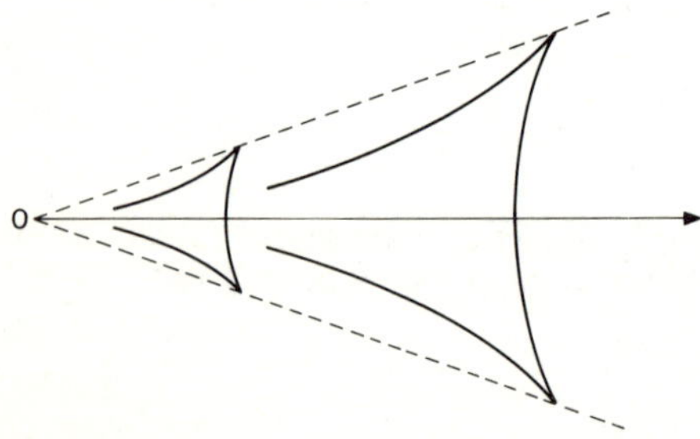

FIGURE 11.6

pair intersecting in a cusp (corresponding to the point
S in Figure 11.5) on the lines making an angle β with
Ox. The ducks and minesweepers of Plate 3 clearly show
this effect.

An alternative (and more simple-minded) way of
looking at this problem is indicated in Appendix 9. The
present method can be extended, however, to calculate
the shape of the wave crests at a large distance from
the original disturbance at O using a purely
geometrical method. To do this is, however, beyond the
scope of this book.

EXERCISES

1 The function $f(x)$ is defined in $0 \leq x \leq b$ by

$$f(x) = \begin{cases} 0, & 0 \leq x \leq x_0 - \tfrac{1}{2}L \\ a \cos[\pi(x - x_0)L], & |x - x_0| \leq L \\ 0, & x_0 + \tfrac{1}{2}L \leq x \leq b. \end{cases}$$

Sketch the graph of f as a function of x, and find
the coefficients in the half-range Fourier sine series

$$f(x) = \sum_{n=1}^{\infty} b_n \sin(n\pi x/b).$$ Be careful to distinguish

between the cases in which b/L is and is not an
integer.
   Evaluate each coefficient to one place of decimals up
to $n = 30$ given that $a = 10$, $x_0 = 4L/3$, $b = 10L$.

2 An observer on a weather ship observes waves of length
5m at a particular time and waves of length 1m at a time
12 hours later. If the waves were all caused by a single
storm at sea, find the time of the storm and its distance
from the weather ship.

3 A gravity wave in a liquid of depth h is described
by the velocity potential

$$\phi = \phi_0 \cosh[k(z + h)]\sin(kx - \omega t),$$

in the region defined by $-h \leq z \leq \zeta(x,t)$, where
$\zeta = a \cos(kx - \omega t)$, $\omega^2 = c^2 k^2 = gk \tanh kh$, $ka \ll 1$
and $a \ll h$. Show (from first principles) that, to the
first order of small quantities,
   (a) the mean potential energy per unit width is
       $\tfrac{1}{4}\rho g a^2$,
   (b) the mean kinetic energy per unit width is $\tfrac{1}{4}\rho g a^2$,
   (c) the mean rate of transport of energy per unit
       width is $\tfrac{1}{4}\rho g a^2 c(1 + 2kh \, \text{cosech} \, 2kh)$.
Hence find the velocity of energy propagation and verify
that it is $d\omega/dk$ in this case.

4 The frequency equation for the propagation of small
disturbances on the surface of deep water is

$$\omega^2 = gk + \sigma k^3/\rho.$$

Show that, if a group of waves travels with speed v
where $\rho v^4 \gg g\sigma$, there are long waves with $k \simeq g/4v^2$
and short waves with $k \simeq 4\rho v^2/9\sigma$.

5   The frequency equation for the propagation of small disturbances on the surface of water of depth h is, when surface tension effects are negligible,

$$(\omega - U\lambda)^2 = gk \tanh kh,$$

where $k = (\lambda^2 + \mu^2)^{\frac{1}{2}}$.   For the case where kh is small, sketch the curves of constant ω in the λ-μ plane, distinguishing between the cases $U^2 < gh$ and $U^2 > gh$.
   Discuss, in each case, how the energy of the waves is propagated.

6   The frequency equation for the propagation of small horizontal disturbances in an incompressible atmosphere moving with uniform velocity U parallel to the x-axis, which is tangential to a line of latitude on a rotating sphere, is

$$(\lambda^2 + \mu^2)(U\lambda - \omega) = \beta\lambda,$$

where β is a positive constant depending on the speed of rotation of the sphere and on the latitude of the origin.   Sketch the curves of constant ω in the λ-μ plane.
   Discuss the way in which energy is propagated, with particular reference to the case $\omega = 0$.

7   The frequency equation for the propagation of small adiabatic disturbances in an atmosphere at constant temperature is

$$\omega^4 - \omega^2[(\lambda^2 + \mu^2 + \nu^2)c^2 + K^2N^2] + (\lambda^2 + \mu^2)c^2N^2 = 0,$$

where c is the speed of sound, N is the Brunt-Väsälä frequency and K is a real constant such that $K > 1$. Sketch, when they exist, cross-sections of the surfaces of constant ω in wave-number space, distinguishing between the three cases $0 < \omega < N$, $N < \omega < KN$ and $\omega > KN$.
   Discuss the energy propagation for waves with frequencies $\omega = 0$, $\omega = N$, $\omega = KN$ and $\omega \to \infty$.

8   The frequency equation for the propagation of small disturbances in a compressible plasma is

$$\omega^4 - \omega^2(a^2 + A^2)(\lambda^2 + \mu^2 + \nu^2) +$$
$$+ a^2A^2\nu^2(\lambda^2 + \mu^2 + \nu^2) = 0,$$

where a is the speed of sound, A is the Alfvén speed
and both are constant. Sketch the cross-sections of the
surfaces of constant ω in wave-number space when
a > A. Discuss how the energy is propagated.
   What happens when a < A ?

STRESSES AND RATES OF STRAIN IN A REAL FLUID

In general, when $\underset{\sim}{u}$ is a general function of $\underset{\sim}{r}$, we proceed as follows. We use a double-suffix notation for the internal stresses: the axes are Ox_1, Ox_2 and Ox_3 and, for the force per unit area (that is, the stress) on a surface perpendicular to Ox_j we write (p_{1j}, p_{2j}, p_{3j}). So there are nine components of stress p_{ij} with $i = 1,2,3$ and $j = 1,2,3$. When all the internal forces are normal ones, of course, $p_{ij} = 0$ if $i \neq j$.

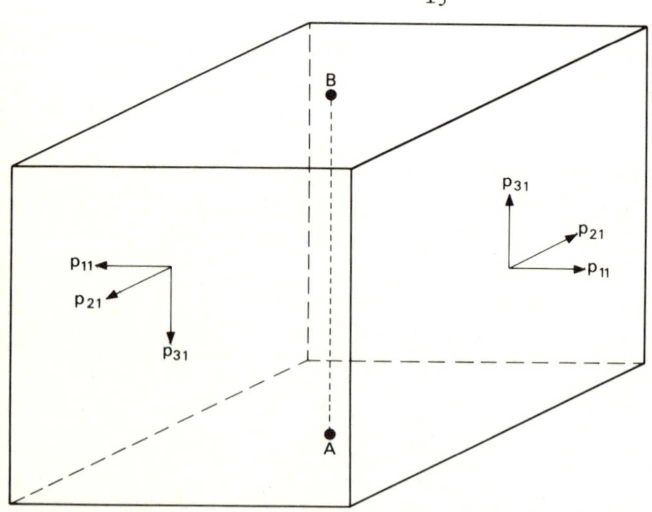

FIGURE A1.1

It is easy to show that $p_{ji} = p_{ij}$ for all values of i, j. We consider the rectangular parallelepiped whose edges are parallel to Ox_1, Ox_2, Ox_3 and of length

ξ, η, ζ respectively. We take moments about the line AB parallel to Ox_3 through the centre of the faces perpendicular to this direction. We know that the rate of change of the angular momentum of the fluid in the box is equal to the moment of the external body force on the fluid plus that of the internal forces. Now the angular momentum about AB is

$$\iiint_V \rho(u_2 x - u_1 y)dV =$$

$$= \iiint_V \left\{ \left[(\rho u_2)_0 + x\frac{\partial(\rho u_2)_0}{\partial x} + y\frac{\partial(\rho u_2)_0}{\partial y} + \ldots \right] x \right.$$

$$\left. - \left[(\rho u_1)_0 + x\frac{\partial(\rho u_1)_0}{\partial x} + y\frac{\partial(\rho u_1)_0}{\partial y} + \ldots \right] y \right\} dV,$$

where V is the volume of the box and the suffix zero denotes evaluation at the mid-point of the box. Now, from symmetry, we have $\iiint_V x\, dV = \iiint_V y\, dV = \iiint_V xy\, dV = 0$; also $\iiint_V x^2 dV = \frac{1}{12}\xi^2 V$, $\iiint_V y^2 dV = \frac{1}{12}\eta^2 V$. It follows that the angular momentum about AB and, therefore, also its rate of change, is proportional to the fifth order of small quantities ξ, η, ζ. Similarly, the moment of the external forces is

$$\iiint_V \rho(F_2 x - F_1 y)dV =$$

$$= \frac{1}{12}\left\{ \xi^2\left[\frac{\partial(\rho F_2)}{\partial x}\right]_0 - \eta^2\left[\frac{\partial(\rho F_1)}{\partial y}\right]_0 \right\}V + \ldots,$$

and this, too, is of the fifth order in the same small quantities. Further, the moment of the internal forces (assuming that they act through the centre of the faces) is

$$(\tfrac{1}{2}p_{21})_+ \xi\eta\zeta + (\tfrac{1}{2}p_{21})_- \xi\eta\zeta - (\tfrac{1}{2}p_{12})_- \xi\eta\zeta - (\tfrac{1}{2}p_{12})_+ \xi\eta\zeta =$$

$$= (p_{21} - p_{12})_0 V + \text{higher-order terms},$$

where the suffices $+$ and $-$ are used to denote opposite faces of the box; this is of the third order of small quantities and must therefore vanish in the limit as ξ, η, $\zeta \to 0$. It follows that $p_{12} = p_{21}$, and by a similar argument we can show that $p_{23} = p_{32}$ and $p_{13} = p_{31}$.

It is not difficult to show that the stress p_{ij} is a second-order tensor, and for a rigorous derivation of the form of the relationship between p_{ij} and the velocity derivatives, we use this fact. Here we shall restrict ourselves to a physical discussion, and in any case we need to make two assumptions - first, that the p_{ij} depends linearly on the velocity derivatives and second, that the form of the relation between them does not depend on the direction (or position) of the axes chosen. This last condition is usually called *isotropy*.

It will be easier to write down the expression which arises and then to explain the physical significance of the various terms. In fact it can be shown that

$$p_{ij} = -p\delta_{ij} + \lambda\Delta\delta_{ij} + \mu\left(\frac{\partial u_i}{\partial x_j} + \frac{\partial u_j}{\partial x_i}\right), \qquad (A1)$$

where

$$\Delta = \sum_{k=1}^{3} \frac{\partial u_k}{\partial x_k} = \frac{\partial u_1}{\partial x_1} + \frac{\partial u_2}{\partial x_2} + \frac{\partial u_3}{\partial x_3},$$

and $\delta_{ij} = 0$ when $i \neq j$ and $\delta_{11} = \delta_{22} = \delta_{33} = 1$. Writing this out in full, we obtain

$$p_{11} = (-p + \lambda\Delta) + 2\mu\frac{\partial u_1}{\partial x_1},$$

$$p_{22} = (-p + \lambda\Delta) + 2\mu\frac{\partial u_2}{\partial x_2},$$

$$p_{33} = (-p + \lambda\Delta) + 2\mu\frac{\partial u_3}{\partial x_3},$$

$$p_{23} = \mu\left(\frac{\partial u_2}{\partial x_3} + \frac{\partial u_3}{\partial x_2}\right),$$

$$p_{31} = \mu\left(\frac{\partial u_3}{\partial x_1} + \frac{\partial u_1}{\partial x_3}\right),$$

$$p_{12} = \mu\left(\frac{\partial u_1}{\partial x_2} + \frac{\partial u_2}{\partial x_1}\right).$$

Note that the mean normal stress is

$$\tfrac{1}{3}(p_{11} + p_{22} + p_{33}) = -p + (\lambda + \tfrac{2}{3}\mu)\Delta.$$

For an incompressible fluid, $\Delta = 0$, and so the mean
normal stress is $-p$ and this does not depend directly
on the velocity field. When the fluid is compressible,
however, $\Delta \neq 0$ and we no longer have that the mean
normal stress is independent of the velocity field
unless $\lambda = -\frac{2}{3}\mu$. In practice it turns out that, to a
very good approximation this relationship does hold
except in very special cases (such as for very high
frequency oscillations and in a discussion of the
attenuation of sound waves). The effect of not putting
$\lambda + \frac{2}{3}\mu$ equal to zero is the same as allowing the fluid
to have some elasticity and it is effectively the bulk
modulus of elasticity of the fluid - this is clear when
we use equation (128) and note that $\Delta = -\dfrac{1}{\rho}\dfrac{D\rho}{Dt} = \dfrac{1}{V}\dfrac{DV}{Dt}$
where $V = \dfrac{1}{\rho}$ is the specific volume. So the mean normal
stress is

$$- p + (\lambda + \tfrac{2}{3}\mu)\frac{1}{V}\frac{DV}{Dt} \, ,$$

and $\dfrac{1}{V}\dfrac{DV}{Dt}$ is the rate of change of volume per unit
volume of the given mass.

Thus in the relation (A1) we see that (in general)
the first term $-p\,\delta_{ij}$ is a normal pressure such as
that defined in Chapter 1. The second term $\lambda\Delta\delta_{ij}$
is related to the compressibility of the fluid: if the
fluid is incompressible, $\Delta = 0$, and so the term
vanishes and if it is compressible then it is a measure
of the normal force required to change its density.
The last term $\mu\left(\dfrac{\partial u_i}{\partial x_j} + \dfrac{\partial u_j}{\partial x_i}\right)$ is the viscous term and is
and obvious generalization of the simple case discussed
in Chapter 5 - it is a measure of the difference in the
i-components of momentum carried by the molecules in
opposite directions across the surface perpendicular to
Ox_j.

THE NAVIER–STOKES EQUATIONS

If we wish to include the tangential components of the internal forces when finding the equations of motion, we need only to alter the expression (133) in Section 7.2. If $\hat{\underset{\sim}{n}} = (\ell_1, \ell_2, \ell_3)$ is the unit normal in the direction of the element of area δS and we use the notation of Appendix 1 and of Section 1.7, we see that the component in the direction Ox_1 is

$$p_{11}\delta S_1 + p_{12}\delta S_2 + p_{13}\delta S_3 = (\ell_1 p_{11} + \ell_2 p_{12} + \ell_3 p_{13})\delta S.$$

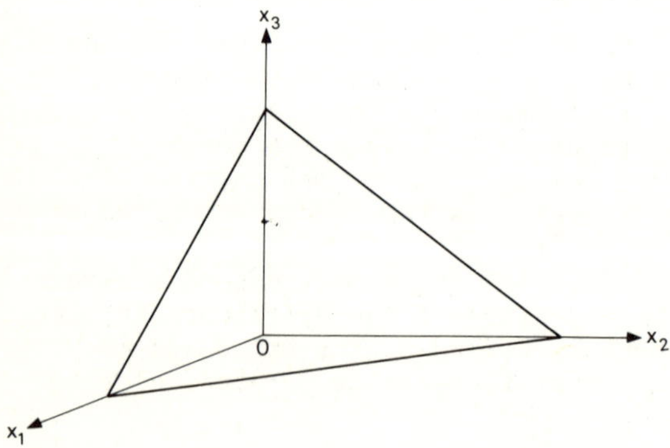

FIGURE A2.1

Hence the component over the whole surface V of the volume V is

$$\iint_{\partial V} (\ell_1 p_{11} + \ell_2 p_{12} + \ell_3 p_{13})\,dS =$$

$$= \iiint_V \left(\frac{\partial p_{11}}{\partial x_1} + \frac{\partial p_{12}}{\partial x_2} + \frac{\partial p_{13}}{\partial x_3} \right) dV$$

and similarly for the other components. So, instead of
$-$ grad p in equation (134) we have the vector whose
components are

$$\frac{\partial p_{i1}}{\partial x_1} + \frac{\partial p_{i2}}{\partial x_2} + \frac{\partial p_{i3}}{\partial x_3}, \qquad i = 1,2,3.$$

So the ith component of equation (134) is

$$\rho\,\frac{Du_i}{Dt} - \rho\,F_i = \frac{\partial p_{i1}}{\partial x_1} + \frac{\partial p_{i2}}{\partial x_2} + \frac{\partial p_{i3}}{\partial x_3} = \sum_{j=1}^{3} \frac{\partial p_{ij}}{\partial x_j}. \qquad (A2)$$

Using equation (A1) we see that the right-hand side of
this expression is

$$\sum_{j=1}^{3} \left\{ \delta_{ij}\left[-\frac{\partial p}{\partial x_j} + \frac{\partial(\lambda\Delta)}{\partial x_j}\right] + \frac{\partial \mu}{\partial x_j}\frac{\partial u_i}{\partial x_j} + \mu\,\frac{\partial u_i}{\partial x_j} + \frac{\partial \mu}{\partial x_j}\frac{\partial u_j}{\partial x_i} + \right.$$

$$\left. + \mu\,\frac{\partial}{\partial x_i}\left(\frac{\partial u_j}{\partial x_j}\right)\right\} = -\frac{\partial p}{\partial x_i} + \frac{\partial}{\partial x_i}(\lambda\Delta) + \mu\nabla^2 u_i + \mu\frac{\partial \Delta}{\partial x_i}$$

$$+ \sum_{j=1}^{3} \frac{\partial \mu}{\partial x_j}\left(\frac{\partial u_i}{\partial x_j} + \frac{\partial u_j}{\partial x_i}\right),$$

where $\Delta = \operatorname{div}\underline{u} = \displaystyle\sum_{j=1}^{3} \frac{\partial u_j}{\partial x_j}$.

For incompressible flow, $\Delta = 0$ (by definition) and
μ is usually taken to be constant $-$ this is undoubtedly
true when there are no temperature gradients in the
fluid (since μ is a function of T only). In this
case, therefore, the right-hand side of equation (A2)
becomes $-\dfrac{\partial p}{\partial x_i} + \mu\nabla^2 u_i$,

and so

$$\frac{Du_i}{Dt} = F_i - \frac{1}{\rho}\frac{\partial p}{\partial x_i} + \frac{\mu}{\rho}\nabla^2 u_i . \qquad (A3)$$

This may be written in the alternative form

$$\frac{D\underline{u}}{Dt} = \underline{F} - \frac{1}{\rho}\operatorname{grad} p + \nu\nabla^2\underline{u}, \qquad (A4)$$

where $\nu = \mu/\rho$ is the kinematic viscosity, since (A3) is
evidently the ith component of (A4).

When the flow is compressible, the expression cannot be simplified in general, and we have

$$\frac{Du_i}{Dt} = F_i - \frac{1}{\rho}\frac{\partial p}{\partial x_i} + \nu\nabla^2 u_i + \left(\frac{\lambda + \mu}{\rho}\right)\frac{\partial \Delta}{\partial x_i} + \frac{\Delta}{\rho}\frac{\partial \lambda}{\partial x_i}$$

$$+ \frac{1}{\rho}\sum_{j=1}^{3}\frac{\partial \mu}{\partial x_j}\left(\frac{\partial u_i}{\partial x_j} + \frac{\partial u_j}{\partial x_i}\right). \qquad (A5)$$

The vector form of this equation is easily shown to be

$$\frac{D\underset{\sim}{u}}{Dt} = \underset{\sim}{F} - \frac{1}{\rho}\,\mathrm{grad}\;p + \nu\nabla^2\underset{\sim}{u} + \frac{\lambda + \mu}{\rho}\,\mathrm{grad}\;\Delta + \frac{1}{\rho}\,\Delta\,\mathrm{grad}\;\lambda +$$

$$+ \frac{2}{\rho}(\mathrm{grad}\;\mu.\nabla)\underset{\sim}{u} + \frac{1}{\rho}\,\mathrm{grad}\,\mu\wedge\mathrm{curl}\;\underset{\sim}{u}. \qquad (A6)$$

Note that for incompressible flow with constant μ, the vorticity equation, obtained by taking the curl of equation (A4), is

$$\frac{D\underset{\sim}{\omega}}{Dt} = (\underset{\sim}{\omega}.\nabla)\underset{\sim}{u} + \nu\nabla^2\underset{\sim}{\omega}, \qquad (A7)$$

and it is no longer true that if $\underset{\sim}{\omega} = \underset{\sim}{0}$ at some instant for an element of fluid, when $\underset{\sim}{\omega} = \underset{\sim}{0}$ for that element at all time. The last term on the right-hand side of equation (A7) is a dissipative term, and in a real fluid the vorticity of an element of fluid will tend to die away with time.

THE TWO-DIMENSIONAL BOUNDARY-LAYER EQUATIONS

The basic assumption of the boundary layer is that $u_3 \ll u_1$ and that

$$\frac{\partial u_1}{\partial x_1} + \frac{\partial u_3}{\partial x_3} = 0 \ . \tag{A8}$$

If we suppose that L is a typical length of the problem and that U is a typical velocity, then we can define a Reynolds number

$$R = \frac{\rho_0 UL}{\mu_0} \ ,$$

where ρ_0 and μ_0 are typical density and viscosity respectively. We also define the non-dimensional variables T, X_1, X_3, U_1, U_3 and P by means of the relations

$$T = \frac{L}{U} t, \quad x_1 = LX_1, \quad x_3 = R^{-\frac{1}{2}}LX_3,$$

$$u_1 = UU_1, \quad u_3 = R^{-\frac{1}{2}}UU_3, \quad p = \rho_0 U^2 P \ .$$

Then x_3/x_1 and u_3/u_1 are of order $R^{-\frac{1}{2}}$, which is small for large R. Also the modified form of equation (A8) does not contain R. If we make these substitutions in equations (A3), assuming that everything is independent of x_2, we have

$$\frac{U^2}{L} DU_1 = F_1 - \frac{\rho_0 U^2}{\rho L} \frac{\partial P}{\partial X_1} + \frac{\nu U}{L^2}\left(\frac{\partial^2 U_1}{\partial X_1^2} + R \frac{\partial^2 U_1}{\partial X_3^2}\right),$$

$$\frac{U^2}{L} R^{-\frac{1}{2}} DU_3 = F_3 - \frac{\rho_0 U^2}{\rho L} \frac{\partial P}{\partial X_3} + \frac{\nu U}{L^2} R^{-\frac{1}{2}} \left(\frac{\partial^2 U_3}{\partial X_1^2} + R \frac{\partial^2 U_3}{\partial X_3^2} \right) ,$$

where $\quad D = \frac{\partial}{\partial T} + U_1 \frac{\partial}{\partial X_1} + U_3 \frac{\partial}{\partial X_3}$.

Hence

$$DU_1 = \frac{L}{U^2} F_1 - \frac{\rho_0}{\rho} \frac{\partial P}{\partial X_1} + \frac{\partial^2 U_1}{\partial X_3^2} + R^{-1} \frac{\partial^2 U_1}{\partial X_1^2} , \tag{A9}$$

and

$$R^{-\frac{1}{2}} DU_3 = \frac{L}{U^2} F_3 - \frac{\rho_0}{\rho} \frac{\partial P}{\partial X_3} + R^{-\frac{1}{2}} \left(\frac{\partial^2 U_3}{\partial X_3^2} + R^{-1} \frac{\partial^2 U_3}{\partial X_1^2} \right) . \tag{A10}$$

So, if we neglect terms multiplied by R^{-1}, the equation (A9) becomes

$$DU_1 = \frac{LF_1}{U^2} - \frac{\rho_0}{\rho} \frac{\partial P}{\partial X_1} + \frac{\partial^2 U_1}{\partial X_3^2} ;$$

and if we neglect terms multiplied by $R^{-\frac{1}{2}}$, equation (A10) becomes

$$0 = \frac{LF_3}{U^2} - \frac{\rho_0}{\rho} \frac{\partial P}{\partial X_3} .$$

Reverting to the original variables, we find that the equations are now

$$\frac{DU_1}{Dt} = F_1 - \frac{1}{\rho} \frac{\partial p}{\partial x_1} + \nu \frac{\partial^2 u_1}{\partial x_3^2} , \tag{A11}$$

$$0 = F_3 - \frac{1}{\rho} \frac{\partial p}{\partial x_3} . \tag{A12}$$

If we carry out the same procedure for equation (A5) and assume that ρ/ρ_0 and μ/μ_0 are both of order unity, it can be shown that the only extra term which need be included to the same order of accuracy is

$\dfrac{1}{\rho}\dfrac{\partial \mu}{\partial x_3}\dfrac{\partial u_1}{\partial x_3}$ in the first equation. This means that (A11) must be modified to give

$$\frac{Du_1}{Dt} = F_1 - \frac{1}{\rho}\frac{\partial p}{\partial x_1} + \frac{1}{\rho}\frac{\partial}{\partial x_3}\left(\mu\frac{\partial u_1}{\partial x_3}\right), \qquad (A13)$$

and equation (A12) is unchanged.

In general, then, the equations of motion in the boundary layer are (A12) and (A13), together with the equation of continuity (A8). The error obtained by using these equations is of order $R^{-\frac{1}{2}}$.

AN INVISCID FLUID WITH ZERO VORTICITY

We have shown that, for an inviscid fluid, equation (144) holds and so we have

$$\frac{D}{Dt}\left(\frac{\underline{\omega}}{\rho}\right) = \left(\frac{\underline{\omega}}{\rho} \cdot \nabla\right)\underline{u} .\tag{A14}$$

We suppose that all velocity gradients within the fluid are bounded (if they are not, we would not expect the inviscid equations to be valid anyway) and so there exists a number M such that

$$\left|\frac{\partial u_i}{\partial x_j}\right| \leq M , \quad i = 1,2,3, \quad j = 1,2,3.\tag{A15}$$

We write $\omega = (\omega_1,\omega_2,\omega_3)$ and $\omega = (\omega_1^2 + \omega_2^2 + \omega_3^2)^{\frac{1}{2}}$, so that

$$\left|\omega_j\right| \leq \omega , \quad j = 1,2,3.\tag{A16}$$

Then we have, using equation (A14),

$$\left|\frac{\omega_1}{\rho}\frac{D}{Dt}\left(\frac{\omega_1}{\rho}\right)\right| =$$

$$= \frac{1}{\rho^2}\left|\omega_1^2\frac{\partial u_1}{\partial x_1} + \omega_1\omega_2\frac{\partial u_1}{\partial x_2} + \omega_1\omega_3\frac{\partial u_1}{\partial x_3}\right|,$$

$$\leq \frac{1}{\rho^2}\left[\left|\omega_1^2\right|\left|\frac{\partial u_1}{\partial x_1}\right| + \left|\omega_1\right|\left|\omega_2\right|\left|\frac{\partial u_1}{\partial x_2}\right| + \left|\omega_1\right|\left|\omega_3\right|\left|\frac{\partial u_1}{\partial x_3}\right|\right],$$

using the triangle inequality

$$\leq \frac{M}{\rho^2}\left[|\omega_1^2| + |\omega_1||\omega_2| + |\omega_1||\omega_3|\right],$$

using inequality (A15)

$$\leq \frac{3M\omega^2}{\rho^2} \,, \quad \text{using inequality (A16)}.$$

Similarly, $\left|\dfrac{\omega_2}{\rho}\dfrac{D}{Dt}\left(\dfrac{\omega_2}{\rho}\right)\right| \leq \dfrac{3M\omega^2}{\rho^2}$ and $\left|\dfrac{\omega_3}{\rho}\dfrac{D}{Dt}\left(\dfrac{\omega_3}{\rho}\right)\right| \leq \dfrac{3M\omega^2}{\rho^2}$.

It follows that

$$\left|\frac{D}{Dt}\left(\frac{\omega^2}{\rho^2}\right)\right| = 2\left|\frac{\omega_1}{\rho}\frac{D}{Dt}\left(\frac{\omega_1}{\rho}\right) + \frac{\omega_2}{\rho}\frac{D}{Dt}\left(\frac{\omega_2}{\rho}\right) + \frac{\omega_3}{\rho}\frac{D}{Dt}\left(\frac{\omega_3}{\rho}\right)\right|$$

$$\leq 2\left|\frac{\omega_1}{\rho}\frac{D}{Dt}\left(\frac{\omega_1}{\rho}\right)\right| + 2\left|\frac{\omega_2}{\rho}\frac{D}{Dt}\left(\frac{\omega_2}{\rho}\right)\right| + 2\left|\frac{\omega_3}{\rho}\frac{D}{Dt}\left(\frac{\omega_3}{\rho}\right)\right|,$$

using the triangle inequality

$$\leq 18M\,\frac{\omega^2}{\rho^2} \,.$$

So we have

$$-18M\left(\frac{\omega}{\rho}\right)^2 \leq \frac{D}{Dt}\left[\left(\frac{\omega}{\rho}\right)^2\right] \leq 18M\left(\frac{\omega}{\rho}\right)^2. \qquad\qquad \text{(A17)}$$

Now the second of these inequalities is

$$\frac{D}{Dt}\left[\left(\frac{\omega}{\rho}\right)^2\right] - 18M\left(\frac{\omega}{\rho}\right)^2 \leq 0,$$

and hence

$$\frac{D}{Dt}\left[\left(\frac{\omega}{\rho}\right)^2 e^{-18Mt}\right] \leq 0.$$

So, if $\omega = \omega_0$ and $\rho = \rho_0$ when $t = 0$, we have

$$\left(\frac{\omega}{\rho}\right)^2 e^{-18Mt} \leq \left(\frac{\omega_0}{\rho_0}\right)^2 \quad \text{for} \quad t > 0,$$

and this may be written in the form

$$\frac{\omega^2}{\rho^2} \leq \frac{\omega_0^2}{\rho_0^2}\, e^{18Mt}.$$

If $\omega_0 = 0$, this means that $\dfrac{\omega^2}{\rho^2} \leq 0$. But $\dfrac{\omega^2}{\rho^2} \geq 0$ (since it is a perfect square). Hence, if $\omega = 0$ when $t = 0$, then

$$\omega = 0 \quad \text{for all} \quad t > 0.$$

Also the first inequality in equation (A17) gives

$$\frac{D}{Dt}\left[\left(\frac{\omega}{\rho}\right)^2 e^{18Mt}\right] \geq 0,$$

and so

$$\left(\frac{\omega}{\rho}\right)^2 e^{18Mt} \geq \left(\frac{\omega_0}{\rho_0}\right)^2 \quad \text{for} \quad t > 0.$$

It follows that

$$\frac{\omega^2}{\rho^2} \geq \frac{\omega_0^2}{\rho_0^2} e^{-18Mt} > 0$$

for all t if $\omega_0 \neq 0$. It follows that, if at any time $t > 0$, the magnitude of the vorticity is zero, then it cannot have been non-zero when $t = 0$.

COUETTE FLOW

The equation of motion for a viscous incompressible
fluid at constant temperature and under the action of no
external forces is found in equation (A4) to be

$$\frac{D\underset{\sim}{u}}{Dt} = -\frac{1}{\rho}\,\text{grad } p + \nu\nabla^2\underset{\sim}{u}\,, \qquad\qquad (A18)$$

and the equation of continuity is

$$\text{div }\underset{\sim}{u} = 0. \qquad\qquad (A19)$$

We consider the motion of such a fluid between two
coaxial cylinders whose radii are $\ a, b\ \ (a < b)\ $ and

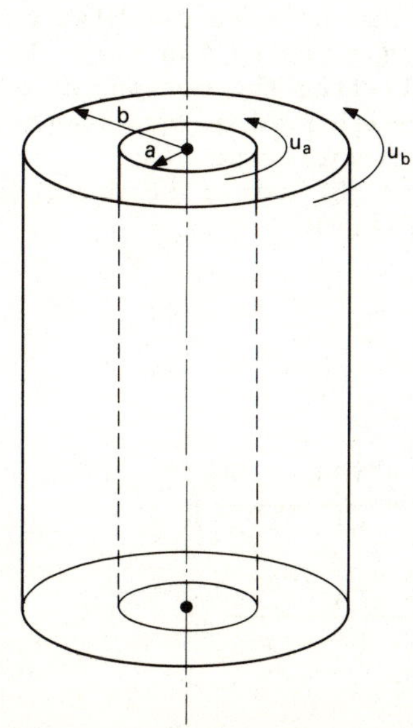

FIGURE A5.1

which are rotating about their axes so that their velocities are u_a, u_b respectively in the azimuthal direction (Figure A5.1). We use cylindrical polar coordinates (r,θ,z) and assume that the motion is perpendicular to the z-axis and that it is axially symmetric. Then the velocity is $(0,v(r),0)$ and equation (A19) is automatically satisfied. The equations (A18) give

$$0 = \nu \left[\frac{1}{r}\frac{d}{dr}\left(r\frac{dv}{dr}\right) - \frac{v}{r^2}\right],$$

and this may be written in the form

$$\frac{d^2v}{dr^2} + \frac{1}{r}\frac{dv}{dr} - \frac{v}{r^2} = 0.$$

This equation has the general solution

$$v = \frac{A}{r} + Br,$$

where A and B are arbitrary constants of integration. We see that the motion is the sum of two terms:

    (i) an irrotational motion given by the term $\frac{A}{r}$, which is exactly like that caused by a rectilinear line vortex along the axis of symmetry; this has uniform circulation $2\pi A$.

    (ii) a solid rotation given by the term Br, which is exactly like the motion of a fluid in a container with no internal boundaries; this has uniform vorticity $\frac{1}{2}B$.

The constants A and B can, of course, be evaluated using the boundary conditions

$$v = u_a \quad \text{when} \quad r = a,$$

$$v = u_b \quad \text{when} \quad r = b.$$

In fact, we find that

$$A = \frac{ab(bu_a - au_b)}{b^2 - a^2}$$

and

$$B = \frac{bu_b - au_a}{b^2 - a^2}.$$

SURFACE TENSION

The molecules of a liquid are sufficiently close
together for intermolecular forces to be relevant to
their motion. It is immediately evident that, whereas
a molecule in the middle of the liquid has similar
molecules all round it, a molecule which is in the
neighbourhood of a free surface (this may be the interface between two liquids or that between a liquid and a
gas) has similar molecules on one side of it and
dissimilar molecules (or similar molecules which are
quite differently spaced) on the other. It is, therefore, under the action of forces which are one-sided
compared with those on a molecule in the middle of the
liquid. On a macroscopic scale the effect of this is
to produce a *surface tension:* it is found that, if a
curve is drawn on the surface of the liquid, there is
a force exerted which is of magnitude σ per unit
length and which is in a direction which is normal to
the curve and tangential to the surface. If $\hat{\underset{\sim}{t}}$ is
the unit vector along the curve and $\hat{\underset{\sim}{n}}$ is that normal
to the surface, we see that the force per unit length
of the curve is $\sigma\hat{\underset{\sim}{t}} \wedge \hat{\underset{\sim}{n}}$, the force being a tension
if σ is positive. (This is equivalent to an increase
in the total free energy of the system of magnitude σA,
where A is the area of the surface.) Values of σ for
various interfaces were given in Table 10.1. At a
liquid/gas interface $\sigma > 0$, but σ may have either
sign at a liquid/liquid interface: only pairs of liquid
for which $\sigma > 0$ are immiscible.

It should be noted that a contaminant changes the
surface tension (it usually tends to reduce the
imbalance of forces described above). The effect
accounts for the usefulness of detergents for which
the effect of reduced σ is to increase wetting ability.

When a contaminant is spread unevenly over the surface, there will be a non-uniformity in σ :   this happens when a toy boat, with a piece of camphor at its stern, is placed in a bath and the effect is, as any child knows, that it is rapidly propelled across the surface of the water.

An important consequence of the surface tension is that, if the surface in which it acts is curved, there is a resultant normal stress across the surface. This can be calculated as follows: consider a small element

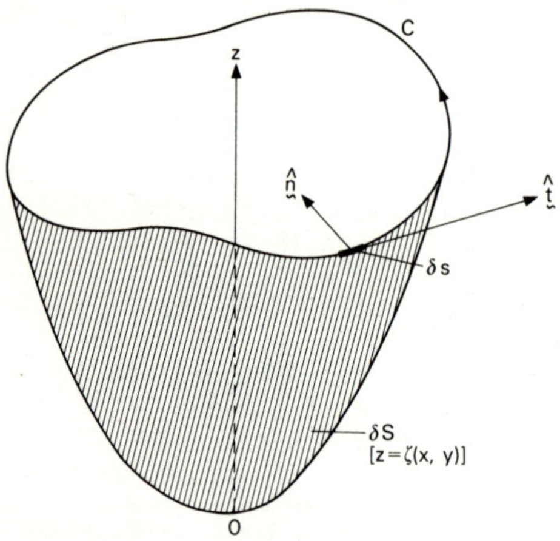

FIGURE A6.1

of surface δS bounded by the curve C (Figure A6.1). We take our origin O in the element of surface (more or less in the middle) and the axis Oz normal to the surface at O, so that the surface is concave towards $z > 0$. We take $\hat{k}$ to be the unit vector in the positive direction of Oz, and choose the sense of the unit normal $\hat{n}$ to be such that $\hat{n} = \hat{k}$ at 0 ; then if the equation of the surface is $z = \zeta(x,y)$, we have

$$\underset{\sim}{n} = \frac{1}{K}\left(-\frac{\partial\zeta}{\partial x}, \ -\frac{\partial\zeta}{\partial y}, \ 1\right), \quad K = \sqrt{1 + \left(\frac{\partial\zeta}{\partial x}\right)^2 + \left(\frac{\partial\zeta}{\partial y}\right)^2}.$$

Then, if $\hat{t}$ is in the sense shown in the figure, the force due to surface tension has as its z-component

$$\sigma\int_C \hat{t}\wedge\hat{n}.\hat{k}\ ds = \sigma\int_C \hat{n}\wedge\hat{k}.\hat{t}\ ds$$

$$= \sigma\iint_{\delta S} \mathrm{curl}\,(\hat{n}\wedge\hat{k}).\hat{n}\ dS, \text{ using Stokes's theorem}$$

$$\to \sigma \iint_{\delta S} \frac{1}{K}\left[\frac{\partial}{\partial x}\left(\frac{\zeta_x}{K}\right) + \frac{\partial}{\partial y}\left(\frac{\zeta_y}{K}\right)\right] dS \quad \text{as} \quad \delta S \to 0.$$

This must be balanced by a disparity in the pressure forces on the opposite sides of the interface: if p_+ is the pressure on the concave side and p_- is that on the convex side, the resultant force due to pressure has z-component

$$\iint_S (p_- - p_+)\underset{\sim}{n}.\underset{\sim}{k}\ dS = \iint_{\delta S} \frac{1}{K}\ (p_- - p_+) dS.$$

It follows at once that

$$\iint_{\delta S} \frac{1}{K}\left\{ (p_- - p_+) + \sigma\left[\frac{\partial}{\partial x}\left(\frac{\zeta_x}{K}\right) + \frac{\partial}{\partial y}\left(\frac{\zeta_y}{K}\right)\right]\right\} dS = 0 .$$

Hence the mean value of the integrand vanishes and, as δS shrinks on to the point 0, the mean value approaches the value of the function at 0 itself. It follows that

$$p_+ - p_- = \sigma\left[\frac{\partial}{\partial x}\left(\frac{\zeta_x}{K}\right) + \frac{\partial}{\partial y}\left(\frac{\zeta_y}{K}\right)\right] \quad \text{evaluated at } 0$$

$$= \sigma\left(\frac{\partial^2 \zeta}{\partial x^2} + \frac{\partial^2 \zeta}{\partial y^2}\right), \quad \text{since } \zeta_x, \zeta_y = 0 \text{ at } 0$$

$$= \left(\frac{1}{R_1} + \frac{1}{R_2}\right),$$

where R_1 and R_2 respectively are the radii of curvature of the curves in which the surface intersects the x-z and y-z planes. It is evident that this form depends neither on the direction nor on the position of the axes, but only on the point 0.

CURVES OF CONSTANT FREQUENCY FOR DEEP-WATER GRAVITY WAVES

We have seen that the frequency equation (256) in this case is

$$g^2(\lambda^2 + \mu^2) = (\omega_0 - U\lambda)^4,$$

and this may be rearranged in the form

$$\mu^2 = -\lambda^2 + \frac{(\omega_0 - U\lambda)^4}{g^2}. \tag{A20}$$

We note first that if we change the sign of μ, the value of λ is unchanged. Also, if we change the sign of ω_0 and of λ, the value of μ remains unchanged. So we need to consider only non-negative values of μ and of ω_0.

Next we note that when $\lambda = 0$, $\mu = \pm \frac{\omega_0^2}{g}$ and that when $\lambda \to \infty$, $\mu \sim \frac{U^2\lambda^2}{g}$.

The special case $\omega_0 = 0$

Here equation (A20) becomes

$$\mu^2 = -\lambda^2 + \frac{U^4\lambda^4}{g^2}. \tag{A21}$$

We can sketch the graph of μ^2 as a function of λ^2 (Figure A7.1) and note that (if we, for the moment, allow negative values of λ^2, μ^2) this is a parabola through the origin and through the point $(g^2/U^4, 0)$.

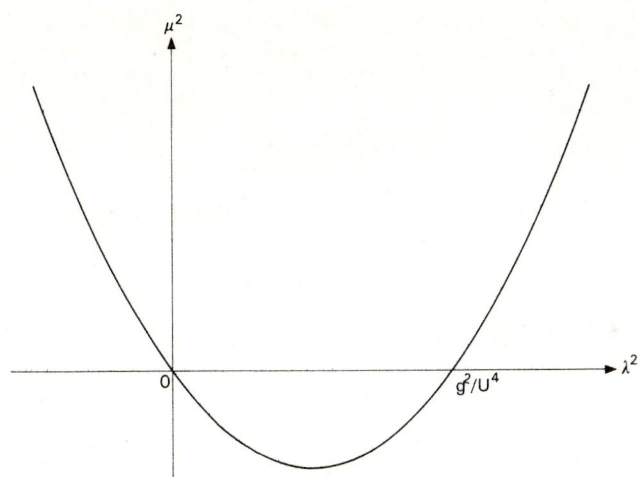

FIGURE A7.1

But it is not permissible for λ^2 and μ^2 to be
negative, since λ and μ are real. So we have real
pairs of numbers (λ,μ) only when $\lambda = \mu = 0$ and
when $\lambda^2 > g^2/U^4$. It is easy to verify that when
$|\lambda| = g/U^2$, then $\mu = 0$ and $|d\mu/d\lambda|$ is infinite.
It follows that the graph of the curve given by equation
(A21) is as shown in Figure A7.2.

The case $\omega_0 > 0$

When $\omega \neq 0$, we have to consider the full equation
(A20). We note that $\mu = 0$ when $(\omega_0 - U\lambda)^4 = g^2\lambda^2$;
that is, when

$$(\omega_0 - U\lambda)^2 = g|\lambda| . \tag{A22}$$

We have to consider separately what happens when $\lambda \geq 0$
and when $\lambda \leq 0$.

When $\lambda \geq 0$, equation (A22) is $(\omega_0 - U\lambda)^2 = g\lambda$
and this may be rearranged in the form

$$\lambda^2 - \left(\frac{2\omega_0}{U} + \frac{g}{U^2}\right)\lambda + \frac{\omega_0^2}{U^2} = 0 .$$

This quadratic in λ has two real, distinct, positive
roots, each of which increases as ω_0 increases.

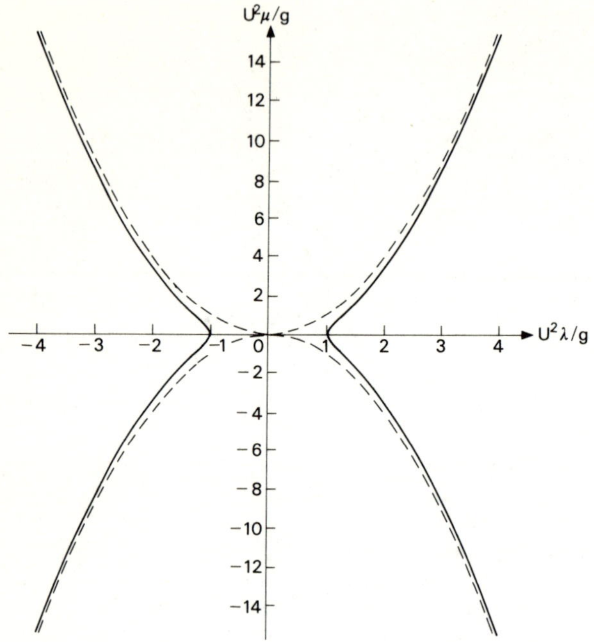

FIGURE A7.2

When $\lambda \leq 0$, equation (A22) is $(\omega_0 - U\lambda)^2 = - g\lambda$ and this may be rearranged in the form

$$\lambda^2 - \left(\frac{2\omega_0}{U} - \frac{g}{U^2}\right)\lambda + \frac{\omega_0^2}{U^2} = 0 .$$

This has two real, distinct, negative roots if $\omega_0 < g/4U$, two real, coincident, negative roots if $\omega_0 = g/4U$ and no real roots if $\omega_0 > g/4U$. We restrict ourselves to the case in which $\omega_0 < g/4U$ and note that then the smaller root (that is the one further from the origin) gets larger as ω_0 increases, whereas the other gets smaller. Clearly these two values approach each other as ω_0 increases, and meet when $\omega_0 = g/4U$. As for the case $\omega_0 = 0$, the value of $|d\mu/d\lambda|$ is infinite at all the points where $\mu = 0$ (when $\omega_0 < g/4U$), and so the graph of μ as a function of λ is as shown in Figure A7.3. The arrows show the direction in which the curves move as ω_0 increases.

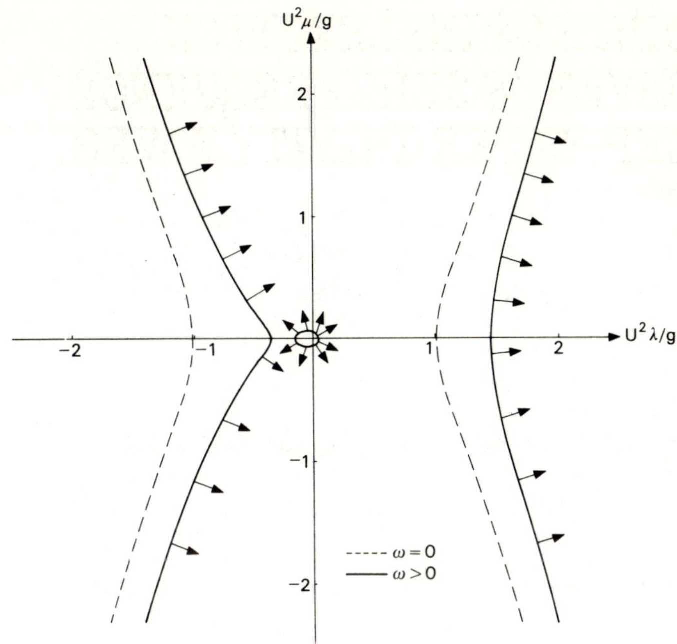

FIGURE A7.3

We can now go back and insert the corresponding arrows for $\omega_0 = 0$ and the result is as shown in Figure 11.4 in the main text.

(The curves for $\omega_0 \geq g/4U$ can be drawn similarly, but are of little practical interest.)

THE POINTS OF INFLEXION ON THE CURVES OF ZERO FREQUENCY FOR DEEP-WATER GRAVITY WAVES

We have (see either equation (256) or equation (A21))

$$\mu^2 = -\lambda^2 + \frac{U^4 \lambda^4}{g^2} \; . \tag{A23}$$

Differentiating this once, we obtain

$$\mu \frac{d\mu}{d\lambda} = -\lambda + \frac{2U^4 \lambda^3}{g^2} \; , \tag{A24}$$

and again

$$\left(\frac{d\mu}{d\lambda} \right)^2 + \mu \frac{d^2\mu}{d\lambda^2} = -1 + \frac{6U^4 \lambda^2}{g^2} \; .$$

At a point of inflexion we have $d^2\mu/d\lambda^2 = 0$, and so

$$\left(\mu \frac{d\mu}{d\lambda} \right)^2 = \left(-1 + \frac{6U^4 \lambda^2}{g^2} \right) \mu^2 \; .$$

Using (A23) and (A24) in this, we obtain

$$\left(-\lambda + \frac{2U^4 \lambda^3}{g^2} \right)^2 = \left(-1 + \frac{6U^4 \lambda^2}{g^2} \right) \left(-\lambda^2 + \frac{U^4 \lambda^4}{g^2} \right) \; ,$$

and this can be expanded and rearranged to give

$$\frac{3U^4 \lambda^4}{g^2} = \frac{2U^8 \lambda^6}{g^4} \; .$$

Now $\lambda \neq 0$ (since $(0,0)$ is an isolated point and so cannot be a point of inflexion). Hence

$$\lambda^2 = \frac{3g^2}{2U^4} \, .$$

This gives

$$|\lambda| = \sqrt{\left(\frac{3}{2}\right)} \frac{g}{U^2} \, , \quad |\mu| = \frac{\sqrt{3}}{2} \frac{g}{U^2} \, , \quad \left|\frac{d\mu}{d\lambda}\right| = 2\sqrt{2} \, ,$$

using equations (A23) and (A24) again.

AN ALTERNATIVE APPROACH TO THE PROBLEM OF BOW WAVES ON THE SURFACE OF DEEP WATER

We consider a point disturbance which, at time t_0, was at A and which moves with uniform velocity U so that

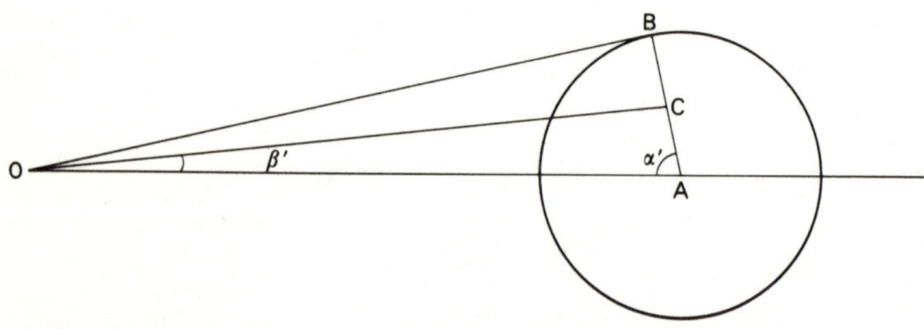

FIGURE A9.1

at time t_1 it is at the point O. At time t_0 it emitted disturbances of all wavelengths and, in particular, one with wavelength $2\pi/k$: by time t_1 a wave crest having this wave number would form a circle centre A and radius $c(t_1 - t_0)$, where $c = \sqrt{(g/k)}$.

We draw the tangent from O to this circle to touch it at B. We note that if at time t_2, where

$t_0 < t_2 < t_1$, the disturbance is at A', a wave crest having wave number $2\pi/k$ would form a circle centre A' and radius $c(t_1 - t_2)$, and that OB is also tangential to this circle. It follows that OB is the envelope at time t_1 of all the wave crests having wavelength $2\pi/k$ which are emitted by the disturbance as it moves along

310

and the effect is that of a plane wave crest of wave-length $2\pi/k$ moving parallel to AB so that it always passes through the present position of the disturbance.

However, the situation is less simple than this because the energy of the disturbance is not propagated with speed c, but with speed $v = c/2$. So, at time t_1, the energy from the disturbance when it was at A will have reached the point C on AB, where $AC = v(t_1 - t_0)$. Further, the straight line OC bounds the region (above the axis) where energy from the disturbance corresponding to wave number $2\pi/k$ can reach.

We write β' for the angle AOC and α' for the angle OAB and note that β' and α' are both functions of k, and hence β' is a function of α'. In fact we have

$$\frac{\sin \beta'}{\sin(\alpha' + \beta')} = \frac{AC}{OA} = \frac{1}{2U}\sqrt{\left(\frac{g}{k}\right)} ,$$

$$\cos \alpha' = \frac{AB}{OA} = \frac{1}{U}\sqrt{\left(\frac{g}{k}\right)} .$$

It follows that

$$\frac{\sin \beta'}{\sin(\alpha' + \beta')} = \tfrac{1}{2} \cos \alpha' ,$$

and this may be rearranged to give

$$\tan \beta' = \frac{\tan \alpha'}{1 + 2 \tan^2\alpha'} . \qquad (A25)$$

We note that

$$\sec^2\beta' \frac{d\beta'}{d\alpha'} = \frac{1 - 2 \tan^2\alpha'}{(1 + 2 \tan^2\alpha')} \sec^2\alpha' ,$$

and so β' has a maximum value β when $\alpha' = \alpha$, where

$$\tan \alpha = 1/\sqrt{2} ,$$

and then

$$|\tan \beta| = 1/2\sqrt{2} ,$$

using equation (A25).

ANSWERS TO EXERCISES

CHAPTER 1

2 $a^2 = \gamma \left\{ [RT - \alpha\rho(1 - \beta\rho)]/(1 - \beta\rho)^2 \right\} \exp(-\alpha\rho/RT).$

3 $S_2 - S_1 = c_v \log_e(T_2/T_1).$

CHAPTER 2

1 Possible examples are $\underset{\sim}{u} = \underset{\sim}{r}t$ and $\underset{\sim}{u} = (\sin t/r^2)\hat{\underset{\sim}{r}}$.

2 $\underset{\sim}{r} = \exp(2t)\underset{\sim}{i} + \exp(-2t)\underset{\sim}{j}$.

3 (a) $ax = by = cz$: straight lines through the origin;

   (b) $x + y + z = b$, $x^2 + y^2 + z^2 = c^2$; circles normal
      to $x = y = z$ and with centres on it;

   (c) $x = c \cos[(z - b)/3]$, $y = c \sin[(z - b)/3]$:
      helices, of constant spacing, on cylinders
      $x^2 + y^2 = c^2$;

   (d) $yz = a$, $x = b + \frac{1}{2}y^2$;

   (e) $x^2 + y^2 + z^2 = c^2$, $x = b + y^2$;

   (f) $x^2 + y^2 = c^2$, $z = b$: circles normal to z-axis and
      with centres on it.

   All except (a) correspond to flow of incompressible
   fluid.

6 $T_0/T = 1/[1 - \frac{1}{2}(\gamma - 1)(u/a_0)^2]$, etc.

10 $p_1 - p_2 = \rho u_2^2 - \rho u_1 u_2$; loss of energy $= (p_2' - p_1)u_1 A_1.$

11 $T \to 0$, $M \to \infty$; the limit is achieved when $p \to 0$ and
   this implies a vacuum.

CHAPTER 3

5 $M^2 = 2(\alpha - 1)^2/\{\gamma\alpha[\alpha(\gamma - 1) + \gamma + 1]\}$,   where

   $\alpha = 1 + (\gamma + 1)\rho_1 V^2/4p_1 +$

       $+ [4(3\gamma + 1)p_1\rho_1 V^2 + (\gamma + 1)^2\rho_1^2 v^4]^{\frac{1}{2}}/4p_1.$

8 $M_2 = 1.$

9 $W/U = 2[(\gamma - 1)\alpha + 1]/[\alpha(\gamma + 1) + \gamma - 1].$

12 $b^2 d^3 \geq 27q^2/8g.$

CHAPTER 4

1 $u = -(p_0/\rho_0 a_0)\sec(\omega b/a_0)\sin(\omega x/a_0)\cos \omega t;$

  $\rho' = (p_0/a_0^2)\sec(\omega b/a_0)\cos(\omega x/a_0)\sin \omega t.$

2 Transmitted wave in phase with incident wave; its
amplitude is $2\rho_1 a_1/(\rho_2 a_2 + \rho_1 a_1)$ times that of
incident wave. Reflected wave in phase with incident
wave if $\rho_2 a_2 > \rho_1 a_1$ and $180°$ out of phase if
$\rho_2 a_2 < \rho_1 a_1;$ its amplitude is $|\rho_2 a_2 - \rho_1 a_1|/$
$(\rho_2 a_2 + \rho_1 a_1)$ times that of incident wave.

CHAPTER 5

2 $U(z) = (\rho g/4\mu)(2hz - z^2)\sin \alpha;\ Q = (\rho g h^3/24\mu)\sin \alpha.$

3 $U_1 = (dp/dx)(z - b)[(\mu_1 + \mu_2)z + 2b\mu_1]/[2\mu_1(\mu_1 + \mu_2)];$

  $U_2 = (dp/dx)(z + b)[(\mu_1 + \mu_2)z - 2b\mu_2]/[2\mu_2(\mu_1 + \mu_2)].$

6 $u = \alpha(\pi\nu t)^{\frac{1}{2}}$ where $\alpha = (- \partial u/\partial z)_{z=0}$

8 $Q = (g\pi a^4/8\nu)\sin \alpha.$

9 $Q = \frac{1}{2}\pi V[- 2a^2 + (b^2 - a^2)/\log_e(b/a)].$

10 $F = - \pi\rho g b^2 + [2\pi\mu/\log_e(a/b)][(g/4\nu)(a^2 - b^2) - W];$

   terminal velocity is $(g/4\nu)(a^2 - b^2).$

11 (a) 6.2×10^7, (b) 0.005, (c) 7.6×10^{-4},

   (d) 3.1×10^6, (e) $3.7 \times 10^6.$

CHAPTER 6

1 $u = U \exp(-z/\delta)\cos(\omega t - z/\delta)$; $z = 4.6$ mm.

3 $u = U[1 - \exp(-z/\delta)\cos(z/\delta)]$, $v = U \exp(-z/\delta)\sin(z/\delta)$.

4 $\delta_1 = 1.83(\nu x/U)^{\frac{1}{2}}$, $\delta_2 = 0.73(\nu x/U)^{\frac{1}{2}}$.

5 (a) $\delta = 4.80(\nu x/U)^{\frac{1}{2}}$, $\delta_1 = 1.74(\nu x/U)^{\frac{1}{2}}$, $\delta_2 = 0.66(\nu x/U)^{\frac{1}{2}}$,

   (b) $\delta = 2.55(\nu x/U)^{\frac{1}{2}}$, $\delta_1 = 1.77(\nu x/U)^{\frac{1}{2}}$, $\delta_2 = 0.78(\nu x/U)^{\frac{1}{2}}$.

6 (a) $x \simeq 360$ mm, (b) $x \simeq 680$ mm.

CHAPTER 7

6 $\phi = Ux - Ua^2x/(x^2 + y^2) + (\kappa/2\pi)\tan^{-1}(y/x)$.

7 $\tan(\omega b/a_0) = \omega b/a_0$.

8 $\rho' \simeq - (\varepsilon\omega^2\rho_0 R^3/a_0^2 r)\cos\{\omega[t - (r - R)/a_0]\}$.

10 $(0, \pm a/\sqrt{3})$.

11 $- \kappa$ at $(-a,b,0)$, $- \kappa$ at $(a,-b,0)$, κ at $(-a,-b,0)$;
   $x^{-2} + y^{-2} = a^{-2} + b^{-2}$.

13 $u = - \tfrac{1}{2}\kappa$ when $y > 0$, $u = \tfrac{1}{2}\kappa$ when $y < 0$; $v = 0$.

CHAPTER 8

2 $(0,0)$.

3 Streamlines are arcs of circles through $(a,0)$, $(-a,0)$ having centres on the y-axis.

4 (a) Line source strength A, (b) line vortex strength $2\pi A$, (c) line doublet strength A.

CHAPTER 9

1 $r^{2n}\sin(2n\theta - \alpha) - 2r^n r_0^n\cos n\theta_0 \sin(n\theta_0 - \alpha)$
$$- r_0^{2n}\sin \alpha = 0.$$

2 $r \sin n\theta = $ constant.

4 Superpose $U \sin \alpha$ parallel to generators.

6 $\kappa = 2\sqrt{2}\pi Ub$.

7   $w = \frac{1}{2}U\{z + z' + [(a^2 + b^2)/(a^2 - b^2)]e^{-2i\alpha}(z - z')\}$

$$- (i\kappa/2\pi)\log_e(z + z'),$$

where $z' = [z^2 - 2(a^2 - b^2)e^{2i\alpha}]^{\frac{1}{2}}.$

8 There is an infinite velocity gradient at at least one point in the flow. (Note that it is inappropriate to use arguments involving viscosity here.)

9 $x/a = [\alpha^2 - 2\alpha/(\pi + 4)]^{\frac{1}{2}}$

$$- [2/(\pi + 4)]\cosh^{-1}[\tfrac{1}{2}(\pi + 4)\alpha]^{\frac{1}{2}};$$

$y/a = \{\pi + 4[\tfrac{1}{2}(\pi + 4)\alpha]^{\frac{1}{2}}/(\pi + 4).$

10 $U\,dz/dw = 1 + 2\exp(-\pi w/Uc')$

$$+ 2[\exp(-2\pi w/Uc') + \exp(-\pi w/Uc')]^{\frac{1}{2}},$$

where $2c'$ is final width of stream.

CHAPTER 10

2 (a) Straight lines (shallow water waves),
   (b) circles (deep-water gravity waves).

4 Amplitude of waves is no longer small and theory must be modified when $kU^2 = g\tanh kh$.

CHAPTER 11

1 $b_n = [4abL/\pi(b^2 - n^2L^2)]\cos(n\pi L/2b)\sin(n\pi x_0/b)$, $n \neq b/L$

   $b_n = (aL/b)\sin(\pi x_0/L)$, $n = b/L$.

In special case, coefficients are 0.1, 0.2, 0.3, 0.3, 0.2, 0.1, 0.0, 0.0, 0.0, 0.0, 0.1, 0.1, 0.2, 0.1, 0.0, -0.3, -0.8, -1.5, -3.2, 0.9, 1.8, 0.3, -0.2, -0.3, -0.3, -0.3, -0.2, -0.1, 0.0, 0.0, etc.

2 Approximately 10 hours before first observation; distance approximately 50 km.

5 $U^2 < gh$: ellipse; energy propagated in all directions.

  $U^2 > gh$: hyperbola; energy propagated within angle $\pm\,\alpha$ of λ-axis, where $\tan^2\alpha = (U^2 - gh)/gh$.

6 Each curve consists of two parts: one is a closed curve which has $\lambda < 0$ and which touches the μ-axis (this corresponds to long waves propagated in all directions); the other is asymptotic to the straight line $\lambda = \omega/U$ (this corresponds to waves of shorter

wavelength propagated forwards in a direction nearly parallel to the λ-axis).

7   $0 < \omega < N$: hyperboloid of revolution (of one sheet); energy propagated outside the asymptotic cone, away from the ν-axis.

   $N \leq \omega < KN$: no waves can propagate in this range of ω.

   $\omega \geq KN$: ellipsoid of revolution; energy propagated in all directions.

8   Each curve consists of three parts: one is a closed curve centred on the origin (this corresponds to long waves propagated in all directions); the other two are asymptotic respectively to $\nu = \pm\omega(a^2 + A^2)^{\frac{1}{2}}/a^2A^2$ (these correspond to waves of shorter wavelength propagated in a direction nearly parallel to the ν-axis).

   When $a > A$, the long waves are propagated with speed of order a and the short waves with speed of order A.

   When $A > a$, the picture is exactly the same, but now the long waves are propagated with speed of order A and the short ones with speed of order a.

INDEX